Crisis Management of Chronic Pollution

Contaminated Soil and Human Health

Urbanization, Industrialization, and the Environment

Series Editor

Ming Hung Wong

Crisis Management of Chronic Pollution: Contaminated Soil and Human Health, *edited by Magalie Lesueur Jannoyer, Philippe Cattan, Thierry Woignier, and Florence Clostre*

Biochar: Production, Characterization, and Applications, *edited by Yong Sik Ok, Sophie M. Uchimiya, Scott X. Chang, Nanthi Bolan*

Crisis Management of Chronic Pollution

Contaminated Soil and Human Health

edited by
Magalie Lesueur Jannoyer
Philippe Cattan
Thierry Woignier
Florence Clostre

CRC Press
Taylor & Francis Group
Boca Raton London New York

CRC Press is an imprint of the
Taylor & Francis Group, an **informa** business

CRC Press
Taylor & Francis Group
6000 Broken Sound Parkway NW, Suite 300
Boca Raton, FL 33487-2742

First issued in paperback 2020

ISBN-13: 978-1-4987-3783-8 (hbk)
ISBN-13: 978-0-367-65837-3 (pbk)

Library of Congress Cataloging-in-Publication Data

Names: Lesueur Jannoyer, Magalie.
Title: Crisis management of chronic pollution : contaminated soil and human health / [edited by] Magalie Lesueur Jannoyer [and three others].
Description: Boca Raton : CRC Press, 2017. | Series: Urbanization, industrialization, and the environment ; 2
Identifiers: LCCN 2016034256 | ISBN 9781498737838
Subjects: LCSH: Soil pollution--West Indies, French. | Soil pollution--West Indies, French--Prevention. | Water--Pollution--West Indies, French. | Water quality management--West Indies, French. | Organochlorine compounds--Environmental aspects--West Indies, French. | Pesticides--Environmental aspects--West Indies, French.
Classification: LCC TD878 .C75 2017 | DDC 363.739/6--dc23
LC record available at https://lccn.loc.gov/2016034256

Visit the Taylor & Francis Web site at
http://www.taylorandfrancis.com

and the CRC Press Web site at
http://www.crcpress.com

We dedicate this book to our colleague and friend Yves Marie Cabidoche, who was present at the very beginning of the story but passed away in 2012. He initiated discussions among scientists and managed working together on this hard and complex task: the characterization, the understanding, and the management of chlordecone pollution in the French West Indies. He motivated a lot of French research teams to work on the subject, whatever the domain. He co-organized the first event on chlordecone pollution, an international workshop on remediation, and was very keen on imparting knowledge and explaining things to people, whoever they were: researchers, politicians, civil society actors, or funders.

Yves Marie was, above all, an engaged researcher and was working for the Guadeloupe Islands he adopted in the early 1980s.

Contents

SECTION III Biotope Contamination and Eco-Toxicological Assessment (Terrestrial and Aquatic Ecosystems)

SECTION IV Pollutant Transfers

SECTION V The Public Health Issue: Exposure and Health Impacts

SECTION VI Remediation

SECTION VII Management Approach

Foreword

Crisis Management of Chronic Pollution: Contaminated Soil and Human Health documents the pollution of the French West Indies (FWI) islands by chlordecone, an organochlorine pesticide used in banana plantations. The initial discovery of chlordecone in spring water from the FWI islands has triggered a collective response of the scientific, political, and societal communities, which has led to a detailed characterization of pollution and the development of (local) knowledge about this pollutant and its effects on the environment and human health. In the final phase, various actions, such as remediation, management of crisis, and information of the local communities, have been undertaken to tackle and contain chlordecone pollution.

Chlordecone, also known as Kepone, is an organochlorine insecticide related to Mirex and DDT. Due to its deleterious effects on the environment and organisms, chlordecone has been classified as a persistent organic pollutant (POP) and was banned globally in 2011 by the Stockholm Convention on POPs. Chlordecone has a high bioaccumulation factor in organisms, is highly lipophilic, persists in the environment and organisms, and has a broad range of toxicities, including neurotoxicity. Sadly, the islands of Martinique and Guadeloupe show the highest prostate cancer diagnosis rates in the world that could be related to pesticides exposure.

After years of unrestricted use on banana plantations, the FWI territories, particularly the islands of Martinique and Guadeloupe, are heavily contaminated with chlordecone. Despite the ban by France in 1990, chlordecone use had been allowed until 1993, because of the unavailability of the alternative pesticide at that time. Since 2003, local authorities have restricted cultivation of crops following severe soil contamination.

Due to the involvement of the FWI territories in the contamination issue, it is appropriate to have a strong French team of researchers to write about this environmental contamination episode. This book aims to review and compile the main developments and knowledge acquired over many years of study and presents the contribution of research to the diagnosis and management of this pollution situation at the beginning of the 2000s. This book is structured into eight sections consisting of 20 chapters, covering a broad range of scientific fields, and answers in an exceptional manner a number of key questions: What is the (source of) pollution and how to characterize it? Which locations and environmental compartments have been contaminated? How long has this contamination be present? What are the most relevant environmental samples to analyze? What are the impacts on public health and the environment? How to tackle and contain the problem? Can it be managed and, if possible, can it be remedied?

This book provides a clear and concise presentation of various aspects of chlordecone pollution. It follows a logical presentation of the key facets associated with chlordecone, beginning with a chronicle to the chlordecone case study in the FWI territories (Section I) and an overview of the crisis emergence. An inventory of the environmental contamination of the natural resources, such as soil, surface water, and groundwater, is documented in Section II. The environmental and ecotoxicological assessments of chlordecone for the terrestrial and aquatic ecosystems of the FWI are further described in Section III. An assessment of chlordecone in marine fish from the FWI related to fishery management concerns is also included in this section. An important section of the environmental assessment (Section IV) involves the investigation of chlordecone transfer from soil to different environmental compartments (water, crops, and animals), with the purpose of further investigating its transfer from locally grown foods to humans. Human exposure via diet and the resulting dietary risk assessment are consequently discussed in Section V, together with a first study on the possible health impacts of chlordecone on pregnancy and child development. This book continues with a very important section on possible remediation (Section VI), taking this approach from a theoretical to a practical level by looking into an organic amendment of soil to decrease pesticide bioavailability and filtration of contaminated water using activated carbon. The management measures taken to contain the contamination at the farm level are discussed in Section VII. Various tools are proposed for healthy gardening, together with a widespread awareness program to help people better understand the situation of environmental contamination caused by chlordecone. Section VIII contains an overview of the crisis management and concludes by questioning our capacity to manage future threats emerging from human activities. Several aspects, such as analytical developments and social analyses, have not been directly included in this book but have been key factors in supporting appropriate environmental crisis monitoring and management.

This book is aimed at serving as a helpful guide to students, researchers, and decision-makers in the assessment and management of chronic pollution, specific for POP-like compounds. Each chapter contains enough relevant references to the literature to help as an effective resource for more detailed information.

This is the most recent (and also the most updated) initiative to compile chemical, environmental, and toxicological information related to chlordecone. The other three monographs are rather outdated since they date from 1978 (*Kepone/Mirex/Hexachlorocyclopentadiene, an Environmental Assessment: A Report* by National Research Council), from 1984 (*Chlordecone, Environmental Health Criteria No. 43* by the World Health Organization), and from 1995 (*Toxicological Profile for Mirex and Chlordecone*, U.S. Department of Health and Human Services). Interestingly, chlordecone has been added in 2009 to the Stockholm Convention list of POPs, following a strong dossier introduced by (almost) the same team of French researchers.

As a side note, the official statistics related to pesticide use on banana plantations in many countries in Latin America, the Caribbean Islands, and Africa indicate that the FWI territories use lesser amount of pesticides. It is just a matter of time until strong stories about environmental contamination with pesticides, as the one presented in the current book, will become publicly available from other locations too. This is a serious matter that we all need to be aware of and in the end take responsibilities for.

Adrian Covaci
Antwerp, Belgium

Acknowledgments

We are very grateful to all the contributors who made this book possible. We thank our coauthors and colleagues for their timely response, patience, and confidence to gather in this book the French expertise on the chlordecone pollution.

The research and ideas developed in this book are the result of more than 10 years of work, most of them in the Caribbean. The work is still in progress with the support of national and local funds as solutions to remediate chlordecone pollution are not effective in real-world conditions and monitoring is still necessary. Financial support from the respective organization is specifically acknowledged at the end of each chapter.

We also want to mention the very good collaboration between research and local authorities, with a special thanks to Eric Godard, who coordinated the National Plan (2008–2012) and did his best to spread knowledge and take measures to protect people and the environment.

We also thank the support of CRC Press for providing us with this book project and for helping with its thorough editing.

Acknowledgments

Editors

Magalie Lesueur Jannoyer earned her PhD in agronomy in the National Agronomical Institute, Paris-Grignon in 1995. Since 1999, she has been working as a senior researcher at CIRAD (International Research Center for Development). She has more than 15 years of experience in research and development projects in tropical horticultural crop management and environmental impact assessment, as well as in the management of fruit quality and safety. She has developed innovative practices, including eco-physiological tools and systemic agronomical analysis, to reduce the use of chemicals and promote the ecological functioning of tropical horticultural systems. For the past 10 years, her main subject of interest has been the assessment and management of soil pollution, specifically the case of organochlorine soil pollution in the French West Indies. Her research focuses on (1) the assessment of the impact of this pollutant on food production and safety (pesticide residues), on agro systems (changes in practices), and on the environment (watershed quality); and (2) the elaboration of efficient tools to manage pollution at different scales (field, farm, watershed, territory). From 2008 to 2010, in collaboration with Y.M. Cabidoche (Inra), she coordinated the research activities of the National Action Plan on chlordecone. She co-organized an international workshop on pollution remediation in 2010. She collaborates with national and international research institutions in French and European Union granted projects. She is recognized as an expert by the French Agency of Food Security and Safety (ANSES) and is a member of the International Society for Horticultural Science council.

Philippe Cattan is an agronomist researcher at the research unit GECO (banana and pineapple systems), CIRAD. He graduated from the National High School of Agronomy of Montpellier (ENSAM 1983), holds a PhD in agronomy (National Agronomical Institute, Paris-Grignon 1996), and received a habilitation to conduct researches in 2010 at the University of the French West Indies and Guyana. After working for 13 years on oilseed crops in Africa (Senegal and Burkina Faso), he joined the research team on banana systems in Guadeloupe at the research station of Neufchâteau. Since 1999, his area of concern has been the impact of agricultural practices on the environment. He first studied hydrological functioning of banana plantation at the plot and watershed scales and the impact of agricultural practices on these states, the properties, and the outgoing flows of soil. Notably, he addressed the issue of pesticide fate. Sincc 2007, hc focused on chlordecone through the coordination of two main projects: the first one (Chlordexco, founded par French National Research Agency) aimed to identify the transfer pathways of chlordecone from soil toward plant and aquatic animals; the second one aimed to implement an observatory of agricultural pollution (OPA-C that became OPALE), notably chlordecone. Nowadays, in partnership with other French institutes (BRGM, INRA, IRD) and the University of the French West Indies and Guyana, he coordinates an integrated project called OPALE, Observatory of Agricultural Pollutrions in the French West Indies about the two-way relationship between environmental states and changes in agricultural practices.

Thierry Woignier is director of research, CNRS and IRD (IMBE-Institut Mediterranéen de Biodiversité et d'Ecologie Marine et continentale). He is a postgraduate engineer in materials sciences (1980), PhD in physics (1984), postdoctorate in Brazil, Universidade de Sao Paolo (1985), and state doctorate in physics (Université de Montpellier 1993). He directed research teams (Materials Science Laboratory in Montpellier CNRS) and was responsible for research programs on composite materials for aerospace applications, nuclear waste containment, and optical properties. Since 2006, he directs the IRD Physical Properties of Soils at The Caribbean Agro Environmental Campus (IRD Martinique), working on the sequestration of greenhouse gases in soils and on the contamination of tropical soils by pesticide (chlordecone). He is responsible for the research programs from the French Overseas Ministry and an expert in the Office Parlementaire Scientifique et Technique. He is interested in the natural sequestration of pesticides in clays and proposes alternative solutions to decontamination by trapping process. He has a wide range of publications (180 scientific papers and 21 book chapters).

Florence Clostre completed her engineering degree in agriculture from AgroSup Dijon, France, in 2002 with a specialization in tropical agriculture at the Institut Supérieur Industriel agronomique Huy-Gembloux, Belgium. She worked for public or semipublic bodies as an expert in pesticide use in agriculture and associated impacts on user health and on the environment. Since 2009, she has been working with CIRAD, a public research institute specializing in agriculture and sustainable development in tropical and subtropical countries. Clostre's research focuses on pollutant uptake by crops, especially tropical crops, soil contamination, enhancement of natural pollutant sequestration in soil with organic matter, and more generally on environmental impact assessment. She has published ten papers and collaborated on two book chapters about organochlorine pollution between 2012 and 2015.

Contributors

Alain Abarnou
Department of biological ressources and Environment
French Research Institute for Exploitation of the Sea
Nantes, France

Patrick Andrieux
ASTRO Research Unit
French National Institute for Agricultural Research
Petit-Bourg, France

Alain Archelas
French National Center of Scientific Research
Institute of Molecular Sciences
University of Aix Marseille
Marseille, France

Luc Arnaud
Water Division
The French Geological Survey
Orléans, France

Abderazak Bazizi
International Center of Agricultural Research for Development
UPR HortSys
Martinique, France

Simon Bellec
Territorial Delegation of Belfort
Bourgogne-Franche-Comté Regional Health Agency
Belfort, France

Jacques A. Bertrand
Department of Biological Resources and Environment
French Research Institute for Exploitation of the Sea
Nantes, France

Xavier Bodiguel
French Research Institute of Exploitation of the Sea
Martinique, France

Olivier Boucher
University Health Centre of Quebec
Laval University
Québec City, Québec, Canada

Jean Pierre Bricquet
UMR HydroSciences Montpellier
French Institute of Research for Development
Montpellier, France

Céline Carles
International Center of Agricultural Research for Development
UPR HortSys
Martinique, France

Philippe Cattan
International Center of Agricultural Research for Development
UPR Geco
Guadeloupe, France

Fanny Caupos
Asconit Consultants
Martinique, France

and

Ecosystem Health and Ecology Lab
University of Rennes
Rennes, France

Jean-Baptiste Charlier
Water Division
The French Geological Survey
Montpellier, France

Florence Clostre
International Center of Agricultural Research for Development
UPR HortSys
Martinique, France

Sylvaine Cordier
Team of Epidemiological Research on Environment, Reproduction and Development
University of Rennes I
Rennes, France

Nathalie Costet
Team of Epidemiological Research on Environment, Reproduction and Development
University of Rennes I
Rennes, France

Renée Dallaire
University Health Centre of Quebec
Laval University
Québec City, Québec, Canada

Pauline Della Rossa
Cirad
UPR HortSys
Martinique, France

Jan Dolfing
School of Civil Engineering and Geosciences
Newcastle University
Newcastle-upon-Tyne, United Kingdom

Carine Dubuisson
French Agency for Food, Environmental and Occupational Health Safety
Maisons-Alfort, France

Laure Ducreux
The French Geological Survey
Guadeloupe, France

Axelle Durimel
COVACHIM M2E Lab
University of French West Indies
Guadeloupe, France

Cyril Feidt
Department of Animal and Functionality of Animal Products
University of Lorraine
Vandoeuvre-lès-Nancy, France

Paula Fernandes
International Center of Agricultural Research for Development
UPR HortSys
Montpellier, France

Agnès Fournier
Department of Animal and Functionality of Animal Products
University of Lorraine
Vandoeuvre-lès-Nancy, France

Marie Fröchen
French Agency for Food, Environmental and Occupational Health Safety
Maisons-Alfort, France

Sarra Gaspard
COVACHIM M2E Lab
University of French West Indies
Guadeloupe, France

Jean-Marie Gaude
International Center of Agricultural Research for Development
UPR HortSys
Martinique, France

Eric Godard
Territorial Animation and Public Health Directorate
Martinique Regional Health Agency
Martinique, France

Julie Gresser
Water Office of Martinique
Martinique, France

Mathilde Guene
International Center of Agricultural Research for Development
UPR HortSys
Martinique, France

Olivier Guyader
French Research Institute of Exploitation of the Sea
Bretagne Center
Pointe du Diable, France

Fanny Héraud
French Agency for Food, Environmental and Occupational Health Safety
Maisons-Alfort, France

Magalie Lesueur Jannoyer
International Center of Agricultural Research for Development
UPR HortSys
Martinique, France

Ulises Jáuregui-Haza
Department of Environment
Higher Institute of Technology and Applied Sciences
La Habana, Cuba

Catherine Jondreville
Department of Animal and Functionality of Animal Products
University of Lorraine
Vandoeuvre-lès-Nancy, France

Stefan Jurjanz
Department of Animal and Functionality of Animal Products
University of Lorraine
Vandoeuvre-lès-Nancy, France

Philippe Kadhel
Department of Obstetrics and Gynaecology
University Health Centre of Pointe à Pitre/Abymes
Guadeloupe, France
and
Department of Pediatry
University of Rennes I
Rennes, France

Marion Labeille
Asconit Consultants
Guadeloupe, France

Yoan Labrousse
Mediterranean Institute of Marine and Terrestrial Biodiversity and Ecology
French Institute of Research for Development
Martinique, France
and
University of Aix Marseille
Marseille, France

Sylvain Lerch
Department of Animal and Functionality of Animal Products
University of Lorraine
Vandoeuvre-lès-Nancy, France

Philippe Letourmy
International Center of Agricultural Research for Development
UPR Agroécologie et intensification durable des cultures annuelles
Montpellier, France

Hervé Macarie
Mediterranean Institute of Marine and Terrestrial Biodiversity and Ecology
French Institute of Research for Development
Martinique, France
and
University of Aix Marseille
Marseille, France

Mathilde Merlo
French Agency for Food, Environmental and Occupational Health Safety
Maisons-Alfort, France

Louise Meylan
International Center of Agricultural Research for Development
UPR HortSys
Le Lamentin, France

Dominique Monti
Dynecar Lab
University of French West Indies
Guadeloupe, France

Charles Mottes
International Center of Agricultural Research for Development
UPR HortSys
Martinique, France

Gina Muckle
University Health Centre of Quebec
Laval University
Québec City, Québec, Canada

Luc Multigner
Team of Epidemiological Research on Environment, Reproduction and Development
University of Rennes I
Rennes, France

Igor Novak
Charles Sturt University
Orange, New South Wales, Australia

Nady Passé-Coutrin
COVACHIM M2E Lab
University of French West Indies,
Guadeloupe, France

Joanne Plet
International Center of Agricultural Research for Development
UPR HortSys
Martinique, France

Guillaume Pompougnac
Regional Forum for Education and Promotion of Health
Guadeloupe, France

Luc Rangon
Mediterranean Instiute of Marine and terrestrial Biodiversity and Ecology
Aix-Marseille University
Martinique, France

Ronald Ranguin
COVACHIM M2E Lab
University of French West Indies
Guadeloupe, France

Philippe Rey
La Drôme Laboratoires
Valence, France

Lionel Reynal
French Research Institute for Exploitation of the Sea
Martinique, France

Serge Robert
French Research Institute for Exploitation of the Sea
L'Houmeau, France

Nicolas Rocle
Institut National de Recherche et Sciences et Technologies pour l'Environnement et l'Agriculture
Cestas, France

Florence Rouget
Team of Epidemiological Research on Environment, Reproduction and Development
Department of Pediatry
University of Rennes I
Rennes, France

Guido Rychen
French National Institute for Agricultural Research
Department of Animal and Functionality of Animal Products
University of Lorraine
Vandoeuvre-lès-Nancy, France

Isabel Sastre-Conde
SEMILLA
Illes Balears, Spain

Alain Soler
International Center of Agricultural Research for Development
UPR GECO
Martinique, France

Anne-Lise Taïlamé
The French Geological Survey
Martinique, France

Jean-Pierre Thomé
Animal Ecology and Ecotoxicology Lab
University of Liege
Sart-Tilman, Belgique

Jean-Philippe Tonneau
International Center of Agricultural Research for Development
UMR Tetis
Montpellier, France

Jean-Luc Volatier
French Agency for Food, Environmental and Occupational Health Safety
Maisons-Alfort, France

Marc Voltz
Soil-Agrosystem-Hydrosystem Interactions Lab
French National Institute for Agricultural Research
Montpellier, France

Thierry Woignier
Mediterranean Institute of Marine and Terrestrial Biodiversity and Ecology
Aix-Marseille University
Marseille, France
and
French Institute of Research for Development
Martinique, France

Section I

Introduction

1 Chlordecone Case Study in the French West Indies

Magalie Lesueur Jannoyer, Florence Clostre, Thierry Woignier, and Philippe Cattan

CONTENTS

1.1 PREAMBLE

In 2000, people of the French West Indies (FWI) discovered, through analyses of spring water, that they had been water that contained chlordecone, an organochlorine pesticide used in banana plantations for tens of years. This was the beginning of a major environmental crisis in the FWI, leading to serious social, economic, and agronomic impacts. This book traces how a collective response emerged through the characterization of pollution, the development of awareness of this pollutant and its effects on human health and environment, and the action plan to manage this pollution. Finally, the issue questions our readiness to manage future threats arising out of human activities.

1.2 GENERAL INTRODUCTION

Chlordecone pollution is now considered an environmental pollution even if it originated from agricultural practices, quite unlike the industrial pollution. This long-term diffuse pollution (with some hot spots) concerns about 20,000 ha in both Guadeloupe and Martinique islands and slowly contaminates all environmental and biological compartments. At this point, the exact polluted areas are still not well known, and the fate and impacts of this molecule on our tropical humid and volcanic context have to be assessed as no reference exists as to when the general pollution was revealed.

1.3 THE FRENCH WEST INDIES LOCATION AND CONTEXT

1.3.1 Geography

Given their latitude, the FWI islands are located in the tropical region, just south of the Tropic of Cancer in the eastern Caribbean islands and the Lesser Antilles (Figure 1.1) (Guadeloupe: latitude 16°15′, longitude: −61°35′; Martinique: latitude: 14°65′ longitude: −61°). These small islands (area of 1628 km^2 for Guadeloupe and 1128 km^2 for Martinique) were volcanically formed and thus have a hilly relief with steep inclines and specific soils, such as nitisol, ferrisol, and andosol (FAO-ISRIC-ISSS 1998). The primary minerals of volcanic rocks are weathered, and these soils contain specific clays: halloysite for nitisol, halloysite, and Fe-oxyhydroxides for ferralsol, and allophane for andosol (Brunet et al. 2009,

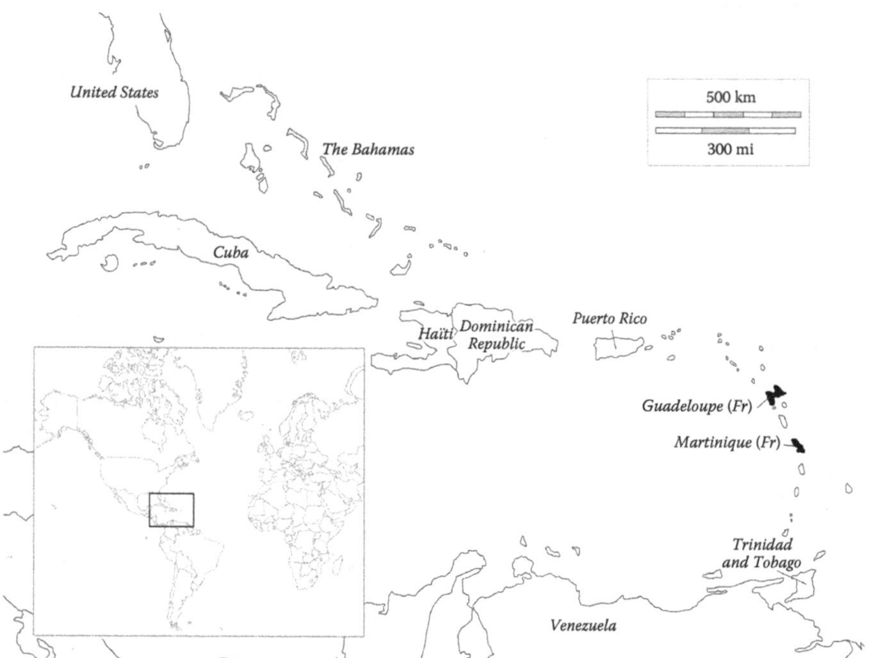

FIGURE 1.1 Map of the Lesser Antilles with Guadeloupe and Martinique in dark. (Adapted from Free Vector Maps, Map of the World—Single Color. WRLD-EPS-01-0006, https://freevectormaps.com/world-maps/WRLD-EPS-01-0006?ref=more_map, 2012; Free Vector Map, Map of the Caribbean Islands with Countries—Single Color. WRLD-CI-01-0001, https://freevectormaps.com/world-maps/caribbean/WRLD-CI-01-0001, 2014.)

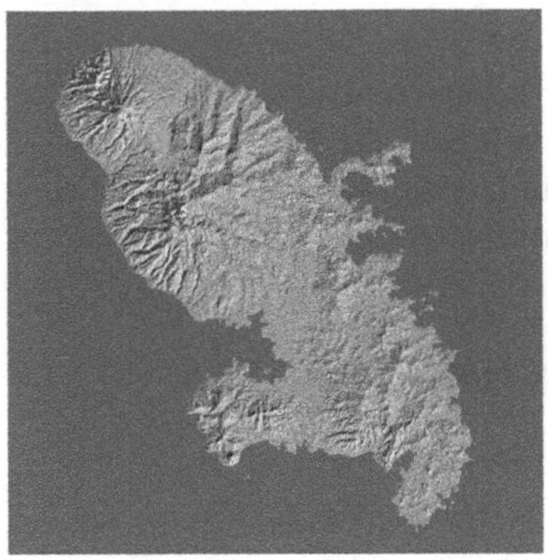
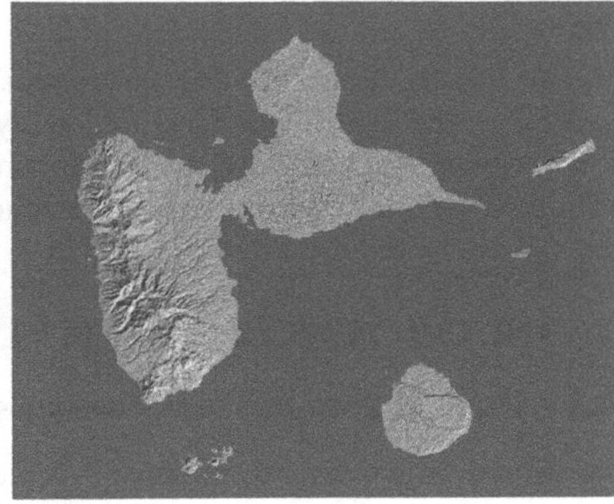

FIGURE 1.2 Elevation map for Guadeloupe and Martinique from NASA Worldview (Courtesy of Global Digital Elevation Map, Terra/Aster).

Cabidoche et al. 2009). The soils have high capacity of infiltration and carbon retention (Chevallier et al. 2010, Dorel et al. 2000, Feller et al. 2001).

The climate is tropical humid, with a mean temperature of 26°C all year long and rainfall between 600 mm to 5400 mm on the higher points (Figure 1.2), marked by a rainy season from June to October (hurricane season) and a dry season from February to April (Météo-France 2015a; Météo-France 2015b).

In 2014, the population was 381,326 in Martinique and 403,750 in Guadeloupe, thus leading to a high population density in the flat and coastal areas (more than 400 inhabitants per km^2).

1.3.2 Economy

The economy is mainly based on service sectors (tourism and administrations) and agriculture. In 2010, the utilized agricultural land was about 24,975 ha in Martinique (22% of the island area) and 31,768 ha in Guadeloupe (20% of the island area) (DAAF Martinique 2011). Agriculture employs 12% of the active population in Guadeloupe and Martinique and contributes to 6% of the gross regional product.

The main crops are banana, sugarcane, root and tubers, and tropical fruits (pineapple, citrus, etc.). Sugarcane, for sugar and rum, and banana, for exportation, are the major crops. In Martinique, banana accounts for 25% of utilized agricultural land, whereas in Guadeloupe, it accounts for 10% of utilized agricultural land (Ministère de l'Agriculture 2011). Banana farms are more specialized for exports to Europe in Martinique, whereas farms with banana cropping systems are diversified in Guadeloupe.

Even though the agricultural sector is accorded priority, the islands are not self-sufficient in terms of food and thus import products such as fruits, vegetables, and tubers to meet more than 50% of their needs; in 2012, Martinique imported 25,000 tons of these products to meet 60% of its needs.

1.3.3 Policy

The two islands have been French overseas departments since 1946. As such, Guadeloupe and Martinique have to comply with both the European Union and French regulations, regarding health (Maximum Limit of Residue), environmental (Water Framework Directive), and socioeconomic (salaries, taxes, etc.) issues.

1.4 THE CHLORDECONE MOLECULE CHARACTERISTICS

1.4.1 Physical Structure and Properties

Chlordecone is an organochlorine molecule made of 10 atoms of carbon and chlorine, and a single oxygen atom (Figure 1.3). The chemical name of chlordecone is 1,1a,3,3a,4,5,5,5a,5b,6-decachlorooctahydro-1,3,4-metheno-2H-cyclobuta[cd] pentalen-2-one according to CAS nomenclature (Dolfing et al. 2012, EPA 2009). It is also known as Kepone®. Its CAS number is 143-50-0 (EPI 2012, EPA 2009).

Chlordecone molecule has a homocubane configuration (Figure 1.3) with high steric hindrance, thus a sort of chemical cage of 0.53–0.65 nm size (Durimel et al. 2013). Due to this peculiar structure, chlordecone molecule has a high stability. The chemical properties are given in Table 1.1, with data from ATSDR (1995) and Risk Assessment Information System (2013).

1.4.2 Toxicological Effects on Animals

The toxicity of chlordecone on the liver and the neuromuscular, endocrine, and reproductive systems has been experimentally ascertained for a range of animals (Guzelian 1982). Moreover, it depressed growth and acted as an endocrine disruptor and a liver carcinogen in rats and mice (Hammond et al. 1979, Larson et al. 1979). On oral lethal dose, at 50%, its values ranged from 71 mg kg^{-1} for male rabbits to 250 mg kg^{-1} for dogs (Larson et al. 1979). Moreover, chlordecone induced various adverse effects (reduced growth, fecundity and fertility, external of poisoning) in fish with the first effects induced for exposure to 0.1 μg L^{-1} of water (Goodman et al. 1982).

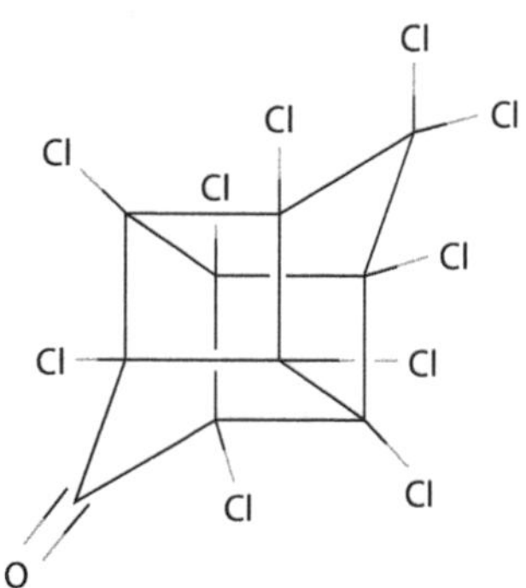

FIGURE 1.3 Structure of the chlordecone molecule, CAS 143-50-0.

TABLE 1.1
Chlordecone Chemical Properties, K_{ow} Octanol-Water Partition Coefficient and K_{oc} Soil Organic Carbon-Water Partitioning Coefficient

Chlordecone Properties	Value	References
Molecular mass M, g mol^{-1}	490.64	EPI (2012)
Molecular surface ($Å^2$)	326	Le Cloirec (2011)
Molecular volume ($Å^3$)	264	Le Cloirec (2011)
Decomposition temperature (°C)	350	Le Cloirec (2011)
Solubility in water S_{water}, mg L^{-1}	2.70–3.00	Kenaga (1980)
Vapor pressure VP, mm Hg	0.225 10^{-6}	IARC (1979)
Henry constant, atm m^3 mol^{-1}	2.2×10^{-6}–2.5×10^{-8}	EPI, Howard (1991)
Log K_{ow}	4.5–5.41	Howard (1991), RAIS (2013)
K_{oc}, L kg^{-1}	2,400–17,500	Kenaga (1980), EPI
Log K_{oc}	3.38–3.415	Howard (1991)
Melting point, °C	350–371	IARC (1979), EPI

Nimmo et al. (1977) reported reduced growth of aquatic invertebrates at similar concentrations.

Bioaccumulation occurs in aquatic and terrestrial environment, especially in animals (Bertrand et al. 2009, Coat et al. 2011, Jondreville et al. 2014). Chlordecone accumulated in estuarine food chains from very low concentrations in water (0.023 µg L^{-1}; Bahner et al. 1977).

1.4.3 Toxicological Effects on Human

The acute toxicity of chlordecone exposure to humans was demonstrated in 1975 following a poisoning episode involving chlordecone plant workers at Hopewell, Virginia (Cannon et al. 1978, Epstein 1978). The exposed workers showed evidence of toxicity, especially with neurological signs—tremor, pleuritic pain, visual disturbances—and reduced quality of sperm (Cannon et al. 1975, Epstein 1978, Taylor 1982). It is also potentially carcinogenic (Clere et al. 2012). In addition, its long-term effects on human health and child development linked to the consumption of polluted food and water are now a serious concern (Boucher et al. 2013, Dallaire et al. 2012, Multigner et al. 2010).

1.5 CHLORDECONE USES

1.5.1 The Kepone® and Curlone® Products

Chlordecone was produced in 1951, patented in 1952, and introduced commercially in the United States by Allied Chemical in 1958 under the trade names Kepone and GC-1189 (Epstein 1978; Figure 1.4). It had both agricultural and industrial uses. In

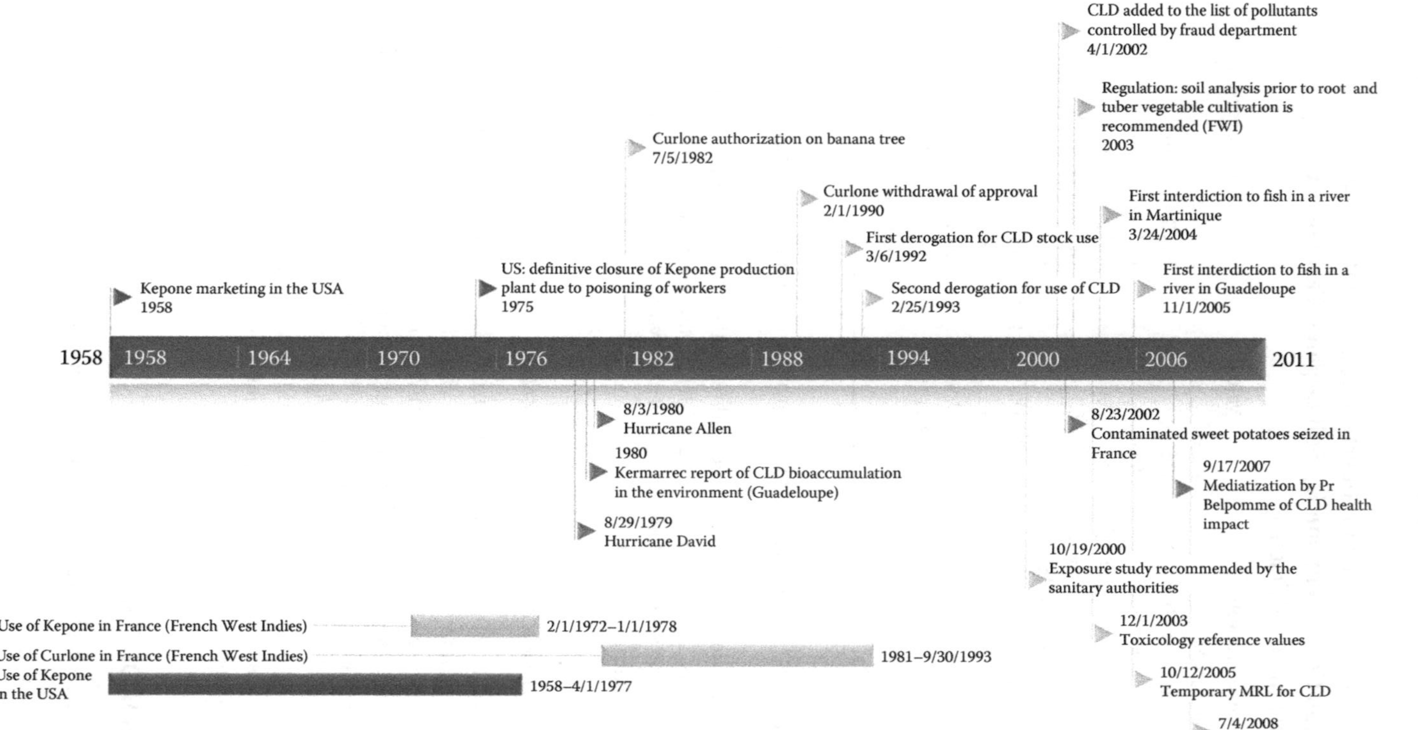

FIGURE 1.4 (See color insert.) Timeline of chlordecone (CLD) use and pollution management.

the plastic industry, it has been used as a fire retardant. In agriculture, it has been used as an insecticide in ant and roach baits in the United States, to fight against mole crickets in Florida, and against weevil in banana in the Caribbean and African countries (Epstein 1978). It has also been converted to Kelevan for use on Colorado potato beetles in eastern European countries (Epstein 1978). In the FWI, the commercial names were first Kepone from Allied Chemical and then Curlone from Lagarrigue Ets (Figure 1.4). These two products contained 5% chlordecone and were presented in the form of white powder.

Chlordecone pollution occurs in many parts of the world (Luellen et al. 2006, Roche et al. 2011, Wei et al. 2006). It is a hydrophobic, nonbiodegradable pesticide that is strongly bound to the organic soil matrix, which is why soils are still contaminated today. Chlordecone is now on the list of persistent organic pollutants prohibited by the Stockholm Convention.

1.5.2 Chlordecone Use in Guadeloupe and Martinique Islands: Practices on Banana Cropping System

As banana tree needs water to grow, banana cropping systems are developed on the humid side of the islands, which is under the influence of trade winds. High temperature and humidity conditions lead to high phytosanitary pressure. The black weevil (*Cosmopolites sordidus*) has been posing a serious problem for the last 30 years. Its larvae attack banana corm and destroy the plant. Damages lead to a high loss of yield and decrease in the sustainability of the banana plantation.

In Guadeloupe and Martinique, chlordecone was used in banana cropping systems from 1972 to 1993 in place of technical-grade lindane (γHCH). Banana fields were theoretically treated according to weevil infestation, which was assessed after a phytosanitary inspection of the field and a count of black weevil larvae in the corm. But in practice, systematic treatments were applied only once a year. The recommended dose was 30 g Kepone/Curlone per banana tree (i.e., 3–5 kg active matter per hectare). This practice led to a diffuse pollution of soil. Chlordecone powder was applied on the soil and around banana trees, in a circle of 30–50 cm width. This practice led to a heterogeneous pollution at the field scale as the treatment was not uniform on the field (Clostre et al. 2014). Some citrus orchards could also have been treated by chlordecone as it was effective against citrus weevil (*Diaprepes* spp.) (Cabidoche and Lesueur Jannoyer 2011). As the citrus areas are limited, in this book we consider banana fields, which are the major contributors of environmental pollution.

1.6 HISTORICAL ELEMENTS OF CHLORDECONE USE AND POLLUTION MANAGEMENT

At the end of the 1970s, Snegaroff (1977) and Kermarrec (1980) noticed contamination of rivers and bioaccumulation in ecosystems, which happened after the use of chlordecone was banned in the United States (Figure 1.4), and it unfortunately coincided with the beginning of a massive use of chlordecone in the FWI until 1993, with a series of authorization and derogations of use. Finally, it was only more than 20 years after Snegaroff's study that the issue had been considered. The revelation of chlordecone pollution began in the early 2000s (see Chapter 2) with a peak media coverage in 2007, which resulted in crisis management followed by a series of measures and plans (Figure 1.4) to tackle it.

Numerous studies were then funded by public administration and funding programs. These studies showed a large chlordecone contamination of groundwater (see Chapter 5), fish, and crustaceans in rivers (Coat et al. 2011; Chapters 6 and 7), and marine fauna (Bertrand et al. 2009; Chapter 8). The studies dealt with (1) the level of soil pollution, which depends on three main determinants: the organic carbon, the clay microstructure, and agricultural practices; and (2) the effects of the pollutant on the ecosystems (Woignier et al. 2015; Chapters 3 and 9 through 11).

This book highlights the contribution of research to diagnoses and management of global pollution situation at the beginning of 2000s. Hence, we attempt to answer some key questions: What is pollution? Where? For how long? How to measure it? What are its impacts? What to do? How to manage it? This book presents an overview of crisis emergence and then it tackles six issues that are of high interest in the management of chronic pollution:

1. The pollution assessment with regard to natural resources
2. The contamination of aquatic ecosystems
3. Transfer of pollutants from soil to different environmental compartments (water, crops, and animals)
4. The exposure and impacts on human health
5. The possible remediation lines
6. The management approach

Some aspects such as analytical progresses (lab investments) (Brunet et al. 2009, Soler et al. 2014) and social analyses (Joly 2010, Ferdinand 2015) have not been directly addressed in this book but are key factors for environmental crisis monitoring and management.

We hope that our book will be helpful to guide students, researchers, and decision makers in the assessment and management of chronic pollution.

REFERENCES

ATSDR, 1995. Toxicological profile for Mirex and Chlordecone. U.S. Department of Health and Human Services, Public Health Service, Agency for Toxic Substances and Disease Registry, Atlanta, GA, 362pp.

Bahner LH, Wilson AJ, Sheppard JM, Patrick JM, Goodman LR, Walsh GE, 1977. Kepone bioconcentration, accumulation, loss, and transfer through estuarine food chains. *Chesapeake Science* 18:299–308.

Bertrand JA, Abarnou A, Bocquené G, Chiffoleau JF, Reynal L, 2009. Diagnostic de la contamination chimique de la faune halieutique des littoraux des Antilles françaises. Campagnes 2008 en Martinique et en Guadeloupe. Ifremer, Martinique, France, 136pp. http://www.ifremer.fr/docelec/doc/2009/rapport-6896.pdf. Accessed May 2015.

Boucher O, Simard M-N, Muckle G, Rouget F, Kadhel P, Bataille H et al., 2013. Exposure to an organochlorine pesticide (chlordecone) and development of 18-month-old infants. *Neurotoxicology* 35:162–168.

Brunet D, Woignier T, Lesueur-Jannoyer M, Rangon L, Achard R, Barthès BG, 2009. Determination of soil content in chlordecone (organochlorine pesticide) using near infrared reflectance spectroscopy (NIRS). *Environmental Pollution* 157:3120–3125.

Cabidoche YM, Achard R, Cattan P, Clermont-Dauphin C, Massat F, Sansoulet J, 2009. Long-term pollution by chlordecone of tropical volcanic soils in the French West Indies: A simple leaching model accounts for current residue. *Environmental Pollution* 157(5):1697–1705.

Cabidoche YM, Lesueur Jannoyer M, 2011. Pollution durable des sols par la chlordécone aux Antilles: Comment la gérer? *Innovations Agronomiques* 16(2011):117–133.

Cannon SB, Veazey JM, Jr., Jackson RS, Burse VW, Hayes C, Straub WE et al., 1978. Epidemic kepone poisoning in chemical workers. *American Journal Epidemiology* 107:529–537.

Chevallier T, Woignier T, Toucet J, Blanchart E, 2010. Carbon sequestration in the fractal porosity of Andosols. *Geoderma* 159:182–188.

Clere N, Lauret E, Malthiery Y, Andriantsitohaina R, Faure S, 2012. Estrogen receptor alpha as a key target of organochlorines to promote angiogenesis. *Angiogenesis* 15(4):1–16.

Clostre F, Lesueur-Jannoyer M, Achard R, Letourmy P, Cabidoche YM, Cattan P, 2014. Decision support tool for soil sampling of heterogeneous pesticide (chlordecone) pollution. *Environmental Science and Pollution Research* 21(3):1980–1992.

Coat S, Monti D, Legendre P, Bouchon C, Massat F, Lepoint G, 2011. Organochlorine pollution in tropical rivers (Guadeloupe): Role of ecological factors in food web bioaccumulation. *Environmental Pollution* 159(6):1692–1701.

DAAF Martinique, 2011. Mémento de la statistique agricole. In *Agreste. Fort-de-France.* Direction de l'alimentation de l'agriculture et de la forêt de la Martinique, Martinique.

Dallaire R, Muckle G, Rouget F, Kadhel P, Bataille H, Guldner L et al., 2012. Cognitive, visual, and motor development of 7-month-old Guadeloupean infants exposed to chlordecone. *Environmental Research* 118:79–85.

Dolfing J, Novak I, Archelas A, Macarie H, 2012. Gibbs free energy of formation of chlordecone and potential degradation products: Implications for remediation strategies and environmental fate. *Environmental Science & Technology* 46(15):8131–8139. doi: 10.1021/es301165p.

Dorel M, Roger-Estrade J, Manichon H, Delvaux B, 2000. Porosity and soil water properties of Caribbean volcanic ash soils. *Soil Use and Management* 16(2):133–140.

Durimel A, Altenor S, Miranda-Quintana R, Couespel Du Mesni P, Jauregui-Haza U, Gadiou R, Gaspard S, 2013. pH dependence of chlordecone adsorption on activated carbons and role of adsorbent physico-chemical properties. *Chemical Engineering Journal* 229:239–249.

EPA, 2009. Toxicological review of chlordecone (KEPONE). EPA/635/R-07/004F, 119pp. + appendixes. U.S. Environmental Protection Agency, Washington, DC. www.epa.gov/iris. Accessed May 2015.

EPI, 2012. Estimation Programs Interface Suite™ for Microsoft® Windows v4.1. U.S. EPI, Washington, DC.

Epstein SS, 1978. Kepone—Hazard evaluation. *The Science of the Total Environment* 9:1–62.

FAO-ISRIC-ISSS (Food and Agriculture Organization of the United Nations, International Soil Reference and Information Centre, International Society for Soil Science), 1998. *World Reference Base for Soil Resources.* FAO, Rome, Italy.

Feller C, Albrecht A, Blanchart E, Cabidoche YM, Chevallier T, Hartmann C, Eschenbrenner V, Larre-Larrouy MC, Ndandou JF, 2001. Soil organic carbon sequestration in tropical areas. General considerations and analysis of some edaphic determinants for Lesser Antilles soils. *Nutrient Cycling in Agroecosystems* 61(1–2):19–31.

Ferdinand M, 2015. De l'usage du chlordécone en Martinique et en Guadeloupe: l'égalité en question. *Revue française des affaires sociales* 1(1–2):163–183.

Free Vector Maps. 2012. Map of the World—Single Color. WRLD-EPS-01-0006. https://freevectormaps.com/world-maps/WRLD-EPS-01-0006?ref=more_map. Accessed July 2016.

Free Vector Map. 2014. Map of the Caribbean Islands with Countries—Single Color. WRLD-CI-01-0001. https://freevectormaps.com/world-maps/caribbean/WRLD-CI-01-0001. Accessed July 2016.

Goodman LR, Hansen DJ, Manning CS, Faas LF, 1982. Effects of Kepone® on the sheepshead minnow in an entire life-cycle toxicity test. *Archives of Environmental Contamination and Toxicology* 11(3):335–342.

Guzelian PS, 1982. Comparative toxicology of chlordecone (Kepone) in humans and experimental-animals. *Annual Review of Pharmacology and Toxicology* 22:89–113.

Hammond B, Katzenellenbogen BS, Krauthammer N, McConnell J, 1979. Estrogenic activity of the insecticide chlordecone (Kepone) and interaction with uterine estrogen receptors. *Proceedings of the National Academy of Sciences of the United States of America* 76(12):6641–6645.

Howard PH, Michalenko EM, Sage GW, Basu DK, Hill A, Aronson D (eds.), 1981. *Handbook of Environmental Fate and Exposure Data for Organic Chemicals.* Lewis Publishers, New York, pp. 110–118.

IARC (International Agency for Research on Cancer), 1979. Chlordecone. *IARC Monographs on the Evaluation of Carcinogenic Risk of Chemicals to Humans* 20:67–81.

Joly P-B, juillet 2010. La saga du chlordécone aux Antilles françaises. Reconstruction chronologique 1968–2008. Inra Science en Société, Paris, France, 82pp. https://www.anses.fr/fr/system/files/SHS2010etInracol01Ra.pdf. Accessed May 2015.

Jondreville C, Lavigne A, Jurjanz S, Dalibard C, Liabeuf JM, Clostre F, Lesueur-Jannoyer M, 2014. Contamination of free-range ducks by chlordecone in Martinique (French West Indies): A field study. *Science of the Total Environment* 493:336–341.

Kenaga EE, 1980. Predicted bioconcentration factors and soil sorption coefficients of pesticides and other chemicals. *Ecotoxicology and Environmental Safety* 4:26–38.

Kermarrec A, 1980. Niveau actuel de la contamination des chaînes biologiques en Guadeloupe: Pesticides et metaux lourds 1979–1980. INRA Antilles-Guyane, Petit-Bourg, Guadeloupe, France.

Larson PS, Egle JL, Jr., Hennigar GR, Lane RW, Borzelleca JF, 1979. Acute, subchronic, and chronic toxicity of chlordecone. *Toxicology and Applied Pharmacology* 48(1, Part 1): 29–41.

Le Cloirec P, 2011. Traitement des eaux—Elimination de la Chlordecone par adsorption sur charbon actif. Etat des lieux et prospectives. *Les Cahiers du PRAM* 9–10:17–19.

Lemarchand J (2016a) http://johan.lemarchand.free.fr/cartes/monde/carteileguadeloupe03.jpg. Accessed April 2016.

Lemarchand J (2016b) http://johan.lemarchand.free.fr/cartes/monde./carteilemartinique05.jpg. Accessed April 2016.

Luellen DR, Vadas GG, Unger MA, 2006. Kepone in James River fish: 1976–2002. *Science of the Total Environment* 358(1–3):286–297.

Météo-France, 2015a. *Le climat en Guadeloupe* [cited 12/26/2015 2015]. http://www.meteofrance.gp/documents/3714888/5579049/Climat971_2pages.pdf/ae75b805-71d7-46de-99de-6f1e39b3068f. Accessed May 2015.

Météo-France, 2015b. *Le climat en Martinique 2014* [cited 12/26/2015 2015]. http://www.meteofrance.gp/documents/3714888/5579049/climat972_2pages.pdf/1bb26ab1-630b-4fd4-9757-9adb5da948a6. Accessed May 2015.

Ministère de l'Agriculture, 2011. La banane en Guadeloupe et en Martinique. *Agreste Primeur* 262:1–4.

Multigner L, Ndong JR, Giusti A, Romana M, Delacroix-Maillard H, Cordier S et al., 2010. Chlordecone exposure and risk of prostate cancer. *Journal of Clinical Oncology* 28(21):3457–3462.

Nimmo DR, Bahner LH, Rigby RA, Sheppard JM, Wilson AJ, Jr., 1977. Mysidopsis bahia: An estuarine species suitable for life cycle toxicity tests to determine the effects of a pollutant. In *Aquatic Toxicology and Hazard Evaluation*, ASTM STP 634, Mayer FL, Hamelink JL (eds.). American Society for Testing and Materials, West Conshohocken, PA, p. 109.

Risk Assessment Information System (RAIS), 2013. Toxicity and properties. The University of Tennessee, Knoxville, TN. http://rais.ornl.gov/cgi-bin/tools/TOX_search. Accessed May 2015.

Roche H, Salvat B, Ramade F, 2011. Assessment of the pesticides pollution of coral reefs communities from French Polynesia. *Revue d'écologie* 66(1):3–10.

Snegaroff J, 1977. Les résidus d'insecticides organochlorés dans les sols et les rivières de la région bananière de Guadeloupe. *Phytiatrie-Phytopharmacie* 26:251–268.

Soler A, Lebrun M, Labrousse Y, Woignier T, 2014. Solid-phase microextraction and gas chromatography–mass spectrometry for quantitative determination of chlordecone in water, plant and soil samples. *Fruits* 69:325–339. doi: 10.1051/fruits/2014020.

Taylor JR, 1982. Neurological manifestations in humans exposed to chlordecone and follow-up results. *Neurotoxicology* 3:9–16.

Wei S, Lau RKF, Fung CN, Zheng GJ, Lam JCW, Connell DW et al., 2006. Trace organic contamination in biota collected from the Pearl River Estuary, China: A preliminary risk assessment. *Marine Pollution Bulletin* 52(12):1682–1694.

Woignier T, Clostre F, Cattan P, Lesueur-Jannoyer M, 2015. Pollution of soils and ecosystems by a permanent toxic organochlorine pesticide: Chlordecone—Numerical simulation of allophane nanoclay microstructure and calculation of its transport properties. *AIMS Environmental Science* 2(3):494–510.

2 From Controversy to Pollution Assessment

A Chronicle of the Chlordecone Crisis Management

Eric Godard and Simon Bellec

CONTENTS

2.1 INTRODUCTION

In the 1990s in Martinique, health and environmental issues were being hotly debated among state services and environmental organizations. The latter accused the local government services responsible for sanitary control of drinking water of not recognizing the issue of pesticide pollution of drinking water and felt that the situation was very alarming.

The Direction of Sanitary and Social Affairs of Martinique (DDASS) (Departmental Direction of Health and Social Development 2001–2010 and since 2010 the Health Regional Agency, ARS) set up sanitary controls in 1991 for pesticide molecules, applying the European Framework of 1975 and 1980, on the abstraction and distribution of drinking water. The analyses revealed only very low and brief pollutions. Nevertheless, the control modalities could have been improved: A DDASS survey conducted in 1996 identified 180 pesticide molecules that were used in Martinique, out of which only 21 were analyzed by the designated laboratory. Since July 1997, 39 new active molecules have been added to the list, increasing the number of searched molecules to 76. Despite these improvements, in May 1999, only nine values were over the regulation threshold of 0.1 $\mu g\ L^{-1}$, with a maximum of 0.26 $\mu g\ L^{-1}$ for trichlorfon (SISE-EAUX database and water quality report of DDASS Martinique). This recurrent debate led to a fact-finding mission of Ministries of the Agriculture and the Environment (Balland et al. 1998). The mission reported an alarming situation: the current state of environmental pollution of drinking water was unreliable and too incomplete; no data were available on the quality of local food; health impacts were not well known for population suspected to be exposed more to pesticides as compared to that of France; the regulation was not complied with; pesticide users were not skilled; and pesticide transfer processes and pollutant diffusion were not well assessed.

Following the release of this report, DDASS of Martinique has been working for 3 years to improve awareness on the presence of pesticides in drinking water and proposed actions to better assess the related risks. In this chapter, we discuss this approach and also the results of our studies.

2.2 POLLUTION OF THE SUBSCRIPTION OF DRINKING WATER BY ORGANOCHLORINE PESTICIDES (1999)

In 1999, the change of analytical laboratory improved the number of searched molecules to 220, out of which 120 were used in Martinique. The La Drôme Departmental Laboratory (LDA26), responsible for this new campaign of analyses, was known for its performance and ability to adapt to the detection of new molecules. Moreover samples were delivered within 48 h (temperature below 10°C).

A campaign was conducted on seven sampling sites for 3 months (from June to August 1999), with weekly samples of raw water, treated water, and sediment sludge on five catchments of surface water, and two weekly samples of ground water collected in wells. The Gradis spring, located in the Northern part of Martinique, was closed since the first results were known because of a very high concentration of βHCH (Table 2.1) and bromacil (used for weeding in pineapple cropping systems).

TABLE 2.1
Detected Molecules in the Water of Gradis Spring

Molecule Name	Number of Measurements	Number of Detections	Number of Detections >0.1 μg L^{-1}	Percentage Detection	Mean Value (μg L^{-1})	Maximum Value (μg L^{-1})
Chlordecone	7	7	7	100	0.47	1.37
βHCH	7	7	7	100	2.12	3.30
Bromacil	7	7	7	100	1.46	1.90
αHCH	7	1	0	14	0.004	0.029

The laboratory also detected a new peak close to HCH in the chromatograms. The molecule related to this peak was chlordecone, but it was quantified only in September 1999 because of difficulties with analysis: no standard was available at the laboratory, with no indication of the molecule type. Twenty-two detections concerning four molecules (chlordecone, bromacil, βHCH, and αHCH) were made, and contents were over the regulatory threshold of 0.1 μg L^{-1} for the first three molecules.

The Interregional Epidemiology Cell (CIRE), the local branch of the French Institute of Health Watch (INVS), was then immediately approached to assess the risks and threshold limits in drinking water for chlordecone and βHCH. They first assessed the contribution to tolerable daily intake (TDI) brought by drinking water according to minimal risk levels of the Agency for Toxic Substances and Disease Registry (ATSDR). Simulations were made using mean and maximum values of the molecule water content (Table 2.2).

TABLE 2.2
Risk Assessment of Gradis Spring—Chlordecone and βHCH Data, Calculated on the Basis on the Maximal Concentration

				Dose Brought by Water Consumption (μg)		Part of TDI Brought by Water (%)	
Molecule	Population	Chronic TDI (μg day^{-1})	Raw Water Consumption (L)	Mean Dose	Max. Dose	Mean Dose	Max. Dose
Chlordecone	Infant	2.5	0.75	0.35	1.42	14.0	57.0
	Child	5	1	0.47	1.90	9.3	38.0
	Adult	30	2	0.94	3.80	3.1	12.7
βHCH	Infant	3	0.75	1.59	2.47	52.9	82.5
	Child	6	1	2.12	3.30	35.3	55.0
	Adult	36	2	4.24	6.60	11.8	18.3

Note: TDI, tolerable daily intake.

The contribution of drinking water to TDI was above the 10% value (all exposures) calculated with mean values of chlordecone and βHCH for infants.

Chlordecone pollution was also revealed for two other water resources: the Monsieur and the Capot Rivers with mean chlordecone content in water of 0.3 μg L^{-1}. Conversely to the case of the Gradis spring, the sanitary risk assessment based on the calculation of the TDI in the worst case (infants with maximum pollutant contents in water), the contribution was less than 10% for chlordecone (4% for βHCH) for the Capot River and 15.3% (11% for βHCH) for the Monsieur River. Using the mean values led to a contribution of 7.3% (chlordecone) and 2.8% (βHCH) for the Monsieur River. Thus, the operators of these water treatment plants maintained their activities without any restriction for these two resources, according to the planned exemption scheme of the drinking water quality (98/83/CE framework, November 3, 1998) (DDASS 1999).

The overall results of 2 years of monitoring (1998 and 1999) of drinking water quality showed the detection of 28 pesticide molecules in all water resources, out of which 15 were banned. Among these 15 banned molecules, 14 were organochlorine compounds. This discovery of water pollution due to historical use ended the acrimonious debate between health services and associations: after years of denial, their alerts were confirmed by public authorities and in October 1999, a press conference was held to educate the population about them. In addition, the Hygiene Board of the Department (CDH) recommendated to intensify controls and also to deepen the risk assessment of dietary intake of pesticides (food contamination).

2.2.1 Extension of the Characterization of Water Pollution (2000)

In June and July 2000, a daily sampling campaign was held on two sub-watersheds supporting mainly agricultural activities, in the Lezarde and Capot Rivers, which was intended to better characterize the pollution close to treated fields and to assess the effect of nematicide inputs. At the same time, raw water samples were taken at the extraction site in the Capot River (Vivé water treatment plant) with the aim to study the relationship between flow rate and pesticide concentration. A search for aldicarb and its metabolites was launched and results showed detections over the regulation thresholds at the outlet of fields, which had been treated 4 months earlier, and in the raw water of the Capot River. The parent compound had not been detected before. Shortly after publication of these results, the use of aldicarb was banned in Martinique.

2.2.2 Review of the Data and Knowledge on Drinking Water in October 2001 in Martinique

Since January 1999, 36 molecules have been detected in drinking water. Detections were not so frequent and were with low levels: Few molecules exceeded the regulatory threshold of 0.1 μg L^{-1}, and maximum values were not so high for the majority of them less than 0.5 μg L^{-1} (Figure 2.1).

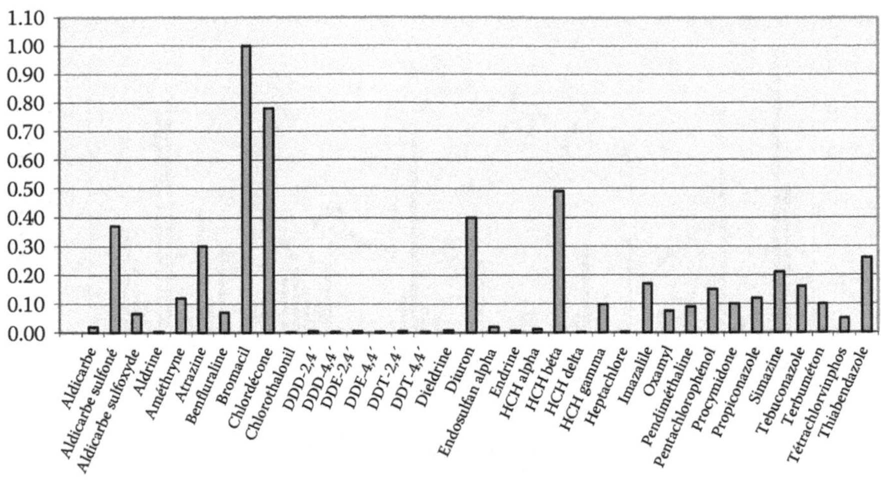

FIGURE 2.1 Maximum pesticide content (μg L^{-1}) in water from January 1999 to June 2001.

Organochlorine compounds from former uses (chlordecone and βHCH, isomer of technical HCH) chronically pollute three water resources: the Gradis spring and Capot and Monsieur Rivers. Contents can exceed 3 μg L^{-1} at the Gradis spring (data not shown), but they are below 0.8 μg L^{-1} for the other two.

For molecules still in use, only 12 molecules exceeded the regulatory threshold of 0.1 μg L^{-1}. The frequency of detection over this limit is low, except for the carbamate derivatives (aldicard). All the maximum contents are below 0.5 μg L^{-1}, except for bromacil that had values close to 1 μg L^{-1}.

2.2.3 Hypotheses on the Origin of Chlordecone and HCH Pollutions

Banned respectively since 1987 and 1993, technical HCH and chlordecone are not in use theoretically and pollution may result from residues still present in soils. This was confirmed by the analysis of a soil sample close to the Gradis spring in August 1999, with contents of 5 and 0.03 mg kg^{-1} dry soil, respectively, for chlordecone and βHCH.

The relationship between pollutant content in water and rainfall showed a decrease in pollutant content after rainfall higher than 30 mm, an increase related to rainfall between 10 and 20 mm, and no variation when low daily rainfall (Figure 2.2). The effect of intense rainfall occured 3 days after. This dilution effect by intense rainfall and the predominance of leaching processes in the diffusion of chlordecone to watercourses were confirmed by recent work conducted in Guadeloupe, in Perou watershed (Crabit et al. 2014).

These data confirmed that the pollution dynamics were primarily due to water infiltration through polluted soils that had received these persistent organochlorine compounds for a long time.

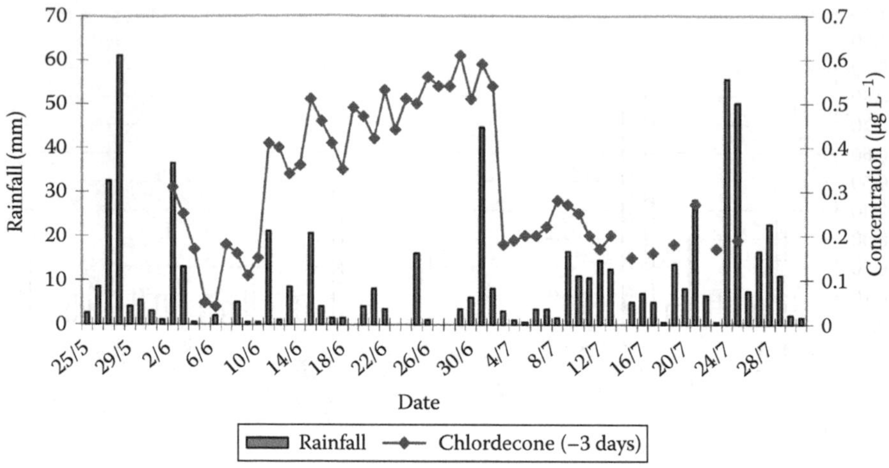

FIGURE 2.2 Evolution of chlordecone content (μg L^{-1}) in Capot raw water (–3 days) according to daily rainfall.

2.2.4 Persistence of Organochlorine Compounds in Soil of Banana Fields and Transfer Risks

Literature data also support pollutant storage in soils and transfer risk for crops:

- In Guadeloupe, Snegaroff (1977) demonstrated environmental pollution (soil, waters, sediment) by organochlorine compounds such as HCH isomers and chlordecone. He also brought the matter of root and tuber crops susceptible to exceed the regulatory thresholds. Nevertheless, a sample of an eggplant grown on a banana field did not reveal a high contamination.
- Soil analyses made by the services of the Departmental Direction of Competition, Consumer Affairs and Prevention of Fraud's (DGCCRF) laboratory showed chlordecone and βHCH levels close to 4 mg kg^{-1} near the polluted Basse-Terre River catchment sites (March 2000).

These organochlorine compounds are still detected in soil and water compartments, even though their use has been banned for more than 10 years, thus bringing to light their persistence and exposure of the population through channels other than drinking water.

2.2.5 Proposal of Sanitary Position to the Hygiene Board of the Department

The proposal of sanitary position (DDASS 2000, written by doctors and engineers from DDASS and CIRE) presented the risk assessment according to the toxicological reference values and put forward a set of recommendations following the results of the sanitary control and monitoring campaign (DDASS 2000):

- Take into account the maximal pollutant content in raw water (0.61 μg L^{-1}) to assess the risks, because treatment plants could not retain all of them.
- In the absence of World Health Organization (WHO) recommendation for chlordecone, consider threshold values from the Environmental Protection Agency (EPA), available in the ATSDR database according to three levels on exposure: acute (10 μg kg^{-1} J^{-1}), intermediary, and chronic (0.5 μg kg^{-1} J^{-1}).
- Consider also other sources of pesticide exposure such as air, food, professional, or household uses.

The proposed sanitary position states the following:

- The excess of regulatory thresholds calls for urgently and specifically treat water.
- The exposure related to maximal content in water remains less than 20% of TDI, which corresponds to a chronic toxicity for people who are more sensitive to pesticide exposure. Thus, it is not imperative to impose restrictions on drinking water supplied from polluted resources. But this tolerance should be limited in time and should not lead to delays in setting up water treatment plants.
- It is imperative to continue to monitor the evolution of pesticide content in water and also to ensure that new molecules are not detected and associated with cumulative effects on human.
- Crops and animal products coming from fields treated by pesticides with a long half-life should be analyzed to assess their possible contribution to exposure to populations.
- It is highly likely that the exposure to organochlorine compounds had been much higher when they were still in use. Regarding their potential for binding on adipose tissues and of the possibility of cumulative effects, assessment of past exposure and its potential consequences is recommended.

2.3 SEARCH OF ORGANOCHLORINE COMPOUNDS IN SOILS AND STAPLE CROPS (2001)

Recommendations to address the sources of food contamination made to the Hygiene Board of Martinique in 1999 and 2000 took time to be implemented; thus, DDASS of Martinique asked for financial support from the French Health Ministry to carry out a study on crop contamination in 2001, and then on aquatic organisms in 2002.

The aim of the first study on crop contamination was to confirm the presence of organochlorine compounds in soils and to assess if they were transferred to crops. This approach was coupled with multiresidue analysis to take into account other possible contamination.

2.3.1 Material and Method

A decision was made to consider tubers that were locally produced, consumed by the local population and likely to have concentrations of organochlorines (dasheen and sweet potato). The selected plots included banana as one of the previous crops, with a large cropping period. Sampling was made with soil-plant couples.

Soil sampling (minimum 500 g fresh weight) was made in a circle of 40 cm around the plant, and according to the root development on 40–50 cm depth for dasheen and 30–40 cm for sweet potato. Five plots were sampled at the harvest stage: four with dasheens, one with sweet potatoes. Two samples of malanga (*Xanthosomas* sp.) were made on another plot. The upper part as well as root and soil particles was removed. The central part was chosen for dasheen, whereas peeled tuber was chosen for sweet potato to avoid soil particle contamination for the analyses. A minimum of 200 g (fresh weight) was required. The La Drôme Departmental Laboratory (LDA26) was chosen for the analyses as it was already accredited by the French Accreditation Committee (COFRAC) for water analysis and was also already involved in water pesticide analyses in Martinique. A 48-hour delay was fixed for sample delivery to preserve samples.

2.3.2 Results

Two major molecules were found in tubers: chlordecone and βHCH. Four were found in the soil: chlordecone, βHCH, Mirex, and dieldrin.

2.3.2.1 Tubers

The results showed that only a small proportion of dasheens were accumulating chlordecone (Table 2.3).

Proportionally sweet potatoes were more sensitive than dasheen, but regarding the low number of samples for Malanga, no conclusion could be made. βHCH was much less frequently detected in tubers and with lower values (Table 2.3). The maximum chlordecone content was 1.6 mg kg^{-1} for dasheen and 1.9 mg kg^{-1} for sweet potato (on a fresh weight basis).

2.3.2.2 Soils

The fifth plot was the most polluted but with a higher dispersion of data (Table 2.4). Most of plots were polluted with βHCH but to a lower extent (Table 2.4).

TABLE 2.3
Contamination of Tubers

		Number of Results > Detection Limit		Percentage of Results > Detection Limit according to the Sample Number	
Tubers	**Number of Samples**	**Chlordecone**	**βHCH**	**Chlordecone**	**βHCH**
Dasheen	63	7	2	11	3
Sweet potato	10	4	0	40	0
Malanga	2	2	0	100	0

TABLE 2.4
Statistical Data of Soil Chlordecone and βHCH Content (mg kg^{-1} Dry Soil)

Molecule	Plot	Mean Value	Standard Deviation	Minimum Value	Maximum Value
Chlordecone	Plot 1	2.67	1.14	0.63	5.10
	Plots 2, 3, 4	3.31	1.46	0.64	6.48
	Plot 5	8.70	2.05	5.80	13.00
βHCH	Plot 1	0.01	0.00	0.01	0.02
	Plots 2, 3, 4	0.15	0.77	0.01	5.00
	Plot 5	0.01	0.01	0.01	0.05

Note: A value of 0.01 mg kg^{-1} (detection limit) had been attributed to the results below this limit.

The majority of samples were polluted with Mirex (perchlordecone), which was surprising because it has not been used in Martinique. The hypothesis was that it could have been included in products made with chlordecone as impurity. Cavelier (in Kermarrec et al. 1980) did not confirm this hypothesis with the analysis of commercial products. Dieldrin, an aldrin metabolite, was analyzed relatively constantly in soils but at a low level: Each plot had at least a sample over the detection limit, except plot 3. With a high number of crop samples under the detection limit, it was not possible to relate plot pollution to tuber contamination. Moreover, pollutant transfer was dependent on soil characteristics and on the chlordecone repartition at plot scale, thus necessitating higher number of samples.

2.3.3 Discussion

Thus, some crops had concentrations of chlordecone. Plots 1 and 2 did not show any dasheen contamination over the detection limit. It appeared necessary to implement agronomical trials to better understand the crop behavior according to the specie, the variety, the soil type, the soil moisture around the tuber, or other factors. Since we learned that organic matter or clays in soil are able to modify the chlordecone migration to crops (Woignier et al. 2013).

2.4 SEARCH OF ORGANOCHLORINE COMPOUNDS IN AQUATIC ORGANISMS (2002)

Between May and August 2002, a study was conducted in collaboration with the French Research Institute for Exploitation of the Sea (IFREMER) with the aim to assess if sea and river products were also contributing to exposure of the population to organochlorine.

The study revealed very high contamination of wild and fish farming species in rivers, and, to a lesser extent, the contamination of marine species coming from costal zones close to the outlet of polluted rivers (Bocquené and Franco 2005, Coat et al. 2006). Tilapias sampled in the Lezarde River had the highest levels of chlordecone (386 and 196 μg kg^{-1} fresh weight). Fish farming products showed chlordecone content of 132 and 23 μg kg^{-1} (fresh weight) for tilapias and crayfishes, grown in freshwater aquaculture ponds located in the northeast of the island. Some of the fish and crustaceans from the sea, especially lobsters, also contained chlordecone. Conversely, unlike organisms from coastal zones or close to polluted outlets, deep sea fish were less contaminated and their chlordecone content was below the detection limit of 1 μg kg^{-1} fresh weight.

2.5 CONCLUSION

The Bellec and Godard report (2002) reviewed the knowledge and data available in 2002 with the aim to raise awareness of the perils of soil and island ecosystem pollutions in population susceptible to such pollutions through food intake.

The significant conclusions were that in Martinique and Guadeloupe, the whole banana areas were probably polluted by organochlorine compounds and that their transfer to tubers and staple crops was possible. Thus, these regions and maybe other areas in the world are in danger of having thousands of hectares of their agricultural land polluted, with resultant contamination of aquatic ecosystems, freshwater, and marine system for years.

People were exposed to these compounds when they were in use, and they are still exposed 10 years after their ban. Thus, global risk assessment should be provided for these departments, including ecosystems and human health, as well as risk management actions to reduce as soon as possible the major sources of exposure and also to protect their populations.

2.5.1 Proposition for Definition of Diet Characteristics

No data were available about diet in tropical French departments. Thus, it was necessary to fill this gap to assess sanitary risks by taking into account food intake. Consumption data could be obtained through a food survey detailed enough to characterize eating habits: 1/by complementing the National Institute of Statistics and Economics Studies (INSEE) survey, a "consumption and household budget," or 2/through a survey on food consumption carried out by the French Agency of Food Sanitary Security (AFSSA, 2003), with the aim not only to characterize the consumption standard of a household for the assessment of mean exposure but also to characterize diets of several population categories, according to the age, the living area, and the socioeconomic level to collect elements for risk management.

2.5.2 Proposition for Improvement of the Knowledge on Food Contamination Levels

Besides the qualitative improvement of search and controls from government services and according to data on consumption, further studies could be conducted. A complete diet approach, implemented in the United States and in some European countries, has been implemented only once on 10 pesticides in France (ENSP 2001). It directly provides the population's mean exposure to pesticides that are detected in foodstuffs.

Crop experiments were necessary to identify the risks involving staple food products and on different soil types by testing the variety and hydric conditions that could have an effect on soil to crop transfer and to draw relevant conclusions to manage these risks. If livestock are located on watersheds with known soil and water pollution, we proposed, in collaboration with veterinary services, analysis of derived food products. Moreover, assessment of the pollution of fishery resources had to be done. Soil analyses showed a major pollution by chlordecone, and with processes of erosion and migration in the saturated soil layer, this molecule could first reach the aquatic environment and then the sea.

To conclude, to reduce food exposure to pesticides as quickly as possible, it was recommended to systematically search for foods susceptible to organochlorine pollution as these compounds were largely spread in the island ecosystem.

2.5.3 Impregnation Status of Living Tissues

After the PhD thesis of Martin (1973), a new study on the organochlorine impregnation status of Martinican people was targeted on some population groups. Breast milk was also a potential source of organochlorine exposure, and residue analyses were proposed to have a better knowledge of infant exposure.

A survey on livestock mammal milk in polluted areas could indicate the concentration risk in fatty tissues as they were potentially exposed through soil, water, or forage intake.

It has been also proposed to survey the wild fauna, as Cavelier and Snegaroff did (Kermarrec et al. 1979, Snegaroff 1977), and complementary to the French National Health Research Institute's (INSERM) work in Guadeloupe on black and gray rats, mice, and mongoose in collaboration with the French Hunting National Office (ONC) and the Inra wild fauna laboratory. These studies could provide information on the bio-concentration risk in terrestrial food webs.

2.5.4 Effects of Pesticides on Populations

According to the recognized role of endocrine disruptor attributed to some pesticides, risks of reproductive disorders were to be considered (Baldi et al. 1998). The French National Health Research Institute (INSERM) was working on human male fertility in Guadeloupe and was foreseeing a study on the potential *in utero* developmental disorders related to pesticides that affect endocrine functions thanks to a cohort of pregnant women. Thus, it was proposed to alert the national bodies on the risk situation in the French West Indies because of the environmental pollution in order to conduct a more relevant study on the health of the local population.

2.5.5 Populations Information

Referring to the European Framework 90/313/CEE of June 7, 1990 that imposes access conditions to information on environment status, and to reduce the risk of conflicts, the report recommended making available to the public the elements taken into account in risk management, as well as references that can allow them to put into perspective this risk situation with other that they experience everyday. This proposal was made upon the hypothesis that self-confidence of the population in public services was reduced if access to information was limited. Thus, it has been proposed to "assume collectively the consequences of past practices in a transparent, constructive and efficient mind facing a pollution situation that was dealing with all components of Martinican society: farmers, fishermen, consumers."

2.5.6 What Actions Have Been Taken Following These Results?

Since the beginning of 2001, the monitoring of pesticide pollution had been discussed in the GREPHY meetings, a new mechanism was put in place in July 2001 and succeded to Hygiene Board of the Department on subject of chlordecone. The study of the DSDS (March 2002) demonstrated a major pollution of Martinican environment. If evidence of water pollution in 1999 did not cause any particular problems and had been publicly published with a lot of media manifestations (presentations, conferences) that provided knowledge and information along time, this was not the same for the data on soil pollution and food contamination.

After a careful analysis of the robustness of the data and animated debates between the defenders of *we don't know enough to communicate the results to people* and those for the communication of results from a study that had been previously announced, the prefect took the decision on their presentation July 1, 2002 during a GREPHY meeting, with a press pack. The communication was rather reassuring arguing that contamination results were not a significant alert element but required analyzing much more scientifically the incidence of the phenomenon as well as more systematic control measures.

Between May and July 2002, the DGCCRF laboratory of Massy developed the chlordecone analysis, and then first results showed quickly the contamination of one-third of the sampled tubers. These first official controls with the quantification of chlordecone molecule led to the seizure of a sweet potato lot from Martinique in the Dunkerque harbor on an alert of the control services of Martinique. The chock occurred when the *Liberation* national newspaper ran the headline "In Martinique, sweet potatoes and strong toxics" in its October 10, 2002 edition (Pons 2002). The situation then came to the national stage, after which the local press seized the news.

Then while the DGCRRF analyses were accumulating on pollution of tubers, GREPHY was informed of the results on aquatic organisms (DSDS and IFREMER) on the November 7, 2002, and it announced the implementation of a series of measures for soils before the planting of crops sensitive to chlordecone transfer. The biggest producers' cooperative (SOCOPMA) had already announced its decision not to accept tubers grown on polluted fields during the GREPHY meeting of July. Some months later, a prefectoral decree on the March 20, 2003, imposed these preventive measures in agriculture.

The DSDS report, which indicated that the situation was foreseeable from the alerts provided by Snegaroff (1977) and Kermarrec (1980), had not been provided to GREPHY members or the population despite several demands from associations (ASSAUPAMAR open letter of October 19, 2002) before its access via the national website (http://www.chlordecone-infos.gouv.fr/) was provided in 2008 by the Pesticides Observatory managed by the French Agency for Environmental and Occupational Health Safety (AFSSET).

At the time of writing, most of the proposals supplied in March 2002 have been implemented even if some measures were not implemented as quickly as was hoped. Several action plans have succeeded each other; the first one was initiated in mid-2002 with a constant objective: the decrease of the population exposure, even if it is still perfectible.

However, the confidence of populations in local food products is not entirely restored, mainly because of reluctance in communication on a subject still considered as a hot topic. The transparency effort on the communication of the public data is a democratic requirement. However, the major difficulty lies in the capacity to consider without shame the situation of pollution, to share information on it, to inform people on the possible failure of the prevention and control systems and on the available tools to make informed choices in their consumption to reduce their exposure and finally to support them on how to better live with the chlordecone risk.

REFERENCES

AFSSA. 2003. Avis de l'AFSSA relatif à l'évaluation des risques liés à la consommation de denrées alimentaires contaminées par la chlordécone en Martinique et en Guadeloupe. AFSSA – Saisines n°2003-SA-0330, 2003-SA-0132, 2003-SA-0091, Maisons-Alfort, France, 8p.

Baldi, I., B. Mohammed-Brahim, P. Brochard, J.F. Dartigues, and R. Salamon. 1998. Effets retardés des pesticides sur la santé: Etat des connaissances épidémiologiques. *Revue d'épidémiologie et de santé publique* 46(2): 134–142.

Balland, P., R. Mestres, and M. Fagot. 1998. Rapport sur l'évaluation des risques liés à l'utilisation de produits phytosanitaires en Guadeloupe et en Martinique. Affaire n 1998-0054-01. Ministère de l'aménagement du territoire et de l'environnement–Ministère de l'agriculture et de la pêche, Paris, France, 47pp.

Bellec, S. and E. Godard. 2002. Contamination par les produits phytosanitaires organochlorés en Martinique. Caractérisation de l'exposition des populations. Direction de la Santé et du Développement Social (DSDS) de la Martinique, Service Santé-Environnement, Fort de France, Martinique, France, 36pp. doi: http://www.observatoire-pesticides.gouv.fr/upload/bibliotheque/501723195631802623768285060734/rapport-organochlores-Godard-Bellec-972-complet.pdf.

Bocquené, G. and A. Franco. 2005. Pesticide contamination of the coastline of Martinique. *Marine Pollution Bulletin* 51(5–7):612–619.

Coat, S., G. Bocquene, and E. Godard. 2006. Contamination of some aquatic species with the organochlorine pesticide chlordecone in Martinique. *Aquatic Living Resources* 19(2):7. doi: 10.1051/alr:2006016.

Crabit, A., P. Cattan, F. Colin, M. Voltz, and L.T. Pak. 2014. Contamination des eaux de rivière d'un bassin versant Guadeloupéen (Pérou, Capesterre Belle-Eau, Guadeloupe). In *44ème Congrès du Groupe français des pesticides "Protection des cultures et santé environnementale: héritages et conceptions nouvelles"*, Schoelcher, Martinique, France, mai 26–29, 2014.

DDASS. 1999. Etude des risques de contamination des eaux de consommation par les pesticides en Martinique; rapport au Conseil Départemental d'Hygiène, séance du 14 octobre 1999. Fort de France, Martinique, France.

DDASS. 2000. Pesticides et Alimentation en Eau Potable en Martinique—Etat des lieux et position sanitaire; rapport au Conseil Départemental d'Hygiène, séance du 19 octobre 2000. Fort de France, Martinique, France.

DDASS. 2001. Pesticides et Alimentation en Eau Potable en Martinique—Etat des lieux et position sanitaire—Bilan actualisé en octobre 2001. Fort de France, Martinique, France.

ENSP. 2001. Bilan des modalités de surveillance de la contamination par les produits phytosanitaires de l'eau et des denrées alimentaires. Laboratoire d'étude et de recherche en environnement et santé (LERES), Rennes, France.

Kermarrec, A., A. Dartenuck, J. Gousseland, A. Villardebo, G. Malato, and N. Cavelier. 1980. Niveau actuel de la contamination des chaînes biologiques en Guadeloupe: pesticides et métaux lourds. 1979–1980. Rapport INRA n 7883. Institut national de la recherche agronomique (INRA), station de zoologie et lutte biologique. Petit Bourg, Guadeloupe, France.

Martin, M. 1973. Les pesticides organochlorés—Recherches des résidus dans le tissu adipeux humain et animal en Martinique. INRA Antilles-Guyane–Université de Bordeaux II, Bordeaux, France.

Pons, F. 2002. In Martinique, sweet potatoes and strong pesticdes. Libération. http://www.liberation.fr/societe/2002/10/12/en-martinique-patates-douces-et-toxiques-durs_418298 (French). Accessed June 2016.

Snegaroff, J. 1977. Les résidus d'insecticides organochlorés dans les sols et les rivières de la région bananière de Guadeloupe. *Phytiatrie-Phytopharmacie* 26:251–268.

Woignier, T., P. Fernandes, A. Soler, F. Clostre, C. Carles, L. Rangon, and M. Lesueur-Jannoyer. 2013. Soil microstructure and organic matter: Keys for chlordecone sequestration. *Journal of Hazardous Materials* 262:357–364.

Section II

Environmental Diagnosis: Pollution Assessment of Natural Resources

3 Heterogeneity of Soil Pollution

Philippe Cattan, Thierry Woignier, Florence Clostre, and Magalie Lesueur Jannoyer

CONTENTS

3.1 INTRODUCTION

Chlordecone is very persistent, sorptive, and highly lipophilic (Epstein 1978). Its persistence in soils is due to its low solubility in water (Dawson et al. 1979), its high affinity for organic matter (Cabidoche et al. 2009, Kenaga 1980), and its poor biodegradability, which is related to its peculiar chemical structure and the high steric hindrance caused by the 10 chlorine atoms (Jablonski et al. 1996). Due to these properties, 20 years after the prohibition of its use (1993), the molecule still persists in soils where it was applied (Brunet et al. 2009, Cabidoche et al. 2009). Moreover, the chlordecone pollution in soils becomes a continuous source of contamination for water resources, crops, and animals (Cabidoche et al. 2009, Gourcy et al. 2009, Jondreville et al. 2013, Luellen et al. 2006).

To manage this pollution, a first step is to delineate polluted areas according to their pollution level. In the case of chlordecone, this should have been a simple task since this molecule was almost exclusively used on banana fields. On this basis, contamination maps have been prepared, which include simple criteria. In Martinique, Desprat et al. (2003) prepared a first map of contamination based on three criteria: rainfall amount accounting for pest pressure, soil retention capacity, and historical banana cultivation. In Guadeloupe, only the banana land use during the application period of chlordecone was accounted for (Tillieut 2005). However, the accuracy of these maps remains poor and they were inadequate to account for spatial and temporal heterogeneity of chlordecone content in soil. One reason is the lack of information about historical land use; another is the inadequate representation about what determines chlordecone content in soil (lack of knowledge about chlordecone fate).

In fact, all the soils are not equivalent in terms of chlordecone contamination and in their capacity to transfer this pollution to water and crops. Two key factors can be put forward: The first is the influence of chemical properties and clay microstructure on the concentrations of chlordecone in soil. The second factor is *farming practices* that determine the amount of chlordecone applied on banana fields. We discuss these two factors next.

3.2 DIFFERENT TYPES OF CONTAMINATED SOILS: INFLUENCE OF CHEMICAL PROPERTIES AND CLAY TYPE

Guadeloupe and Martinique, French West Indies, are volcanic islands in humid tropics. They present a high variability in terms of soil types, ranging from regosols and andosols (volcanic young soils formed from basalt rocks and ashes) to nitisols, ferralsols, and vertisols (soils containing classical phyllosilicate clays) (Colmet-Daage et al. 1965, FAO-ISRIC-ISSS 1998). However, banana crops are mainly cultivated in the vicinity of the volcanoes and the three main contaminated soil types are andosol, nitisol, and ferralsol.

Andosols are globally more polluted by chlordecone than nitisols and ferralsols (Table 3.1) despite similar initial inputs (Brunet et al. 2009, Cabidoche et al. 2009, Levillain et al. 2012). This aspect needs to be considered in light of their chemical properties and clay composition.

TABLE 3.1
Mean Soil Chemical Properties and Chlordecone Content according to Soil Type

	pH in Water	CEC (meq %g)	OC (%)	Chlordecone (mg kg^{-1})
Andosol (n = 30)	5.6 (5.2–6.0)	16.0 (10.5–29.2)	4.2 (1.9–7.6)	5.6 (0.2–17.2)
Ferralsol (n = 40)	5.1 (4.0–6.4)	31.3 (15.5–58.5)	1.6 (1.1–3.6)	1.2 (0.1–3.5)
Nitisol (n = 30)	5.5 (4.8–6.3)	26.7 (15.2–38.5)	1.6 (1.1–2.9)	2.6 (0.1–5.8)

Note: CEC, cation exchange capacity; OC, organic carbon.

3.2.1 Chemical Properties Differ according to Soil Type

Soil chemical properties, especially the organic carbon content, but also the pH and cation exchange capacity, influence the sorption of pesticides in soils (Baskaran et al. 1996, Kumar and Philip 2006, Olvera-Velona et al. 2008). Recently, adsorption of chlordecone on activated carbons was shown to be pH-dependent (Durimel et al. 2013) and sorption on soil was correlated with the organic carbon content but not with the pH, while relationship with cation exchange capacity was unclear (Fernandez Bayo et al. 2013).

Among the three main types of contaminated soils found in the French West Indies, andosols differ from ferralsols and nitisols by their chemical properties, especially their particularly high carbon content (Table 3.1). In the literature, the most widely accepted explanation for the higher retention of chlordecone in andosols (Cabidoche et al. 2009, Levillain et al. 2012) is the latter's high organic carbon content and the high affinity of pesticides for soil organic carbon (Kenaga 1980). At watershed scale, Levillain et al. (2012) found that chlordecone content increased significantly with the organic carbon content but the relationship was poor. Likewise at regional scale, soil organic carbon has been positively correlated to pollutant content for a range of organochlorine pesticides (Gong et al. 2004, Mishra et al. 2012).

3.2.2 Allophane Structure and Physical Properties Enhance Chlordecone Trapping in Andosols

Soils also differ from each other by their clays: halloysite for nitisol, halloysite and Fe-oxyhydroxides for ferralsol, and allophane for andosol. Allophanes are amorphous aluminosilicate with Al/Si ranging between 1 and 2; the unit cell appears as spheres with diameter between 3 and 5 nm; these clays have a bulk density close to 0.5 g cm^{-3} and develop a specific surface area as high as 700 m^2 g^{-1} (Woignier et al. 2006).

We have demonstrated (Woignier et al. 2012) that the chlordecone content in soil is directly related to the allophane content and Figure 3.1 confirms that chlordecone and allophane contents are well correlated ($R^2 = 0.7067$). One objective of this section is to show the influence of the clay microstructure on the retention of chlordecone in soils.

3.2.2.1 Allophane Microstructure

To understand why the allophanes contain more pesticides than the other clays, it is necessary to characterize the peculiar pore structure of the allophane aggregates. The pore volume and the specific surface area are well correlated with allophane content ($P < 0.0001$ for both) (Woignier et al. 2006). The specific surface area of andosols is as high as 180 m^2 g^{-1} and pore volume close to 2.5 cm^3 g^{-1} (Woignier et al. 2006). These data show that allophane clay favors larger porous features. Moreover, the combination of high specific surface area and large pore volume suggests that the porous structure is made up of both micropores and mesopores.

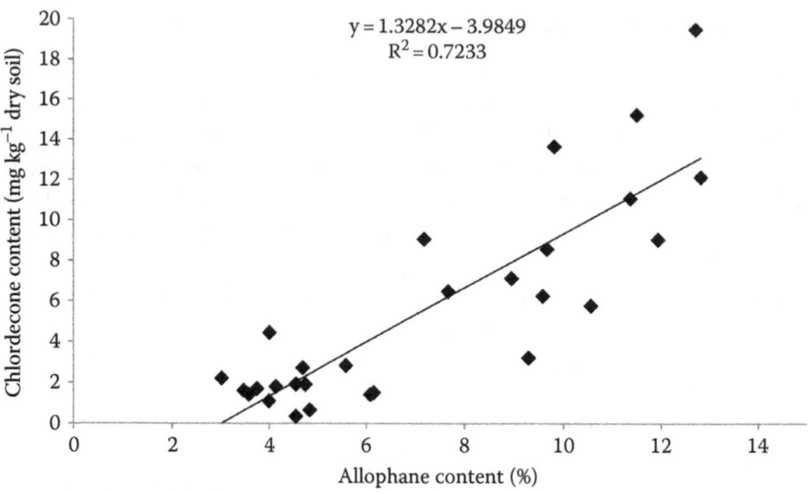

FIGURE 3.1 Chlordecone content in soils is strongly correlated with soil allophane content.

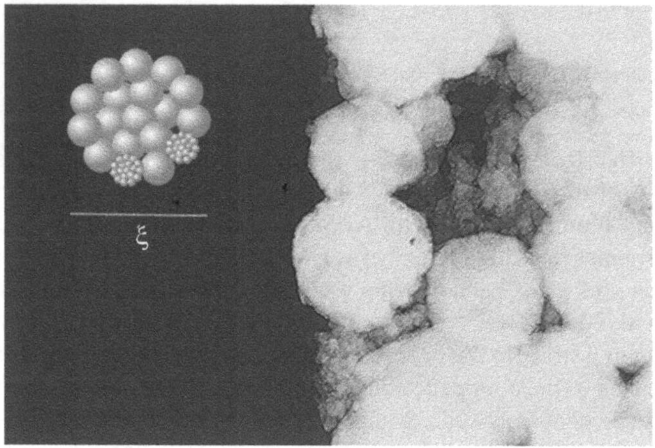

FIGURE 3.2 Transmission electron microscopy micrograph and model of the allophane aggregates (figure width: 1 μm).

Scanning and transmission electron micrograph of andosols demonstrates the spongy structure of allophane compared to the plate-like structure of classical phyllosilicate clays: kaolinite or halloysite (Woignier et al. 2012). The allophane structure is thus a very open structure (Figure 3.2) made of aggregated small particles (≤3 nm) building clusters with radius close to 10 nm. These clusters can stick and form larger and larger aggregates. This description is in agreement with the results in the literature (Adachi and Karube 1999, Chevallier et al. 2010), and with a qualitative fractal description.

To precisely describe the fine structure of the soils samples, we performed small-angle x-ray scattering experiments on allophane soils. Our results showed that the size of the fractal labyrinth ξ significantly increased with an increase in allophane content. This labyrinth morphology suggests that accessibility inside the clay microstructure is reduced (Woignier et al. 2012).

3.2.2.2 Different Trapping Mechanisms in Allophane

The spongy structure influences the transport inside the allophane aggregates. At the scale of the allophane aggregates, accessibility is difficult because of the tortuous structure and small pore size of allophane clay. Chlordecone transfers within the soil depend on hydraulic conductivity and diffusion processes in the porous microstructure. With fractal approach (Woignier et al. 2012) and numerical simulation (Woignier et al. 2015), we find that, at the scale of the allophane aggregates, the calculated transport properties are hindered by the allophane microstructure. Hydraulic conductivity decreased by four orders of magnitude and diffusion process decreased 20-fold. These results could be important in explaining chlordecone sequestration in these soils. The tortuous and inaccessible microstructure likely partly explains its large capacity to immobilize and trap pesticides. In these fractal structures, possible reactions with chemical or biological species that could extract the pesticides are thus hindered; the pesticide remains trapped inside allophane clay and thus cannot be extracted.

Conversely to usual clays, after a strong drying, the andosol's porosity irreversibly collapses (Bartoli et al. 2007). So during the cycle of dry and wet seasons, the allophane aggregates could collapse because of the drying and rewetting processes. This shrinkage, which increases with the allophane content, can contribute to the whole pesticide trapping in andosols by closing the porosity.

As in Woignier et al. 2012, the more accepted explanation for the retention effect reported in the literature is the high affinity of pesticides for soil organic carbon. However, physical parameters like transport properties inside clay aggregates and porosity closing can contribute to pesticide sequestration.

3.3 HIGH DIVERSITY OF FARMING PRACTICES PREVENTS PREDICTION OF CHLORDECONE CONTENT IN SOIL ON LARGE AREAS

Agricultural practices depend on the farmer and on-farm working conditions (Levillain et al. 2012). This is particularly true for practices like plant health treatments, which depend, according to Houdart (2006), on the farmer's strategy, the type of farming system, and land use. For organochlorine pesticide treatments, land use and agricultural practices explain a large part of soil contamination (Boul et al. 1994, Spencer et al. 1996, Wang et al. 2007). By studying the distribution of chlordecone contamination in soil in Guadeloupe (Levillain et al. 2012), we tried to characterize the link with the agricultural activity.

We first examined banana land-use duration with the hypothesis that the longer the period of banana cultivation, the higher the contamination of soil by chlordecone. Overall, we found that plot contamination increased with the banana land-use duration in andosols and ferralsols. Short-duration plots were, on average, less contaminated than medium- and long-duration plots. Although the differences were significant, they were almost of the same order of magnitude (e.g., for andosols, 1.39 mg kg^{-1} for short duration, 2.16 mg kg^{-1} for the two other classes). Besides this effect for andosols and ferralsols, we found no differences according to duration classes for nitisols. An explanation lies in the lower retention capacity of nitisols, unlike the other two soil types, which could mask the impact of input variations related to banana land-use frequency. Overall, the length of time between application and assessment of pollution caused a great uncertainty about various assessments like the duration of banana cultivation for periods dating back nearly 50 years.

Second, we examined the link between chlordecone contamination in soil and the farm types. We found that large farms were more contaminated than small farms. Indeed, Figure 3.3 (we selected farms with more than 10 plots to assess the intra-farm variation of chlordecone concentration) shows that inter-farm variation of

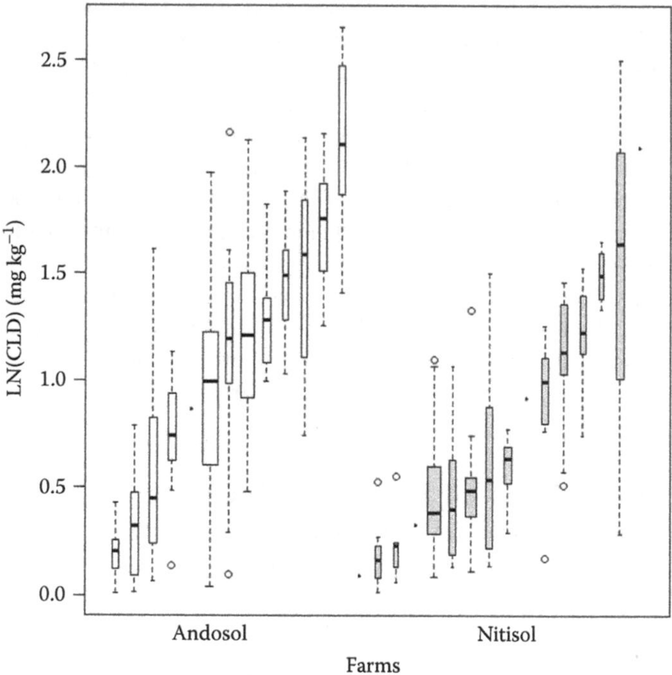

FIGURE 3.3 Distribution of chlordecone concentration on farms located on andosols and nitisols with more than 10 banana plots. Each box represents a farm. The bottoms and tops of the boxes represent the 25th and 75th percentiles; the band inside the box is the median; the whiskers extend to the most extreme data point, which is no more than 1.5 times the interquartile range of the box.

chlordecone concentration was far higher than the intra-farm one, irrespective of the soil type. Regardless of the soil type, the treatment strategy associated with a given farm was an essential component with respect to soil contamination.

Explaining the differences of contamination between farms required us to give attention to the conditions of banana cultivation during the period of chlordecone application.

In Guadeloupe, banana plantations with high productivity appeared in the 1930s. Between 1960 and 1980, the socioeconomic setting (sugar factory shutdowns, closure of Basse-Terre harbor, etc.) caused banana plantations to shift from the western to the eastern coastal region. Mountain banana plantations thus gave way to lowland banana plantations cultivated by banana companies on former mechanized sugarcane fields. In the 1980s, urban growth led to the disappearance of lowland plots. New plots appeared on cleared forestland or on former food crop fields.

In the 1970s–1980s, the intensification-based modernization policy and implementation of banana quality standards led to the disappearance of unprofitable plantations: smallholdings, mountain plantations, and large extensive farms (Zebus 1982). Many farmers became members of banana industry organizations. This required them to meet tonnage quotas and caused widespread intensification of chemical control to minimize the yield loss risk.

Chlordecone was used in this setting. There were two distinct treatment periods, depending on the commercial brand used: the Kepone period (1972–1978) and the Curlone period (1982–1993). The Kepone period (1972–1978) was a transition from the use of old products (HCH and Aldrin) to chlordecone. During this period, Joly (2010) estimated that 25% of the plots had been sprayed. Chlordecone was then used on a frequent, but not systematic, basis. The Curlone period began in May 1982 when chlordecone was reintroduced and subsidized. During this period (1982–1993), treatments were consistently undertaken. Overall, at the farm level, the number of treatments increased with the size of banana-growing areas, and was modulated according to cash flow (Zebus 1982).

These conditions resulted in a high diversity of farming system and types of farms. Some of those farms appeared or disappeared during the period of chlordecone application, whereas others continuously intensified their production. Equally, treatment strategy varied between the 1970s and the 1980s as well as the extent of the treated areas. These variations, combined with a multitude of conditions of chlordecone dispersion in the environment, resulted in a high variability of chlordecone concentration in soils. Consequently, because of the major farm effect, contamination may vary spatially and so it does not allow us to predict the direction of change (farms with high treatment strategy may be close to farms with low treatment strategy). Thus, spatial interpolation seems difficult and prevents mapping the chlordecone contamination accurately. Currently, the lack of information on historical activity of agriculture prevents the determination of contaminated areas without resorting to an intensive and expensive analysis campaign. For the future, observatories of agricultural practices are necessary to better characterize the spatial distribution of agricultural activity.

3.4 CHLORDECONE DISTRIBUTION AT WITHIN-PLOT SCALE DEPENDS MORE ON ANTHROPOGENIC FACTORS THAN ON CHEMICAL PARAMETERS

At plot scale, the heterogeneity of chlordecone pollution is determined by three main factors: the mode of application, the modification of the spatial arrangement management of plantation (linked with the mode of cultivation), and tillage. Chlordecone was a powder applied on soil in rings around banana mats. Such a localized mode of application implies an initially high heterogeneity of the pollutant distribution at soil surface. Afterward, with successive crop cycles, spatial shift of (1) banana rows and (2) mats along the row could occur, causing spatial shifts of localized applications throughout the period of chlordecone use. This contributed to modulate the variability of each application. Finally, the intensity of tillage modifies the pollution distribution both horizontally and vertically. The outcome of the interactions between these three factors on pollution distribution at within-plot scale is difficult to predict.

3.4.1 Horizontal Heterogeneity

We studied chlordecone content at different sampling points in each selected plot using a systematic grid design and 20 samples per plot at two depths, 0–30 and 30–60 cm (Clostre et al. 2014). Horizontal heterogeneity could be very high with chlordecone contents differing by an order of magnitude (4.5–54.1 mg kg^{-1} dry soil, plot F, Figure 3.4) at within-field scale in the 0–30 cm soil layer in the case of an andosol plot in no-tillage farming. Even in less heterogeneous plots, the range of values remained large with maximum contents twofolds and fourfolds higher than

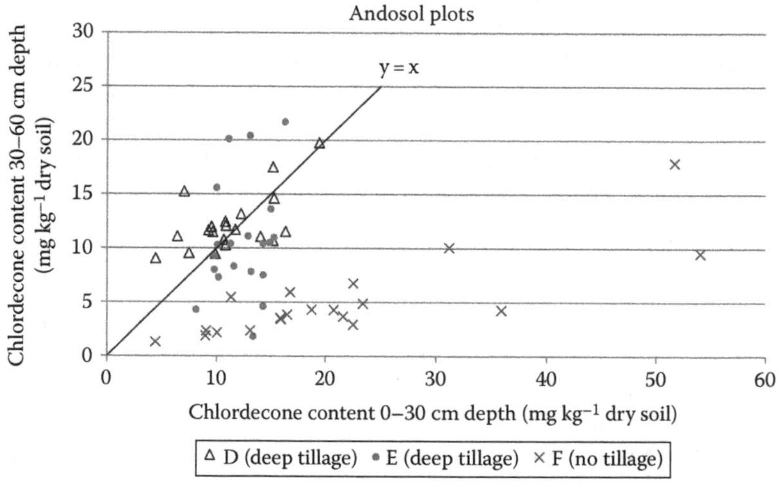

FIGURE 3.4 Chlordecone content in 0–30 and 30–60 cm soil layers for three andosol plots (D–F) with different tillage practices (no tillage; deep tillage: >40 cm deep). (Adapted from Clostre, F. et al., *Environ. Sci. Pollut. Res.*, 21, 1980, 2014.)

minimum contents. Thus, localized application of chlordecone led to highly heterogeneous pollution at within-field scale.

The precision of chlordecone content obtained through sampling tends to vary with tillage depth: the shallower the tillage, the higher the horizontal variability and hence the poorer the precision of sampling for a same number of subsamples. Deep and intensive tillage attenuated the initially high variability of pollutant distribution through a homogenization of the tilled soil layer.

Adsorption and mobility of pesticides are known to be influenced by soil properties, organic matter being the main soil property affecting pesticide fate (Arienzo et al. 1994, Huang et al. 2013). Nevertheless, at within-field scale, the organic carbon content poorly explains the variation of chlordecone content (Clostre et al. 2014).

Topography and erosion affect the organic carbon distribution by driving the top soil down (De Gryze et al. 2008, Jian-Bing et al. 2006). In plots with steep slopes, the erosion of top soil, with higher organic matter and chlordecone contents than deeper soil layers, could create an increasing gradient of chlordecone content from slope top to bottom.

3.4.2 Heterogeneity along the Soil Profile

Regardless of the soil type, chlordecone can be found even in deep soil layers and its content decreases with depth along the soil profile (Figure 3.5). This distribution is partly explained not only by the mode of application of chlordecone and subsequent slow leaching (Cabidoche et al. 2009). The decreasing organic carbon content, and hence the decreasing adsorption potential, with increasing soil depth could also play a role (Baskaran et al. 1996, Dorel et al. 2000). Similarly, organochlorine contents

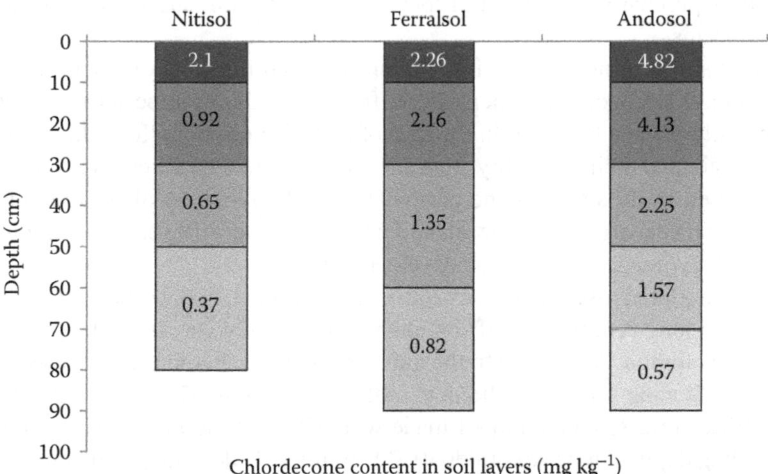

FIGURE 3.5 Mean chlordecone content (mg kg^{-1} dry soil) at different depths along the soil profile (in cm), in composite sample at plot scale (n = 10 for nitisol, n = 20 for ferralsol and andosol).

decreases with increasing depth in soil profile, with a positive correlation with the organic carbon content for some compounds (Wang et al. 2006, Zhang et al. 2006).

In andosol plots, chlordecone contents were either close in the two layers (plots D and E, deep tillage, Figure 3.4), or lower in the 30–60 cm soil layer than in the 0–30 cm soil layer (plot F, no tillage, Figure 3.4). Thus, in andosols, deep tillage led to a dilution and a homogenization of chlordecone in the 0–60 cm layer. A similar trend was observed in ferralsol plots (Clostre et al. 2014).

It should be noted that, after a deep tillage, a profile inversion (interchange of topsoil with subsoil) may occur in some cases, leading to higher contents in the lowest plough layer than in the upper plough layer.

To conclude, chlordecone distribution in the soil profile is influenced by the interactions between the initial input of chlordecone at soil surface, the higher organic carbon content in soil surface layer, and tillage practices.

3.5 CONCLUSION: IMPLICATIONS FOR POLLUTION ASSESSMENT

Soil pollution from agricultural practices sometimes reveals high variations in pollutant content as it is the case for chlordecone at the field scale, as well as at the territory scale. The assessment of the mean pollution and its variability requires specific attention.

This is a great challenge since soil contamination has agronomic consequences (preventing cultivation of some vegetables) or economic consequences (value of land). For a better land-use management, specific soil diagnosis may be introduced at the island scale for both agricultural activities and individual home gardening. The French Accès au Logement et Urbanisme Rénové (access to housing and renovated urban planning) law recommends inclusion of chlordecone soil diagnosis at the time of purchase of residential lots. But soil contamination assessment faces many problems.

First, to assess agricultural soil pollution, we need information about the former practices of chlordecone treatment (dose, frequency, duration/period) and also the former yearly banana land use. In the case of historic pollution, these data are not so easy to obtain and their reliability from surveys is poor, because cultivation registers were not used in the chlordecone period (1972–1993) and aerial photography was scarce and expensive (made at the island scale almost every 10 years) as remote sensing technology/imagery was not so developed then.

Second, if global assessment is effective (Desprat et al. 2003), updating soil pollution map is a long process even if soil analysis is the only way today to characterize mean field pollution. According to the aims of pollution characterization, we recommend the following scheme at the field scale (Figure 3.6): The sampling method is not modified (grid spacing, aliquot made with 20 sampling points per hectare) but the sampling depth can be simplified (0–30 cm for crop decision; 0–60 cm for stock evaluation/estimate with two modalities according to the tillage practices).

Third, total zoning scheme implies a high cost both with regard to soil analyses and time for the society to assess pollutant contents at the regional scale because

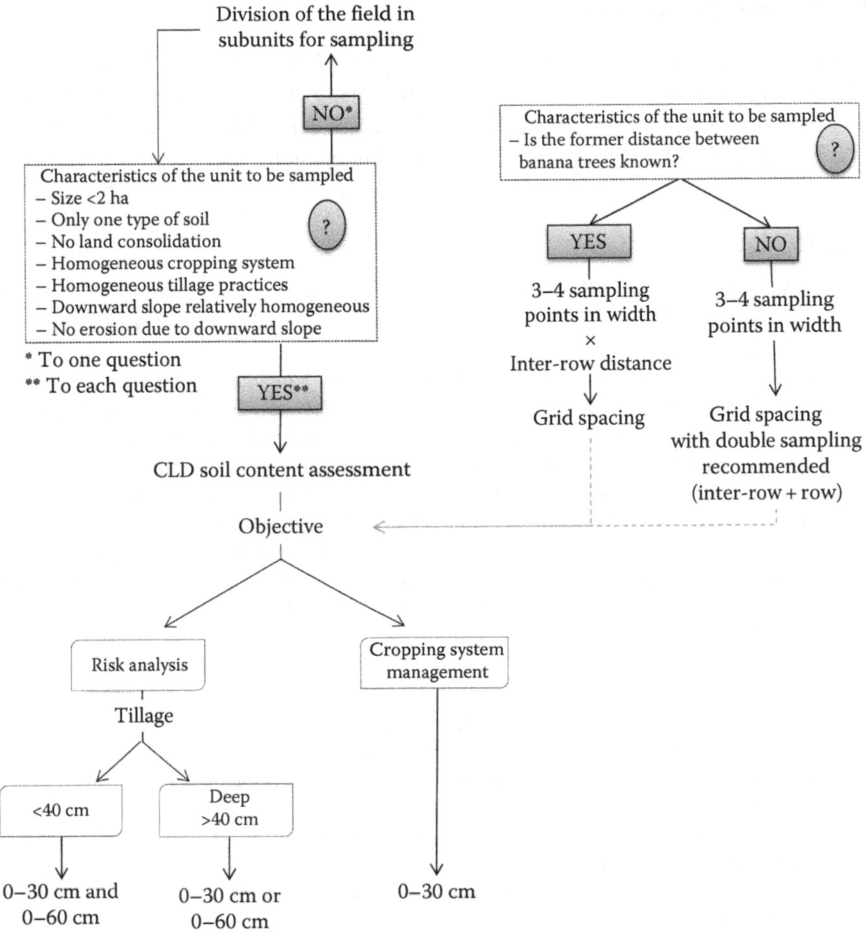

FIGURE 3.6 Decision tree for characterization of the sampling procedure.

dense sampling is needed at the field scale. Further studies will have to develop methods to orientate soil sampling campaigns, using previous results and including the farm effect (Clostre et al. 2014, Levillain et al. 2012), or relating water to soil pollution as water measurements are integrated at the watershed outlet scale (Mottes 2013).

Lastly, to simulate soil pollution evolution over time at different scales, as sampling is expensive and pollutant movement is relatively low, updated data are also needed on pollutant fate in specific soil conditions (physical properties and organic matter content, rainfall, etc.) (Mottes 2013). The variability in the value of some variables can be high: K_{oc} estimates vary from 1 order of magnitude between 2,500 and 17,500 kgL^{-1} (Cabidoche et al. 2009, Fernandez Bayo et al. 2013, Gourcy et al. 2009).

ACKNOWLEDGMENTS

Funding was provided by the French Chlordecone National Plan ("JAFA" project), the French National Research Agency ("Chlordexco" project), the French Ministry for Overseas Development (MOM), and the 2007–2013 Martinique FEDER fund ("GIT—Integrated Land Use and Territorial Management" project).

REFERENCES

Adachi, Y. and J. Karube. 1999. Application of a scaling law to the analysis of allophane aggregates. *Colloids and Surfaces A: Physicochemical and Engineering Aspects* 151(1–2):43–47.

Arienzo, M., T. Crisanto, M.J. Sanchez-Martin, and M. Sanchez-Camazano. 1994. Effect of soil characteristics on adsorption and mobility of (14C)Diazinon. *Journal of Agricultural and Food Chemistry* 42(8):1803–1808. doi: 10.1021/jf00044a044.

Bartoli, F., J.C. Begin, G. Burtin, and E. Schouller. 2007. Shrinkage of initially very wet soil blocks, cores and clods from a range of European andosol horizons. *European Journal of Soil Science* 58(2):378–392. doi: 10.1111/j.1365-2389.2006.00889.x.

Baskaran, S., N.S. Bolan, A. Rahman, and R.W. Tillman. 1996. Pesticide sorption by allophanic and non-allophanic soils of New Zealand. *New Zealand Journal of Agricultural Research* 39(2):297–310. doi: 10.1080/00288233.1996.9513189.

Boul, H.L., M.L. Garnham, D. Hucker, D. Baird, and J. Aislabie. 1994. Influence of agricultural practices on the levels of DDT and its residues in soil. *Environmental Science & Technology* 28(8):1397–1402. doi: 10.1021/es00057a004.

Brunet, D., T. Woignier, M. Lesueur-Jannoyer, R. Achard, L. Rangon, and B.G. Barthès. 2009. Determination of soil content in chlordecone (organochlorine pesticide) using near infrared reflectance spectroscopy (NIRS). *Environmental Pollution* 157 (11): 3120–3125. doi: 10.1016/j.envpol.2009.05.026.

Cabidoche, Y.M., R. Achard, P. Cattan, C. Clermont-Dauphin, F. Massat, and J. Sansoulet. 2009. Long-term pollution by chlordecone of tropical volcanic soils in the French West Indies: A simple leaching model accounts for current residue. *Environmental Pollution* 157(5):1697–1705. doi: 10.1016/j.envpol.2008.12.015.

Chevallier, T., T. Woignier, J. Toucet, and E. Blanchart. 2010. Organic carbon stabilization in the fractal pore structure of andosols. *Geoderma* 159(1–2):182–188.

Clostre, F., M. Lesueur-Jannoyer, R. Achard, P. Letourmy, Y.-M. Cabidoche, and P. Cattan. 2014. Decision support tool for soil sampling of heterogeneous pesticide (chlordecone) pollution. *Environmental Science and Pollution Research* 21(3):1980–1992. doi: 10.1007/s11356-013-2095-x.

Colmet-Daage, F., P. Lagache, J. de Crécy, J. Gautheyrou, M. Gautheyrou, and M. de Lannoy. 1965. Caractéristiques de quelques groupes de sols dérivés de roches volcaniques aux Antilles françaises. *Cahiers ORSTOM. Série Pédologie III* 3(2):91–121.

Dawson, G.W., W.C. Weimer, and S.J. Shupe. 1979. Kepone: A case study of a persistent material. *Water American Institute of Chemical Engineers Symposium Series* 75(190):366–374.

De Gryze, S., J. Six, H. Bossuyt, K. Van Oost, and R. Merckx. 2008. The relationship between landform and the distribution of soil C, N and P under conventional and minimum tillage. *Geoderma* 144(1–2):180–188. doi: http://dx.doi.org/10.1016/j.geoderma.2007.11.013.

Desprat, J.F., J.P. Comte, and G. Perian. 2003. Cartographie par analyse multicritère des sols potentiellement pollués par organochlorés en Martinique, rapport phase 2. BRGM, Montpellier, France.

Dorel, M., J. Roger-Estrade, H. Manichon, and B. Delvaux. 2000. Porosity and soil water properties of Caribbean volcanic ash soils. *Soil Use and Management* 16(2):133–140.

Durimel, A., S. Altenor, R. Miranda-Quintana, P. Couespel Du Mesnil, U. Jauregui-Haza, R. Gadiou, and S. Gaspard. 2013. pH dependence of chlordecone adsorption on activated carbons and role of adsorbent physico-chemical properties. *Chemical Engineering Journal* 229:239–249. doi: http://dx.doi.org/10.1016/j.cej.2013.03.036.

Epstein, S.S. 1978. Kepone-hazard evaluation. *Science of the Total Environment* 9(1):1–62. doi: 10.1016/0048-9697(78)90002-5.

FAO-ISRIC-ISSS. 1998. *World Reference Base for Soil Resources*. Rome, Italy: FAO.

Fernandez Bayo, J., C. Saison, C. Geniez, M. Voltz, H. Vereecken, and A.E. Berns. 2013. Sorption characteristics of chlordecone and cadusafos in tropical agricultural soils. *Current Organic Chemistry* 17(24):2976–2984. doi: 10.2174/13852728113179990121.

Gong, Z.M., F.L. Xu, R. Dawson, J. Cao, W.X. Liu, B.G. Li, W.R. Shen et al. 2004. Residues of hexachlorocyclohexane isomers and their distribution characteristics in soils in the Tianjin Area, China. *Archives of Environmental Contamination and Toxicology* 46(4):432–437. doi: 10.1007/s00244-003-2301-9.

Gourcy, L., N. Baran, and B. Vittecoq. 2009. Improving the knowledge of pesticide and nitrate transfer processes using age-dating tools (CFC, SF6, 3H) in a volcanic island (Martinique, French West Indies). *Journal of Contaminant Hydrology* 108(3–4):107–117.

Houdart, M. 2006. Spatial organisation of agricultural activities and water pollution by pesticides. Modelling in Capot's watershed, Martinique (French West Indies), Geography, UAG, Martinique, France.

Huang, Y., Z. Liu, Y. He, F. Zeng, and R. Wang. 2013. Quantifying effects of primary parameters on adsorption–desorption of atrazine in soils. *Journal of Soils and Sediments* 13(1):82–93. doi: 10.1007/s11368-012-0572-3.

Jablonski, P.E., D.J. Pheasant, and J.G. Ferry. 1996. Conversion of Kepone by *Methanosarcina thermophila*. *FEMS Microbiology Letters* 139(2–3):169–173. doi: http://dx.doi.org/10.1016/0378-1097(96)00137-1.

Jian-Bing, W., X. Du-Ning, Z. Xing-Yi, L. Xiu-Zhen, and L. Xiao-Yu. 2006. Spatial variability of soil organic carbon in relation to environmental factors of a typical small watershed in the black soil region, Northeast China. *Environmental Monitoring and Assessment* 121(1–3):597–613. doi: 10.1007/s10661-005-9158-5.

Joly, P.B. 2010. *La saga du Chlordécone aux Antilles françaises. Reconstruction chronologique 1968–2008*, INRA-Sens, Versailles-Grignon, France.

Jondreville, C., C. Bouveret, M. Lesueur-Jannoyer, G. Rychen, and C. Feidt. 2013. Relative bioavailability of tropical volcanic soil-bound chlordecone in laying hens (*Gallus domesticus*). *Environmental Science and Pollution Research International* 20(1):292–299.

Kenaga, E.E. 1980. Predicted bioconcentration factors and soil sorption coefficients of pesticides and other chemicals. *Ecotoxicology and Environmental Safety* 4(1):26–38.

Kumar, M. and L. Philip. 2006. Adsorption and desorption characteristics of hydrophobic pesticide endosulfan in four Indian soils. *Chemosphere* 62(7):1064–1077. doi: http://dx.doi.org/10.1016/j.chemosphere.2005.05.009.

Levillain, J., P. Cattan, F. Colin, M. Voltz, and Y.-M. Cabidoche. 2012. Analysis of environmental and farming factors of soil contamination by a persistent organic pollutant, chlordecone, in a banana production area of French West Indies. *Agriculture, Ecosystems & Environment* 159:123–132. doi: 10.1016/j.agee.2012.07.005.

Luellen, D.R., G.G. Vadas, and M.A. Unger. 2006. Kepone in James River fish: 1976–2002. *Science of the Total Environment* 358(1–3):286–297. doi: 10.1016/j.scitotenv.2005.08.046.

Mishra, K., R.C. Sharma, and S. Kumar. 2012. Contamination levels and spatial distribution of organochlorine pesticides in soils from India. *Ecotoxicology and Environmental Safety* 76:215–225. doi: http://dx.doi.org/10.1016/j.ecoenv.2011.09.014.

Mottes, C. 2013. *Evaluation des effets des systèmes de culture sur l'exposition aux pesticides des eaux à l'exutoire d'un bassin versant. Proposition d'une méthodologie d'analyse appliquée au cas de l'horticulture en Martinique.* AgroParisTech, Paris, France.

Olvera-Velona, A., P. Benoit, E. Barriuso, and L. Ortiz-Hernandez. 2008. Sorption and desorption of organophosphate pesticides, parathion and cadusafos, on tropical agricultural soils. *Agronomy for Sustainable Development* 28(2):231–238. doi: 10.1051/agro:2008009.

Spencer, W.F., G. Singh, C.D. Taylor, R.A. LeMert, M.M. Cliath, and W.J. Farmer. 1996. DDT persistence and volatility as affected by management practices after 23 years. *Journal of Environmental Quality* 25(4):815–821. doi: 10.2134/jeq1996.00472425002500040024x.

Tillieut, O. 2005. Cartographie de la pollution des sols de Guadeloupe par la chlordécone, rapport technique. DAF-SPV, Basse-Terre, Guadeloupe.

Wang, F., X. Jiang, Y.R. Bian, F.X. Yao, H.J. Gao, G.F. Yu, J.C. Munch, and R. Schroll. 2007. Organochlorine pesticides in soils under different land usage in the Taihu Lake region, China. *Journal of Environmental Sciences (China)* 19(5):584–590.

Wang, X., X. Piao, J. Chen, J. Hu, F. Xu, and S. Tao. 2006. Organochlorine pesticides in soil profiles from Tianjin, China. *Chemosphere* 64(9):1514–1520. doi: http://dx.doi.org/10.1016/j.chemosphere.2005.12.052.

Woignier, T., F. Clostre, P. Cattan, and M. Lesueur-Jannoyer. 2015. Pollution of soils and ecosystems by a permanent toxic organochlorine pesticide: Chlordecone—Numerical simulation of allophane nanoclay microstructure and calculation of its transport properties. *AIMS Environmental Science* 2(3):494–510. doi: 10.3934/environsci.2015.3.494.

Woignier, T., F. Clostre, H. Macarie, and M. Jannoyer. 2012. Chlordecone retention in the fractal structure of volcanic clay. *Journal of Hazardous Materials* 241–242:224–230. doi: 10.1016/j.jhazmat.2012.09.034.

Woignier, T., J. Primera, and A. Hashmy. 2006. Application of the DLCA model to natural gels: The allophanic soils. *Journal of Sol Gel Science and Technology* 40(2–3):201–207. doi: 10.1007/s10971-006-7593-6.

Zebus, M.F. 1982. Bases économiques et sociales de la production de banane en Guadeloupe. Mémoire de DAA-INAPG, Paris, France.

Zhang, H.B., Y.M. Luo, Q.G. Zhao, M.H. Wong, and G.L. Zhang. 2006. Residues of organochlorine pesticides in Hong Kong soils. *Chemosphere* 63(4):633–641. doi: http://dx.doi.org/10.1016/j.chemosphere.2005.08.006.

4 Characterization of River Pollution at the Watershed Scale

Magalie Lesueur Jannoyer, Charles Mottes, Florence Clostre, Céline Carles, Mathilde Guene, Joanne Plet, Pauline Della Rossa, Abderazak Bazizi, and Philippe Cattan

CONTENTS

4.1 INTRODUCTION

Water pollution is a major problem in the French West Indies, notably caused by chlordecone, although it has not been used for 30 years. Then, according to the Water Office of Martinique (2013), most rivers are polluted over the regulation thresholds imposed by the European Water Framework Directive (WFD—2000/60/EC). Three limits are set by EU regulation: (1) A river is considered to be in a "good chemical status" if pesticide concentration is not more than 0.1 μg L^{-1}; (2) a river is considered to be in a "poor chemical status" when the sum of pesticide concentrations exceeds 0.5 μg L^{-1}; and (3) a river is considered untreatable for water consumption if pesticide concentration exceeds 2 μg L^{-1}.

A watershed scale, which is a functional hydrological unit for river and pollutant fluxes, is used to monitor water pollution (Finizio and Villa 2002). The outlet of watersheds integrates their hydrogeological functioning (FAO; Dooge 1986; Zehe et al. 2014). Surrogate analysis of drinking water samples is usually done at the watershed outlet. However, if such measurements are usefull to provide information on the current situation, they will be limited to aid future decision-making on water quality. A better understanding of the whole agro-hydrosystem complexity is necessary to achieve that objective. For this reason, the impact and functioning of physical, biological, and decisional processes, as well as human activities at the watershed scale and their effects on water pollution by pesticides, need to be better understood.

This may strongly affect the fate of pesticides, and several questions emerge: Where and when to measure river pollution? What are the contributing factors to water pollution? How to monitor, assess, and interpret changes over time or space in polluted water content?

These are important considerations to set up and implement a monitoring and management program.

In this chapter, in answer to these questions, we present two case studies and the methods and tools that were applied to characterize water pollution caused by chlordecone.

1. *The Galion watershed*: A river watershed of 45 km^2 and 20.5 km of watercourse, located in the center of Martinique island. Due to its large size and diversity of land use, this watershed was selected to tackle the question related to variations in pollution with space.
2. *The Ravine watershed*: An order 1 watershed of 1.3 km^2, a 3 km watercourse, located in the Atlantic northern side of Martinique, near the La Montagne Pelée volcano. Due to its small size and timely transfers or fast transfers of water and pollutant fluxes due to volcanic soil and geological formations (Charlier 2007), this watershed was selected to tackle the question related to variations in pollution with time.

4.2 CHARACTERIZING THE VARIABILITY OF THE RIVER POLLUTION TO GUIDE POLLUTION MANAGEMENT

Initial assessments were generally elaborated with one-off samples, made at the outlet, due to integrated information and cost consideration. Nevertheless, the integrated pollution concentration at the outlet may be low, whereas the pollutant content can be high in some river segments. These values vary according to the contributing area, whether from point or nonpoint sources. Characterizing the spatial variation of chlordecone pollution according to these segments is an important aspect in elaborating a pollution management plan. The case of the Galion River is explained here.

4.2.1 MEASUREMENTS AND SAMPLING AT THE WATERSHED SCALE

To characterize river pollution, we sampled 35 points on the Galion River and its tributaries according to confluence and former banana cropping areas. The former banana cropping areas were mapped according to the duration of banana land use from 1970 to 1993 (simplified from Desprats et al. 2004).

Sampling was made during the dry period with the hypothesis that chlordecone content in water would be at its highest (Gourcy et al. 2009), and with the aim to characterize the worst case of river pollution due to sanitary issues. Water was sampled one-off in a glass bottle to avoid chlordecone adsorption on the plastic container. Water samples were analyzed at the CIRAD laboratory in Martinique using the solid phase microextraction technique (SPME, Soler et al. 2014), which was both quick and cost-effective.

4.2.2 The Measurement at the Outlet Does Not Provide an Efficient Representation of Pollution Processes

The Galion River's profile of chlordecone pollution (see Table 4.1) shows that chlordecone content in water decreases from the spring to the mid-watercourse and decreases from the mid-watercourse to the outlet. In the upstream, water is free of chlordecone. In the midstream, in the middle of the watershed area, the chlordecone content quickly exceeds the 0.1 μg L^{-1} threshold, from the first positive sampling point. The maximum chlordecone content measured in the river is about 10-fold the

TABLE 4.1
Chlordecone Content in Water of the Galion River during the Dry Period and Distance from the Outlet of Each Sampling Point

Sampling Point in the River (S), and in Tributary (T)	Distance to the Outlet (m)	Chlordecone Content in Water (μg L^{-1})
S1	10328	<LOD
S3	8248	<LOD
S4	7476	0.19
T1	6717	0.42
S6	6499	0.53
T2	6373	3.57
S10	6239	1.02
S11	5445	0.72
T3	5400	0.09
T4	4781	22.98
S13	4775	1.23
S21	4621	0.85
T5	4454	0.51
S22	3198	0.87
S34	2386	0.13
S35	2097	0.60

Note: S on the Galion river; T from tributary; LOD, limit of detection.

first EU threshold (0.1 μg L^{-1}) and close to the third one (2 μg L^{-1}). In the midstream, the river is thus considered to be in a bad chemical status according to the regulatory EU thresholds. Downstream, the chlordecone content decreases with quite high variations, with its value at the outlet being close to the second EU threshold (0.5 μg L^{-1}). Globally, water pollution levels were over the EU thresholds for the major downstream part and for more than 7.5 km from the outlet.

Table 4.1 shows a high variability of chlordecone content between tributaries. There is no general trend, and no apparent and clear logic of behavior relating the levels of pollution of tributaries to those of the main river, which may be explained by the lack of information related to tributaries' discharge or other hydrological processes (e.g., contribution of water table). Consequently, the measurement at the outlet does not provide an efficient representation of pollution processes in the watershed.

4.2.3 Soil Pollution Explains Water Pollution

Figure 4.1 shows our mapping of river pollution at the watershed scale; we used the river segment method (van Gils et al. 2009, Koç 2010). The length of a segment is the distance between two sampling sites. We mapped the chlordecone content in water with ArcGIS 10.0 software. For rivers, via an SQL (Structured Query Language) request, we selected each entity included in one class of chlordecone content in water. The map of soil pollution was built according to the banana land use duration for the period 1970–1993 and five classes of risk. The risk classes were assessed and validated by 35 soil analyses at the field scale (Della Rossa et al. submitted).

Then, we overlapped the simplified pollution soil risk map with the river pollution map during the dry season (Figure 4.1). We observed connections between the soil risk map and the river pollution map. When the river crossed a risky area, chlordecone content in water increased: the more risky the soils nearby the river, the more polluted the river. Thus, the contributing areas were well identified by comparing different sampling points in the watershed. Historically, banana fields are the reservoir of chlordecone pollution, which that slowly diffuses to the river at the watershed scale. This area is of major interest for monitoring and management measures. It primarily concerns the midstream and downstream areas of the watershed.

4.2.4 Orientation for Pollution Management

Fortunately, in this volcanic island, headwaters are usually located in forested areas; thus, they are chlordecone-free and can be used for producing drinking water. Measures for managing the pollution will have to focus on former banana cropping areas, most of the time located in the river midstream (here the central part of Galion watershed) due to the favorable agro-ecological cropping conditions (soil, slope, rainfall, temperature). Some tributaries are highly polluted and thus have to be primarily considered: La Digue River and two tributaries of La Tracée River (Figure 4.1) even if their pollution is diluted downstream in the mainstream.

This general scheme is close to the James River situation in the United States, the first pollution case by chlordecone (Kepone®) in the 1980s: upstream the point

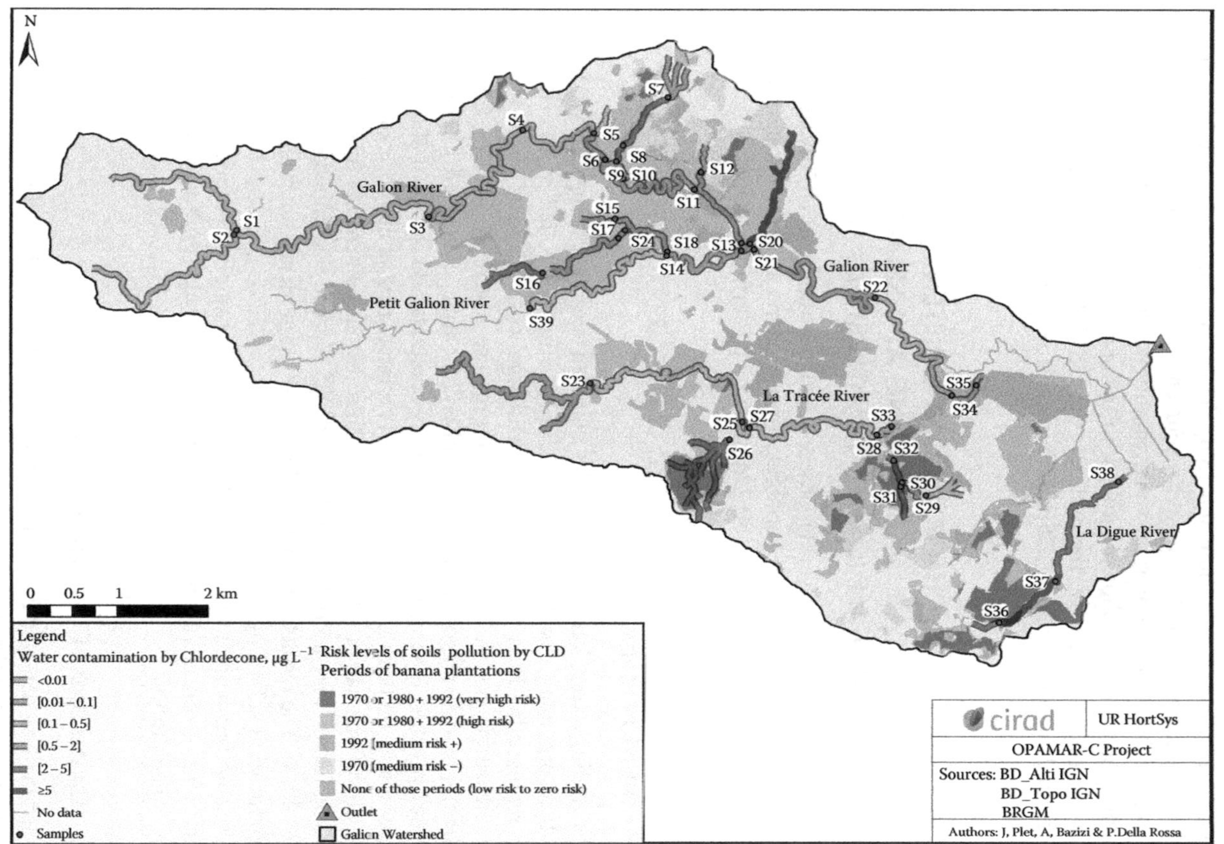

FIGURE 4.1 **(See color insert.)** Map of soil pollution risk and of polluted river segments for the Galion watershed, Martinique.

source pollution, the river is not polluted; from Hopewell to the bay, the chlordecone content in water increases and then decreases after the outlet in the Chesapeake Bay (Nichols 1990).

The study of the Galion River highlights the relationship between water and soil pollution, the latter reflecting human activity. This makes us to consider the spatial distribution of historical land use as well as current water uses to position monitoring points. Then if we need to monitor water quality during chlordecone applications, we ought to consider the following points: in the upper central part (S3–S6) where agriculture was less intensive; the mid part (S11, S13, S20) where most of banana fields are; and nearby the outlet (S35) for an integrative measurement and to quantify exports toward the sea.

4.3 VARIABILITY OF CHLORDECONE CONTENT IN WATER OVER TIME: MONITORING AT A SMALL WATERSHED SCALE (ORDER 1 WATERSHED)

As we saw for the Galion River, chlordecone concentration variability over space can be high. We also studied the variability over time, on an order 1 watershed: the Ravine River. We avoided the one-off sampling by performing integrative sampling for 67 weeks.

4.3.1 The Monitoring System

Water samplings were done in glass bottles using an automatic sampler (ISCO 6712, ISCO Incorporation). The frequency of sampling was controlled by the river discharge representing the river fluxes. The discharge was calculated based on water height measurements in the stream by a pressure sensor (CS420-L, Campbell Scientific). The data logger (CR1000, Campbell Scientific) controlled 100 mL water sampling by the ISCO sampler following every cumulated discharge of 300 m^3 in the dry season and 1800 m^3 in the wet season. The integrated sample was obtained by mixing all 100 mL samples from the week. Chlordecone content was analyzed at the La Drôme Laboratory according to the NF17025 method accredited by COFRAC (Mottes 2013). The results were obtained with 30% confidence interval, with a limit of quantification at 0.01 $\mu g\ L^{-1}$.

This sampling method was preferred to the one-off sampling because it integrates the variations of concentrations that occur during the week, resulting in an averaged weekly concentration.

4.3.2 Chlordecone Concentrations over Time

Chlordecone was detected during the 67 weeks (100%) at the outlet, which means that the river was polluted in a permanent manner (Figure 4.2c). All samples were over the quantification limit threshold of 0.01 $\mu g\ L^{-1}$. The mean chlordecone content in water for 67 weeks was 0.28 (Mottes et al, 2015) thus almost three times over the European Water Framework Directive threshold of 0.1 $\mu g\ L^{-1}$

(Figure 4.2d). Figure 4.2 shows weekly chlordecone concentrations varying from 0.07 to 0.77 μg L^{-1} and there is neither a clear trend in chlordecone concentrations over time (Figure 4.2c) nor a clear relationship between chlordecone and weekly discharges (Figure 4.2b and e) nor rainfall (Figure 4.2a).

Two significant lessons can be drawn from Figure 4.2. The first one deals with the reliability of water measurements based on one-off sampling plans. In our case, the probability to have obtained a value lower than 0.1 and 0.5 μg L^{-1} should have been 7.5%; and 92.5% respectively. Thus, the first regulatory threshold is almost always overpassed; this probability prevents us from clearly formulating a pollution management plan.

The second one deals with the difficulty to evidence positive or negative evolution of pollutant concentration since a same global pollution state of the environment (good or bad according to the 0.1 μg L^{-1} threshold) may result in pollutant concentrations in rivers varying from onefold to tenfold. Then, modeling pollution is probably of a great interest to embed variation factors of pollutant concentration and also to assess the evolution of the environmental pollution state. Nevertheless, assessing river pollution does not seem to be sufficient to characterize pollution and so additional variables must be accounted for (soil, water table pollution, etc.). Thus, integrated model helps identify areas and periods to focus on for pollution monitoring and management, as well as the major levers for action (Mottes 2013; Mottes et al. 2015).

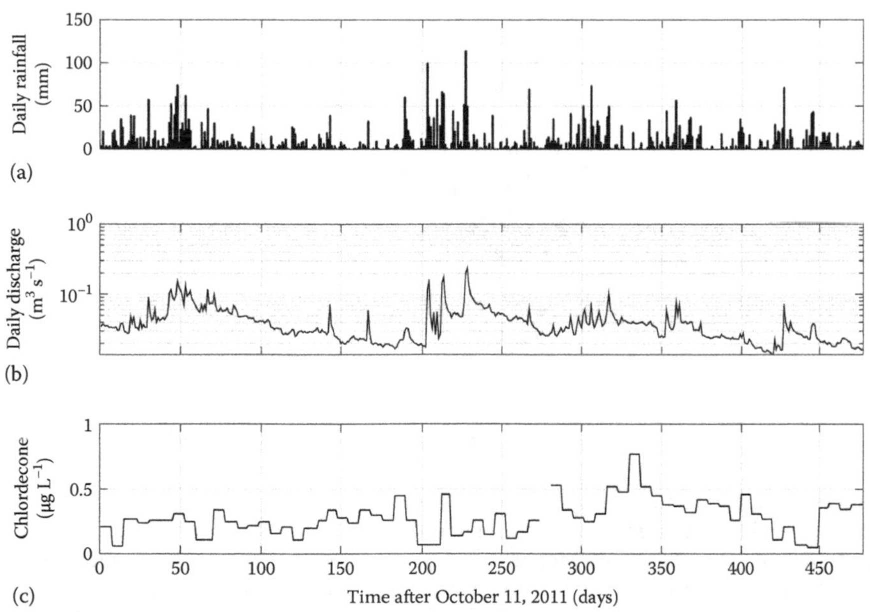

FIGURE 4.2 Time series of daily rainfall (a), time series of daily discharge (b), time series of weekly chlordecone concentrations (c). *(Continued)*

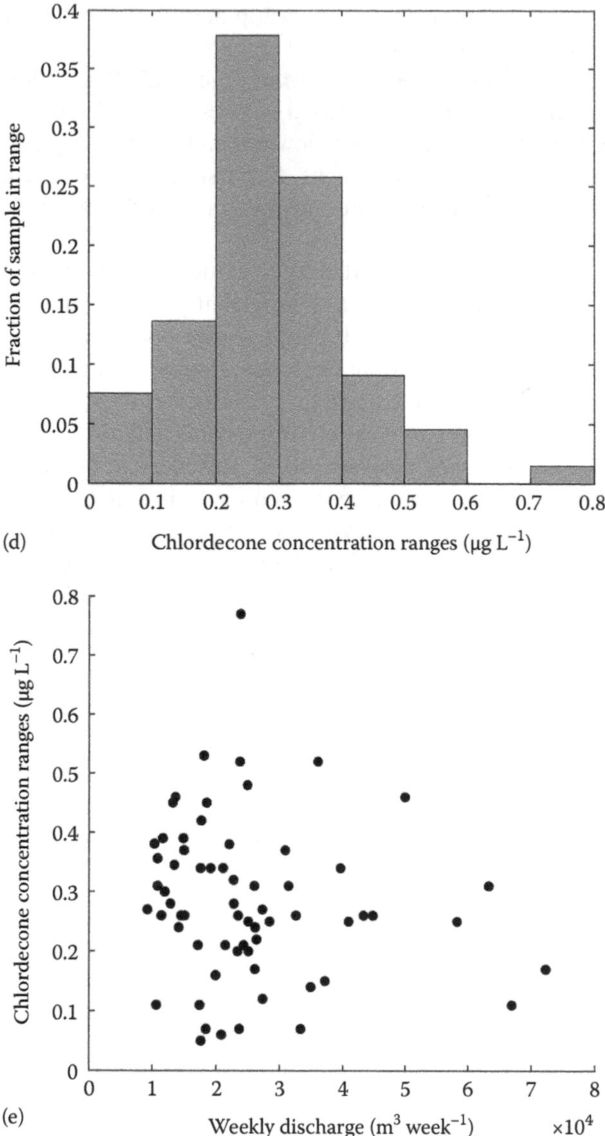

FIGURE 4.2 (*Continued*) Chlordecone concentration distribution (d), and relationship between weekly discharge and chlordecone concentrations (e) at the outlet of Ravine watershed, Martinique. (Data from Mottes, C., Assessment of the impact of cropping systems on water pesticides exposure at the watershed outlet. Proposal of an analysis methodology applied to horticulture cropping systems in Martinique, PhD dissertation, AgroParisTech, Paris, France, 2013, 257pp.)

4.4 CONCLUSION

More than 30 years after the ban on the use of chlordecone in the French West Indies, chlordecone contents in water are still very high and fishing is still forbidden in most of the rivers, unlike the case of the James River (Virginia, USA), where as soon as chlordecone contents in water fell below 0.02 μg L^{-1}, recreational fishing was authorized in 1980, 13 years after the ban (Luellen et al. 2006). The major differences are the *pollution source*, point source in the case of Hopewell *versus* diffuse source in the case of the French West Indies and the *hydro-system functioning* and pollution processes, leaching from the plant and high transport of chlordecone-free sediment from upstream in the case of the James River (Nichols 1990) *versus* high soil drainage and primarily contribution of subsurface ground water in the French West Indies (Charlier et al. 2011; Mottes et al. 2015). As pollutant transfers are complex, modelling at the scale of a small watershed is useful to understand the variations over time and to test hypotheses in pollution contribution and evolution.

At the watershed scale, even if a few fields are polluted, rivers are polluted over the European Water Framework Directive thresholds. The variability of the river segments is high, as well as in time or space. A more refined analysis, using maps from Geographic Information System, helps to focus the pollution monitoring and management on specific contributing areas. In our case, the targeted areas mainly depend on the historical banana land use and farm strategies concerning chlordecone use. Remediation measures will have to focus primarily on these areas, located mainly in the midstream of the watershed. Thus, feasibility studies will have to take into account the techniques to be employed, while keeping in mind the assessment of the environmental impact on aquatic ecosystems and on the agronomical properties of soils. Conclusion based on one-off sampling (increase or decrease in river pollutant concentrations) must be drawn carefully, and sampling should also take into account the high variation with time of the pollutant river concentrations.

ACKNOWLEDGMENTS

These works were part of the OPA-C project (Chlordecone Pollution Observatory in the French West Indies) and the PhD thesis of C Mottes (2010–2013). They benefited from financial support from the second Chlordecone National Action Plan, Martinique Water Office, EU Martinique FEDER funds (Territory Integrated management project), and the French Ministry of Overseas. We also thank the departmental laboratories (LDA972 and LDA26) and the CIRAD and IRD laboratory teams, Alain Soler and Yoan Labrousse, for their technical support.

REFERENCES

Charlier J.B. 2007. Fonctionnement et modélisation hydrologique d'un petit bassin versant cultivé en milieu volcanique tropical. PhD dissertation. Université des Sciences et Techniques du Languedoc, Montpellier II, Montpellier, France.

Charlier J.B., Lassachagne P., Ladouche P., Cattan P., Moussa R., Voltz M. 2011. Structure and hydrogeological functioning of an insular tropical humid andesitic volcanic watershed: A multi-disciplinary experimental approach. *Journal of Hydrology* 398: 155–170.

Della Rossa P., Lesueur Jannoyer M., Mottes C., Plet J., Bazizi A., Aranud L., Jestin A., Woignier T., Gaude J.M., Cattan P. Relating soil pollution to river pollution in a tropical environment. What lessons to draw from the contamination of a watershed with a persistent organic pollutant, the chlordecone. *Environment and Ecosystems*. Submitted to Agriculture.

Desprats J.F., Comte J.P., Chabrier C. 2004. Cartographie du risque de pollution des sols de Martinique par les organochlorés. Rapport Phase 3. BRGM RP 53262 FR, 23pp.

Dooge J.C.I. 1986. Looking for hydrological laws. *Water Resources Research* 22(9): 2246–2258. doi:10.1029/WR022i09Sp0046S.

FAO. Watershed management field manual. FAO Forestry Department, Roma, Italia. http://www.fao.org/docrep/006/t0165e/t0165e01.htm. Accessed May 2015.

Finizio A., Villa S. 2002. Environmental risk assessment for pesticides: A tool for decision making. *Environmental Impact Assessment Review* 22(3): 235–248. doi:10.1016/S0195-9255(02)00002-1.

Gourcy L., Baran N., Vittecoq B. 2009. Improving the knowledge of pesticide and nitrate transfer processes using age-dating tools (CFC, SF6, 3H) in a volcanic island (Martinique, French West Indies). *Journal of Contaminant Hydrology* 108: 107–117.

Koç C. 2010. A study on the pollution and water quality modeling of the river Buyuk Menderes, Turkey. *Clean-Soil, Air, Water* 38(12): 1169–1176.

Luellen D.R., Vadas G.G., Unger M.A. 2006. Keponde in James River fish: 1976–2002. *Science of the Total Environment* 358: 286–297.

Mottes C. 2013. Evaluation des effets des systèmes de culture sur l'exposition aux pesticides des eaux à l'exutoire d'un bassin versant. Proposition d'une méthodologie d'analyse appliquée au cas de l'horticulture en Martinique. PhD dissertation, AgroParisTech, Paris, France, 257pp.

Mottes C., Lesueur Jannoyer M., Charlier J.-B., Carles C., Guene M., Le Bail M., Malézieux E. 2015. Hydrological and pesticide transfer modeling in a tropical volcanic watershed with the WATPPASS model. *Journal of Hydrology* 529(3): 909–927. doi: 10.1016/j.jhydrol.2015.09.007.

Nichols M.N. 1990. Sedimentologic fate and cycling of Kepone in an estuarine system: Example from the James River estuary. *The Science of the Total Environment* 97/98: 407–440.

Soler A., Lebrun M., Labrousse Y., Woignier T. 2014. Solid-phase microextraction and gas chromatography-mass spectrometry for quantitative determination of chlordecone in water, plant and soil samples. *Fruits* 69(4): 325–339. doi: 10.1051/fruits/2014021.

van Gils J., van Hattum B., Westrich B. 2009. Exposure modeling on a river basin scale in support of risk assessment for chemicals in European river basins. *Integrated Environmental Assessment and Management* 5(1): 80–85.

Zehe E., Ehret U., Pfister L., Blume T., Schroder B., Westhoff M., Jackisch C. et al. 2014 HESS opinions: From response units to functional units: A thermodynamic reinterpretation of the HRU concept to link spatial organization and functioning of the intermediate scale catchments. *Hydrology and Earth System Sciences* 18: 4635–4655. doi: 10.5194/hess-18-4635-2104.

5 Groundwater Quality Assessment

Luc Arnaud, Jean-Baptiste Charlier, Laure Ducreux, and Anne-Lise Taïlamé

CONTENTS

5.1 INTRODUCTION

The application of chlordecone, an organochlorine pesticide, on banana plantations in the French West Indies, has led to a chronic contamination of all environmental compartments: soil (Cabidoche et al. 2009; Levillain et al. 2012; Clostre et al. 2015a; Devault et al. 2016), water (Gourcy et al. 2009), plants (Cabidoche and Lesueur-Jannoyer 2012; Clostre et al. 2015b), and marine wildlife (Coat et al. 2006, 2011).

Bananas are grown on highly permeable volcanic soils (Cattan et al. 2006), making infiltration one of the predominant processes in the water cycle at the drainage basin scale (Charlier et al. 2008, 2011). This results in intensive leaching of pesticides toward groundwater and also the contamination of the rivers draining contaminated groundwater (Charlier et al. 2009, 2015). All of these are very important to consider, especially in a tropical context with heavy rainfall, resulting in abundant groundwater recharge. Evaluating groundwater contamination is therefore a fundamental step in our understanding of the processes controlling the contamination of all of the environmental compartments.

Some studies have enabled researchers to characterize small-scale groundwater and river water contamination to that of an aquifer or a drainage basin (Charlier et al. 2009; Mottes et al. 2015). However, we might ask whether conceptual models of the contamination of volcanic aquifers—built in those small scales—are representative of larger scales. Indeed, volcanic aquifers develop in geological formations that are very heterogeneous due to their mode of deposition. We proposed the use of the European Water Framework Directive's (WFD) groundwater quality monitoring network managed by BRGM to determine the nature and extent of groundwater contamination. This work is drawn on studies carried out recently by BRGM in the French West Indies. First, we determined the spatial distribution of chlordecone contamination in groundwater on the islands of Martinique and Guadeloupe. Then, we characterized the biannual and monthly evolution of the contamination. Finally, we investigated the role of groundwater in river water contamination.

5.2 HYDROGEOLOGICAL SETTING OF THE FRENCH WEST INDIES

5.2.1 Geology

The Lesser Antilles Island arc extends 850 km from eastern Venezuela in the south to the Anegada Passage in the north, with a radius of curvature of about 450 km. In the north of Martinique island, the Lesser Antilles split into two distinct arcs (Samper et al. 2007; Figure 5.1). The double arc is clearly visible in the Guadeloupe archipelago, where there are two types of geology: to the east, Grande Terre Island is a vast limestone plateau (the older, outer arc that was active from the Eocene (–54 My) to the Oligocene (–34 My) and, to the west, Basse-Terre Island is a volcanic island (the more recent, inner arc created by active volcanism from the Miocene [–23 My] to the present). Martinique is located at the spot where the arc splits in two. To the east, the outer arc is represented by the Caravelle and Sainte-Anne peninsulas and, to the west, the inner arc is represented by the Trois-Îlets peninsula, the Carbet pitons, Mount Pelée, and the Piton Mont Conil.

In Martinique, eight volcanic units have been identified, from the Basal Complex and Saint Anne series (old arc), dated between 24.8 ± 0.4 and 20.8 ± 0.4 My (Germa et al. 2010) to the active Mount Pelée volcano (Figure 5.1).

Located in the northern part of the recent arc, Basse-Terre Island (Guadeloupe archipelago) is made of a cluster of andesite-dominated composite volcanoes and it results from the activity of six main volcanic massifs, showing an overall southward migration. In the north, the Basal Complex and the Septentrional Chain have been identified as the remnants of the oldest volcanic massifs of the island. Southern Basse-Terre is composed of the following volcanoes: the Axial Chain, the Grande Découverte massif, and the Mont Caraïbe. The Soufrière volcano is the currently active stage of the composite volcano of La Grande Découverte (CVGD) and is younger than 0.2 My. A simplified geological map of Basse-Terre Island is shown in Figure 5.1.

5.2.2 Main Characteristics of Volcanic Aquifers in the Lesser Antilles

The volcanism on both Guadeloupe and Martinique is andesitic with predominantly explosive volcanoes. Aboveground, the andesitic volcanism alternates with the deposition of domes and successive sectorial collapse in the composite volcanos, ash flows, lava flows, and reworked formations (e.g., lahars and debris flows) channeled

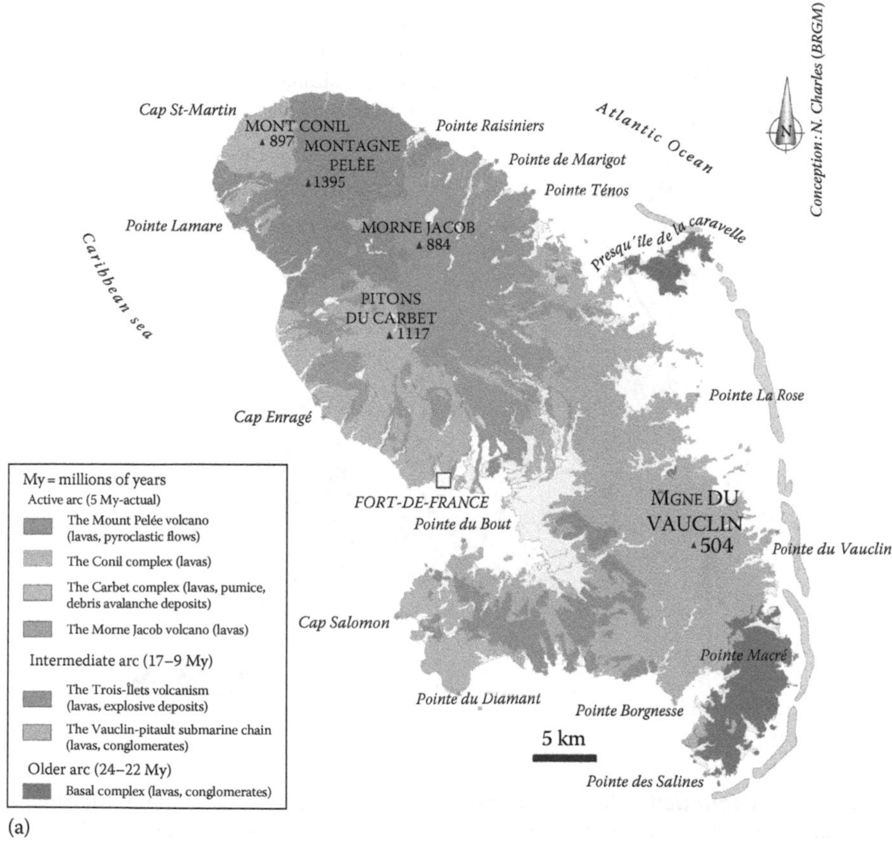

FIGURE 5.1 **(See color insert.)** (a, b) Geology and structure of Martinique and Guadeloupe. *(Continued)*

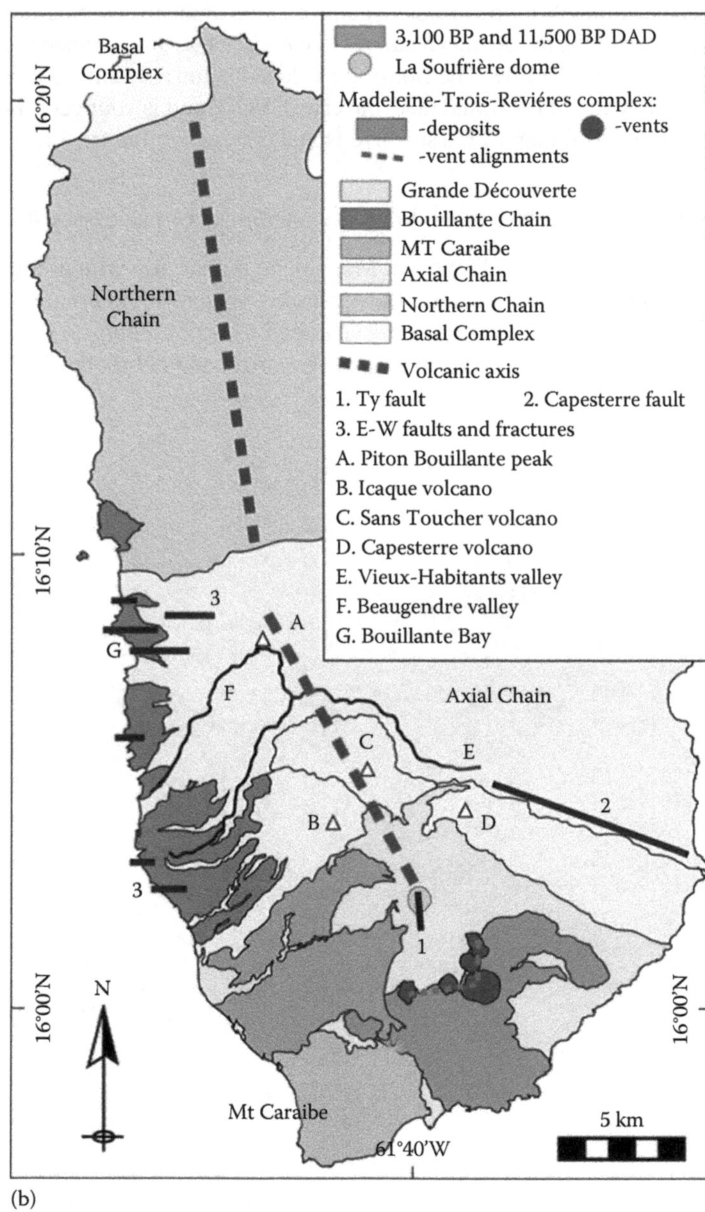

(b)

FIGURE 5.1 (*Continued*) **(See color insert.)** (a, b) Geology and structure of Martinique and Guadeloupe. ([a]: Data from Audru et al., 2015; Modified from Westercamp, D. et al., 1989 Carte géologique de la France (1/50, 000), feuille Martinique. BRGM. Notice explicative par W.D., Andreieff P., 1990; [b]: Adapted from Mathieu, L. et al., *Bull. Volcanol.* 75:700, 2013; Modified from Dagain 1981; Boudon et al. 1988; Gadalia et al. 1988; Samper et al., 2007.)

in peripheral valleys, and atmospheric fallout on a larger scale. The heterogeneity of volcanic aquifers can be observed at various scales.

- At the local scale, between, for example, interstitial and fracture porosity in ash and lava flows, respectively. These two types of porosity can coexist within the same aquifer.
- At the flow scale, interbedded ash layers (e.g., paleosols) and debris formations (lahars, breccia) can act as locally impermeable barriers, whereas scoriated lava flows can create preferential flow pathways.
- At the scale of the entire aquifer, preferential flow can occur along the axes of paleo-valleys, where the results of weathering of geological formations can decrease the overall permeability as the age of the deposits increases.

These heterogeneities therefore create a significant vertical and horizontal compartmentalization of volcanic aquifers. This results in a highly varied lithology that leads to a functioning at the scale of small basins (Foster et al. 1985; Lachassagne 2006; Charlier et al. 2008, 2011). Most of the aquifers are small, a few square kilometers at most, and have very heterogeneous hydrodynamic properties (Charlier et al. 2011, 2014; Vittecoq et al. 2015). A few of them, those that are the most permeable, are porous and are found mostly in recent unweathered pyroclastic formations. Most of the others, in lavas and hyaloclastites, have a fissure and/or a fracture porosity/hydraulic conductivity, which can vary highly according to the fracturation level and the deposit age.

5.3 MONITORING GROUNDWATER QUALITY

5.3.1 The European Water Framework Directive

Article 8 of the European WFD (Directive 2000/60/EC) stipulates that "Member States shall ensure the establishment of programs for the monitoring of water status in order to establish a coherent and comprehensive overview of water status within each river basin district," and that "for groundwater, such programs shall cover monitoring of the chemical and quantitative status."

To this end, the WFD introduced the notion of groundwater bodies as units for the assessment of chemical and quantitative status.

Six groundwater bodies have been identified in Martinique (North, North Atlantic, North Caribbean, Middle, South Caribbean, and South Atlantic) and five in Guadeloupe (Grande-Terre, North Basse-Terre, South Basse-Terre, Marie Galante, and La Désirade) (see gray outlines in Figures 5.2 and 5.3).

The Guadeloupe and Martinique Water Offices and BRGM have financed a groundwater quality monitoring network, managed by BRGM, since 2007 in Martinique and since 2008 in Guadeloupe.

5.3.2 Groundwater Quality Monitoring Network

The groundwater quality monitoring networks are made up of 17 wells and 3 springs in Martinique, and 10 wells and 2 springs in Guadeloupe (Figures 5.2 and 5.3).

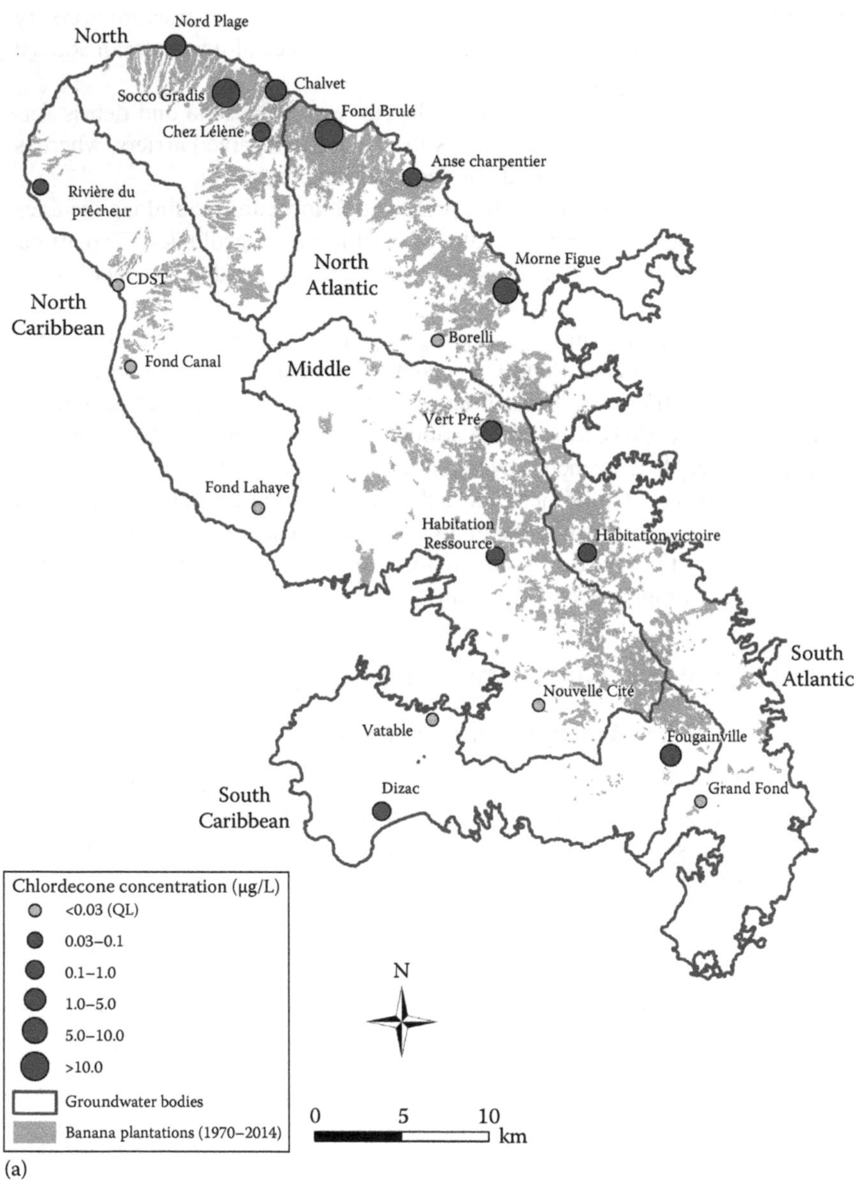

FIGURE 5.2 (a) Location of sampling points in Martinique Island and chlordecone. *(Continued)*

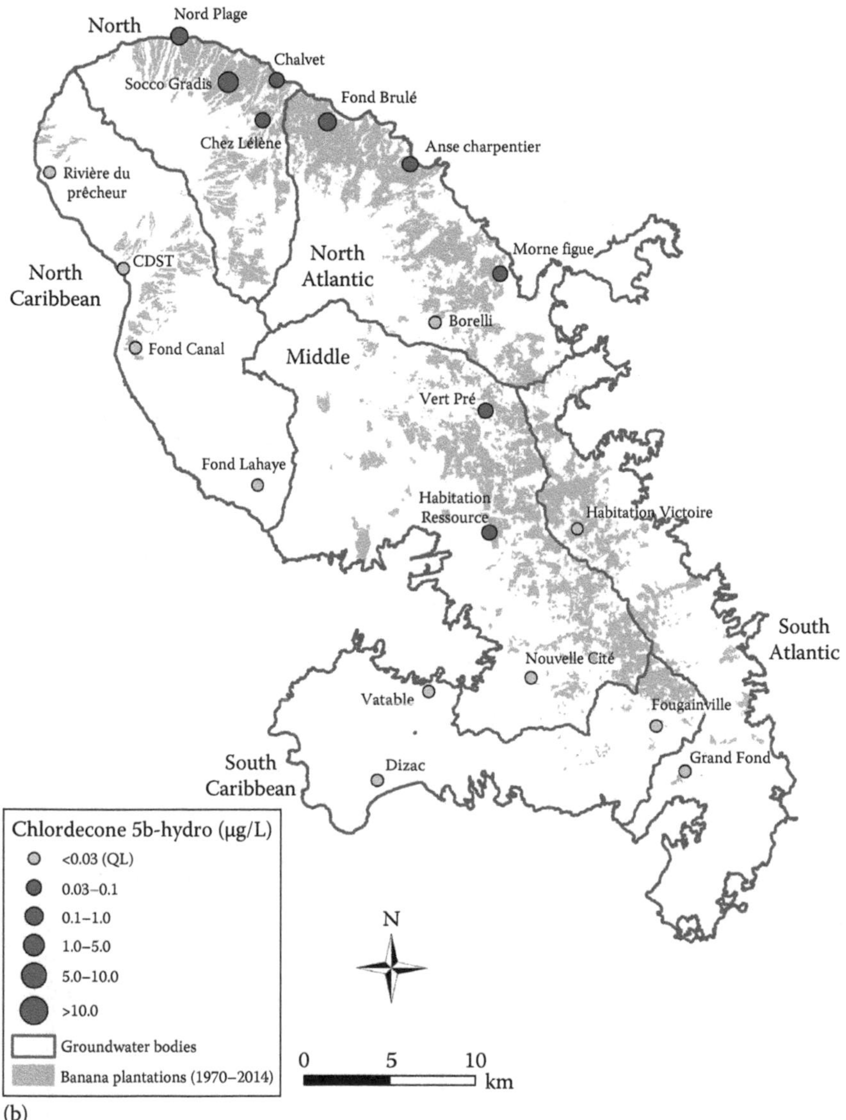

FIGURE 5.2 *(Continued)* (b) 5b-Hydrochlordecone mean concentrations over the period 2008–2014.

The groundwater quality monitoring program includes two sampling campaigns each year, one in April for the dry season and one in October/November for the rainy season. In Martinique, samples have also been collected monthly at two monitoring points since December 2008. One of the objectives of this monthly monitoring program was to improve our understanding of the pesticide transport toward groundwater. The results of these analyses are available to the public at the French website http://www.ades.eaufrance.fr (ADES 2016).

FIGURE 5.3 Location of sampling points on Guadeloupe Island and chlordecone mean concentrations over the period 2008–2014.

5.3.3 Sampling Procedure and Analytical Methods

BRGM complies with the technical recommendations of the AQUAREF groundwater sampling guidelines (http://www.aquaref.fr/). In addition to organic pollutants (organophosphorus and organochlorinated pesticides, PCB, phenoxy acids), physicochemical parameters are measured *in situ*, and major elements—organic matter, nitrogen compounds, and mineral micropollutants—are also analyzed. Here, we focus on the two molecules most commonly detected in water in the French West Indies: chlordecone and 5b-hydrochlordecone ($C_{10}Cl_9HO$; CAS number 53308-47-7), which is a dechlorinated product of chlordecone (Clostre et al. 2015a; Devault et al. 2015).

All pesticides are analyzed in the BRGM laboratory, which is accredited by COFRAC, the French accreditation body (http://www.cofrac.fr/). Chlordecone and 5b-hydrochlordecone are determined in water samples by LLE (liquid/liquid extraction) followed by gas chromatographic separation (GC) and mass spectrometric (MS/MS) identification. The chlordecone and 5b-hydrochlordecone limit of quantification (LOQ), determined by spiking natural water samples, was 0.03 μg L^{-1} for both compounds.

5.4 GROUNDWATER CONTAMINATION

5.4.1 Spatial Distribution of Chlordecone Contamination

5.4.1.1 Martinique Island

Figure 5.2 shows the average concentrations of chlordecone during the period 2008–2014 at the monitoring points in the WFD monitoring network in Martinique. Chlordecone is detected in 13 of the 20 monitoring points. It has been detected systematically since 2008, during each sampling campaign, barring three wells (Rivière du Prêcheur, Habitation Ressource and Dizac), where the contamination seems to be episodic.

Groundwater contamination seems to be generalized near and downstream from areas where chlordecone is known to have been applied. The Dizac well, the impluvium of which is not known to have ever been used for growing bananas, reveals episodic groundwater contamination. This might be explained by the illegal use of this molecule (marketed solely for banana crops) during the period when it was on the market. At the scale of the entire island, the average concentrations between 2008 and 2014 vary between 0.04 (Rivière du Prêcheur) and 32.1 μg L^{-1} (Fond Brûlé).

Figure 5.2 shows average concentrations of 5b-hydrochlordecone during the period 2008–2014. The molecule is not always detected along with chlordecone and its concentrations are systematically lower. For Martinique, it is quantified in 10 of the 20 monitoring points, and is systematically detected in the island's most contaminated wells and springs (Socco Gradis, Nord Plage, Fond Brulé, and Morne Figue).

5.4.1.2 Guadeloupe Island

Figure 5.3 shows the average concentrations of chlordecone during the period 2008–2014 at the monitoring points in the WFD network and the drinking water wells monitored by the Guadeloupe Regional Public Health Agency (ARS). Chlordecone was detected in 7 of the 35 monitoring points in the south of Basse-Terre Island, near and downstream from areas where it is known to have been applied, an area known locally as the Basse-Terre "Banana Crescent." At five monitoring points (La Plaine, Belleterre, Soldat, Tabacco 1, and Marquisat springs), chlordecone has been systematically detected since 2008, during each sampling campaign. Available data for Saint-Denis 1 and Fromager, two wells drilled in 2014, are still too recent to enable us to draw conclusions concerning water contamination.

Neither the groundwater monitored in the northern part of Basse-Terre nor any of the other islands in the archipelago seems to be contaminated by chlordecone. However, a few isolated detections have been reported in Nord Basse-Terre, Grande-Terre, and recently in La Désirade. Two hypotheses have been proposed to explain this observation: (1) the former, illegal use of chlordecone outside of the major banana growing areas and (2) the contamination of groundwater by pipes carrying raw water to Grande-Terre and La Désirade from the southern part of Basse-Terre. At the scale of the entire island, the average concentrations vary between 0.03 (detection limit used as a reference) and 1.2 µg L^{-1} (Soldat spring) between 2008 and 2014.

5.4.1.3 Summary of the Level of Groundwater Contamination and Classification of the Chemical Status (WFD) of Groundwater in the French West Indies

The most contaminated groundwater bodies are located in areas where pesticide use was greatest in terms of duration and extent: the northern and northeastern Martinique and the Banana Crescent in southern Guadeloupe. Average concentrations exceed the drinking water standard (0.1 µg L^{-1}) almost systematically, and are more than 100 times greater than the standard in some areas. In some wells, chlordecone concentrations are, at times, extremely high. For example, maximum concentrations of 108 µg L^{-1} at Lorrain and 44 µg L^{-1} at Basse Pointe-Socco Gradis in November 2009 in Martinique, and 26 µg L^{-1} of chlordecone, were measured at Fromager in March 2014 in Guadeloupe.

Other pesticides are detected in the groundwater in Martinique and in Basse-Terre on Guadeloupe, but always at lower concentrations. Those most often quantified are propiconazole, bromacil, dieldrin, β-HCH and metalaxyl. Atrazine (as well as its metabolites) is also detected in the Grande-Terre groundwater body, as is glyphosate in Marie-Galante groundwater body.

In terms of the WFD, the contamination of groundwater by chlordecone is, therefore, in large part responsible for the poor chemical status of three groundwater bodies in Martinique and one body in the southern part of Basse-Terre.

5.4.2 Evolution of Chlordecone Contamination

5.4.2.1 Multiannual Variations

The evolution of chlordecone concentrations at biannual frequency for wells in Martinique is shown in Figure 5.4. Similar trends are clearly observed from one monitoring point to another, with:

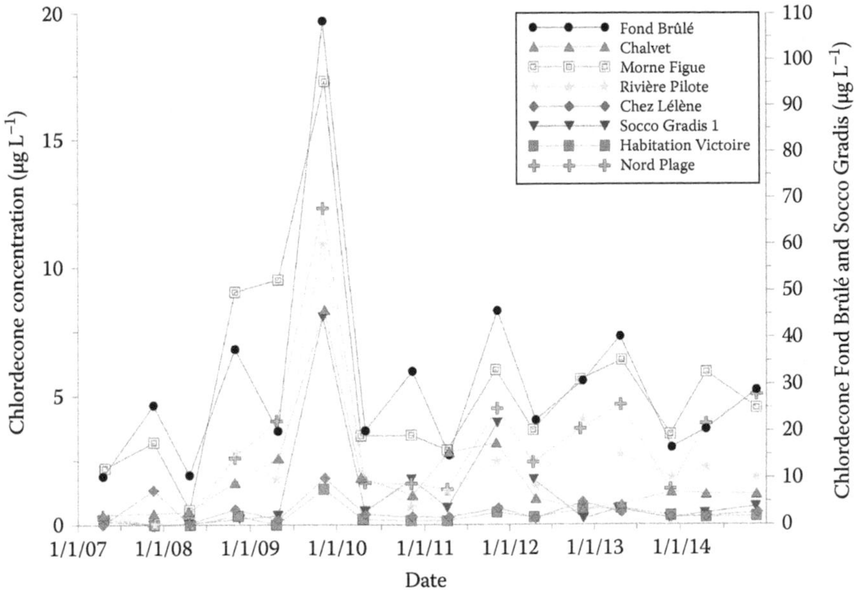

FIGURE 5.4 Evolution of chlordecone concentration with a biannual time step for the monitoring network in Martinique.

- A nearly systematic increase in concentrations between the end of the dry season (April) and the end of the rainy season (November) between 2007 and 2012, which indicates an increase in chlordecone concentration during rainy periods and a decrease during low water levels. Beginning in 2013, these variations are no longer stable and seem to reverse for the most contaminated monitoring points (increase in concentrations during the 2013 and 2014 dry seasons). For the monitoring points with the least contamination, the evolution of concentrations is stationary for the most part after the end of 2012. This change comes after an exceptionally rainy month of April 2013.
- An unusual rainy season in 2009 during which a very major increase in the concentration is recorded. For each point, these are the maximum concentrations recorded since monitoring began.

To conclude, apart from the concentration peak observed in 2009 in Martinique, the overall evolution of groundwater contamination in the French West Indies seems to be stable for the most part since 2008. Seasonal variations (dry season vs. rainy season) seem to explain the alternation of the peaks at the biannual time step.

5.4.2.2 Monthly Variations

In spite of a rainfall pattern strongly controlled by alternating dry and wet periods, the monthly monitoring done since December 2008 (Figure 5.5) for two wells in Martinique (Chalvet and Chez Lélène) provides a satisfactory resolution to the

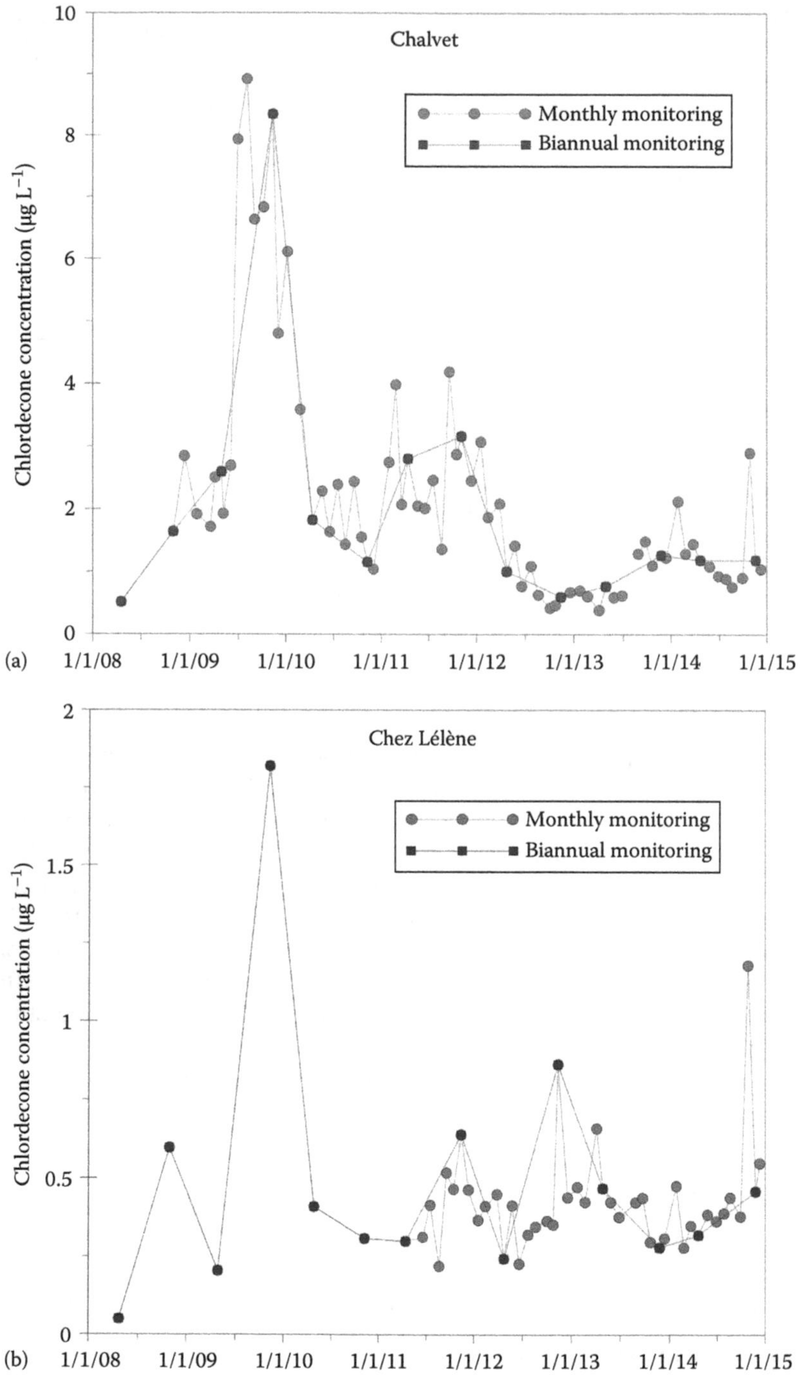

FIGURE 5.5 Evolution of chlordecone concentrations at the monthly and biannual time steps for the (a) Chalvet and (b) Chez Lélène wells (2007–2014).

study of the dynamics of contamination. This monitoring shows concentration peaks that are not revealed by biannual monitoring, in particular the peaks measured during summer 2009, summer 2010, and February and September 2011 for the Chalvet well, and in January and October 2014 for the two monitoring points. Although the dynamics seem to be fairly similar, with high absolute values in both cases (much higher than the drinking water standard), the measured concentrations are seven times greater in the Chalvet well. Using the results of this monthly monitoring, we were able to study the factors that might control the variability of the concentrations (Arnaud and Tailame 2011).

5.4.2.3 Correlation between Water Table Fluctuations and Groundwater Contamination

The water table fluctuations recorded in the Chalvet well (Figure 5.6) reveal a predominantly seasonal pattern to which a multiannual pattern is added. This type of groundwater-level fluctuation, with a strong inertia, indicates that the aquifer is both transmissive and has a storage function consistent with the hydrogeological properties of the aquifer (interstitial porosity of pumice and pyroclastic flows). For the

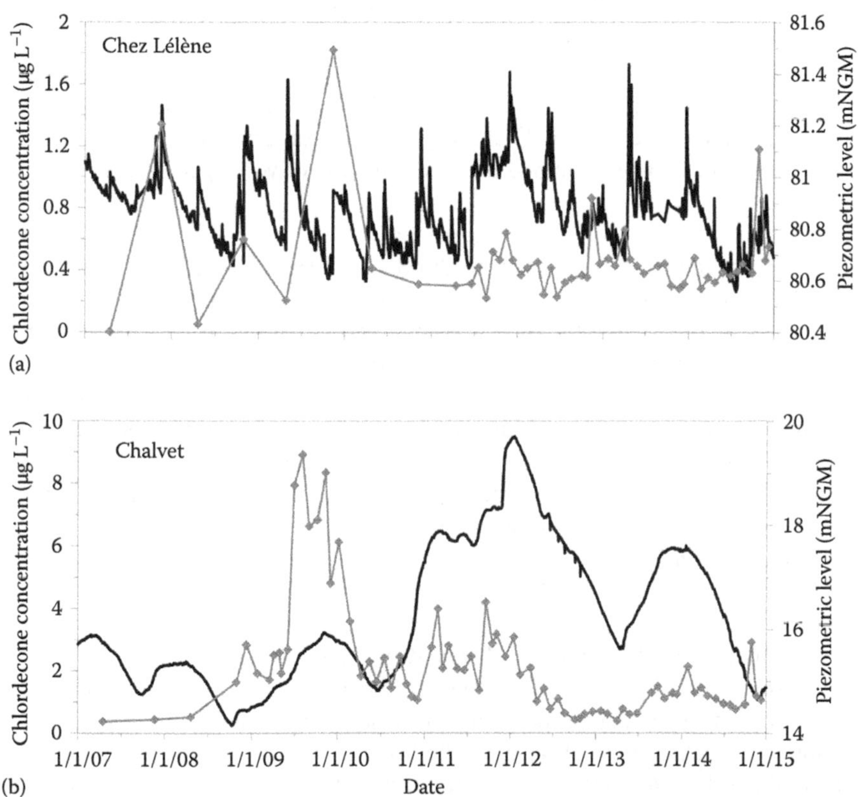

FIGURE 5.6 Evolution of chlordecone concentrations and groundwater levels in the (a) Chez Lélène and (b) Chalvet wells (2007–2014).

Chez Lélène well, the water table fluctuations are much more rapid and lower in amplitude (maximum annual fluctuation 0.8 m). This can be explained by a more transmissive pumice aquifer and the proximity of the Falaise River, which controls the hydraulic head and buffers water table fluctuations. The water table reacts to major rainfall events within 1–2 h.

The increase in chlordecone concentrations is often observed during periods of rising groundwater levels, at both the Chalvet and Chez Lélène wells (Figure 5.6). There was, however, one exception—involving the Chalvet well in October 2014, when an increase in chlordecone concentration was observed at the end of the period of aquifer discharge.

On the other hand, the intensity of water table fluctuations does not seem to correlate with the amplitude of variations in concentration. It is also interesting to note that the lowest concentrations in the Chalvet well, measured between June 2012 and June 2013, were recorded following a long dry period (i.e., with no aquifer recharge) during the 2012 rainy season (Figure 5.6).

To conclude, periods of high and low contamination seem to be controlled by periods of rising and falling groundwater levels, and therefore by aquifer recharge.

5.4.2.4 Correlation between Aquifer Recharge and Groundwater Contamination

A statistical analysis of monthly infiltration was done to characterize infiltration conditions compared to inter-annual means. The high chlordecone concentrations measured between June 2009 and February 2010 and observed for all of the islands' monitoring points can be explained by a combination of very specific hydroclimatic conditions:

- Deficient recharge from 2006 to 2008 that resulted in a very low water table. During this period, chlordecone concentrations were generally low.
- Excess recharge in 2008/2009 accompanied by an initial increase in concentrations during the second half of 2008.
- A very humid beginning of the rainy season in 2009 with two remarkable tropical storms in May and June given large amounts of rainfalls, and thus of recharge. A major increase in concentrations was then observed during the month of June. During the exceptional tropical storm in May 2009, around 300 mm of rain was experienced in 3 days.
- A dry season in 2009/2010 accompanied by a decrease in concentrations beginning in December 2009.

The very high concentration peak observed in the second semester in the Chalvet well (as well as in all of the wells in the WFD network in Martinique) seems to be the result of a major recharge following a dry period that lasted several years. This resumption of recharge might have resulted in a remobilization of the chlordecone stored in the soil and/or in the aquifer's unsaturated zone, a remobilization that is all the more pronounced due to the fact that the water table was previously particularly low. Such a high level of contamination might therefore reoccur when very rainy episodes follow several years of deficient rainfall.

To conclude, the variations in chlordecone concentrations cannot be explained by the variations of the hydrologic cycle (at the annual scale) alone. It is highly probable that the multiannual piezometric cycles play a major role in the level of groundwater contamination. Due to the inertia of certain aquifers, the validation of this hypothesis will require long-term data acquisition.

5.4.3 Contribution of Groundwater to River Contamination

Surface water–groundwater interaction was analyzed in the Falaise River catchment in 2011–2012, where the Chez Lélène well is located (a distance of 10 m between well and river). By monitoring the physicochemistry along the river channel from upstream to downstream parts in the catchment, we were able to confirm the groundwater contributions to the stream, this being greatest during low water level periods (Arnaud et al. 2012). The comparison of the evolution of chlordecone concentrations in surface water and groundwater (Figure 5.7) shows that during the dry season the evolutions are similar, although concentrations are four times greater in the groundwater. On the other hand, during the rainy season, the fluctuations seem to be the opposite: The increases in concentration in the groundwater are concomitant with dilution of the concentration in the river, and vice versa. This can be explained by analyzing the hydro(geo)logical processes occurring during the two periods. During the dry period, the aquifer recharges the river and contributes to surface water contamination. The lower contamination of surface water is due to the contribution of other, uncontaminated groundwater bodies, notably those originating above the banana plantation zone (a similar process has been observed in Guadeloupe,

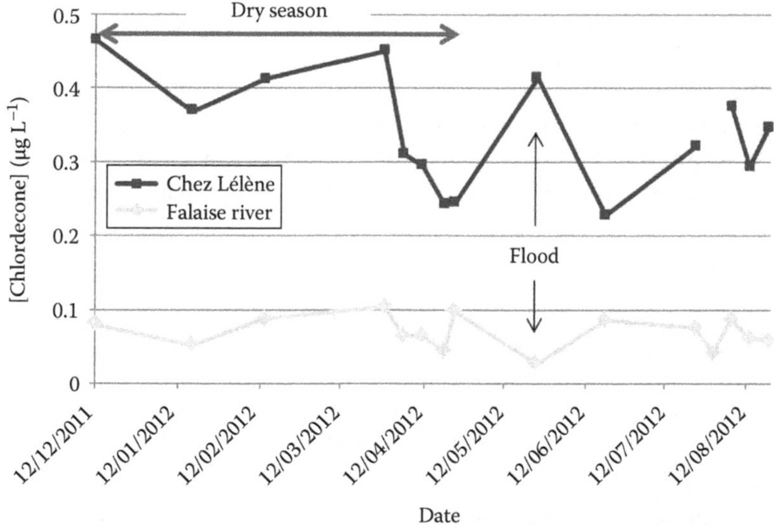

FIGURE 5.7 Chlordecone concentrations in the Chez Lélène well and in the Falaise River between December 2011 and August 2012.

Charlier et al. 2015). During the rainy season, infiltrating water near the well leaches chlordecone, increasing the concentration in the groundwater, whereas surface runoff that is potentially less concentrated due to the contribution of the uncontaminated head of the basin dilutes the river. To conclude, groundwater contributes to the degradation of the quality of the river water during low water level periods, whereas its influence can be diminished during rainy periods when runoff from uncontaminated areas is predominant.

5.5 CONCLUSION

In Guadeloupe and Martinique, groundwater contamination by chlordecone, a highly residual organochlorine pesticide, is widespread in areas where the molecule has been applied. Very high concentrations are observed, usually exceeding the drinking water standard (0.1 $\mu g\ L^{-1}$), and concentrations more than 100 times the standard have been detected in Martinique. Groundwater contamination therefore seems to be generally greater than what is cited in the literature for surface water. With the exception of seasonal variations, the groundwater contamination appears to be stable since the quality monitoring program began (2008). To date, no decreasing trend has been observed. Given what we know today, it is reasonable to think that the evolution of chlordecone concentrations is governed by both the hydrologic cycle during periods of aquifer recharge and discharge, and by the multiannual piezometric cycles, due to the high inertia of transport in the aquifers.

Although the biannual dynamics seem to be similar in all of the monitoring points due to a certain homogeneity of the regional rainfall pattern, the measured concentrations are spatially highly variable. The level of contamination of each aquifer is certainly strongly influenced by specific local conditions: soil and hydrogeological contexts, history of agricultural practices, and so on. The role that these play in groundwater contamination raises questions and they would be well worth studying more precisely at the catchment scale.

Finally, this study shows that in certain contexts, where infiltration is assumed to be predominant, groundwater contributes to chlordecone contamination of surface water. This review of the contamination of groundwater by chlordecone highlights the significant role played by aquifers in the degradation of river water quality.

REFERENCES

ADES. July 2016. National Portal to Groundwater Data (Portail National d'Acces aux données sur les eaux souterraines). French public information system on water. French Ministry of Environment and Sustainable Development. http://www.ades.eaufrance.fr/ConsultationPEResultatRecherche.aspx (consulted July 2016). Accessed June 2016.

Arnaud L., Baran N., Gourcy L., Taïlamé A.-L., Senergues, M. 2012. Étude du transfert de la chlordécone vers les eaux souterraines en Martinique. BRGM report. BRGM/RP-61767-FR.

Arnaud L., Tailame A.-L. 2011. Contrôle de surveillance et contrôle opérationnel de la qualité des masses d'eau souterraine de la Martinique. Rapport annuel 2010. BRGM report. BRGM/RP-60232-FR.

Audru J.-C., Le Roy M., Rançon J.-P. 2015. Curiosités géologiques de la Martinique. BRGM Editions. Editions Orphie.

Boudon G., Dagain J., Semet M., Westercamp D. 1988. Le massif volcanique de la Soufrière (département de la Guadeloupe, Petites Antilles). Carte géologique au 1/20 000ème, BRGM Editions.

Cabidoche Y.M., Achard R., Cattan P., Clermont-Dauphin C., Massat F., Sansoulet J. 2009. Long-term pollution by chlordecone of tropical volcanic soils in the French West Indies: A simple leaching model accounts for current residue. *Environ. Pollut.* 157:1697–1705.

Cabidoche Y.M., Lesueur-Jannoyer M. 2012. Contamination of harvested organs in root crops grown on chlordecone-polluted soils. *Pedosphere* 22(4):562–571.

Cattan P., Cabidoche Y., Lacas J., Voltz M., 2006. Effects of tillage and mulching on run-off under banana (*Musa* spp.) on a tropical Andosol. *Soil Tillage Res.* 86:38–51. doi: 10.1016/j.still.2005.02.002.

Charlier J.-B., Arnaud L., Ducreux L., Ladouche B., Dewandel B. 2015. CHLOR-EAU-SOL – volet EAU: Caractérisation de la contamination par la chlordécone des eaux et des sols des bassins versants pilotes guadeloupéen et martiniquais. BRGM report. BRGM/RP-64142-FR.

Charlier J.-B., Cattan P., Moussa R., Voltz M. 2008. Hydrological behaviour and modelling of a volcanic tropical cultivated catchment. *Hydrol. Process.* 22(22):4355–4370. doi: 10.1002/hyp.7040.

Charlier J.-B., Cattan P., Voltz M., Moussa R. 2009. Transport of a nematicide in surface and groundwaters in a tropical volcanic catchment. *J. Environ. Qual.* 38(3):1031–1041. doi: 10.2134/jeq2008.0355.

Charlier J.-B., Lachassagne P., Ladouche B., Cattan P., Moussa R., Voltz M. 2011. Structure and hydrogeological functioning of an insular tropical humid andesitic volcanic watershed: A multi-disciplinary experimental approach. *J. Hydrol.* 398(3–4):155–170. doi: 10.1016/j.jhydrol.2010.10.006.

Clostre F., Cattan Ph., Gaude J.-M., Carles C., Letourmy Ph., Lesueur-Jannoyer M. 2015a. Comparative fate of an organochlorine, chlordecone, and a related compound, chlordecone-5b-hydro, in soils and plants. *Sci. Total Environ.* 532:292–300. doi: 10.1016/j.scitotenv.2015.06.026.

Clostre F., Letourmy Ph., Lesueur-Jannoyer M. 2015b. Organochlorine (chlordecone) uptake by root vegetables. *Chemosphere* 118(2015):96–102. doi: 10.1016/j.chemosphere.2014.06.076.

Coat S., Bocquené G., Godard E. 2006. Contamination of some aquatic species with the organochlorine pesticide chlordecone in Martinique. *Aquat. Living Resour.* 19:181–187.

Coat S., Monti D., Legendre P., Bouchon C., Massat F., Lepoint G. 2011. Organochlorine pollution in tropical rivers (Guadeloupe): Role of ecological factors in food web bioaccumulation. *Environ. Pollut.* 159:1692–1701.

Dagain J. 1981. La mise en place du massif volcanique Madeleine-Soufrière, Basse-Terre de Guadeloupe, Antilles. Thèse de 3ème cycle de l'Université de Paris Sud.

Devault D.A., Laplanche C., Pascaline H., Bristeau S., Mouvet C., Macarie H. 2016. Natural transformation of chlordecone 5h-hydrochlordecone in French West Indies soils: statistical evidence for investigating long-term persistence of organic pollutants. *Environ. Sci. Pollut. Res.* 23(1):81–97.

DIRECTIVE 2000/60/CE (DCE). 2000. Du parlement européen et du conseil du 23 octobre 2000 établissant un cadre pour une politique communautaire dans le domaine de l'eau.

Foster S.S.D., Ellis A.T., Losilla-Penon M., Rodriguez-Estrada H.V. 1985. Role of volcanic tuffs in ground-water regime of Valle Central, Costa Rica. *Ground Water* 23(6):795–801.

Gadalia A., Gstalter N., Westercamp D. 1988. La chaîne volcanique de Bouillante, Basse-Terre de Guadeloupe: identité pétrographique, volcanologique et géodynamique. Géologie de la France, n° 2–3, pp 101–130.

Germa A. 2010. Évolution volcano-tectonique de l'île de la Martinique (arc insulaire des petites Antilles): Nouvelles contraintes géochronologiques et géomorphologiques. UFR scientifique d'Orsay. Thèse de l'Université Paris XI, Paris, France.

Gourcy L., Baran N., Vittecoq B. 2009. Improving the knowledge of pesticide and nitrate transfer processes using age dating tools (CFC, SF_6, ^3H) in a volcanic island. *J. Contam. Hydrol.* 108(3–4):107–117.

Lachassagne P. 2006. Chapitre: XIII. DOM-TOM. 1. Martinique. In *Aquifères et Eaux souterraines en France*, Roux J.-Cl. (ed.), AIH-IAH, BRGM Editions, Tome 2, pp. 768–781.

Levillain J., Cattan P., Colin F., Voltz M., Cabidoche Y.-M. 2012. Analysis of environmental and farming factors of soil contamination by a persistent organic pollutant, chlordecone, in a banana production area of French West Indies. *Agric. Ecosyst. Environ.* 159:123–132.

Mathieu L., Wyk de Vries B., Mannessiez C., Mazzoni N., Savry C., Troll V.R. 2013. The structure and morphology of the Basse Terre Island, Lesser Antilles volcanic arc. *Bull. Volcanol.* 75:700.

Mottes C., Lesueur-Jannoyer M., Charlier J.-B., Carles C., Guéné M., Le Bail M., Malézieux E. 2015. Hydrological and pesticide transfer modeling in a tropical volcanic watershed with the WATPPASS model. *J. Hydrol.* 529:909–927. doi: 10.1016/j.jhydrol.2015.09.007.

Samper A., Quidelleur X., Lahitte P., Mollex D. 2007. Timing of effusive volcanism and collapse events within an oceanic arc island: Basse-Terre, Guadeloupe archipelago (Lesser Antilles Arc). *Earth and Planetary Science Letters.* 258(1–2):175–191.

Vittecoq B., Reninger P.A., Violette S., Martelet G., Dewandel B., Audru J.-C. 2015. Heterogeneity of hydrodynamic properties and groundwater circulation of a coastal andesitic aquifer controlled by tectonic induced faults and rock fracturing—Martinique island (Lesser Antilles—FWI). *J. Hydrol.* 529(2015):1041–1059.

Westercamp D., Andreieff P., Bouysse P., Cottez S., Battistini R. 1989. Notice explicative de la carte géologique au 1/50 000 de la Martinique. BRGM, Orléans.

Westercamp D., Pelletier B., Thibault P.A., Traineau H. 1990. Carte géologique de la France (1/50 000), feuille Martinique. BRGM, Orléans.

Section III

Biotope Contamination and Eco-Toxicological Assessment (Terrestrial and Aquatic Ecosystems)

6 Environmental Assessment of Rivers

What Have We Learnt from Implementation and Results of the First Chlordecone National Action Plan?

Marion Labeille, Fanny Caupos, and Julie Gresser

CONTENTS

6.1 INTRODUCTION

A National Action Plan (Ministry of Public Health) specific to chlordecone was put forward in 2008 to manage this pollution. It recommends two actions regarding aquatic habitats: action 2, monitoring contamination of river waters, and action 5, assessing the contamination of aquatic fauna, fishes and macro-crustaceans, in Martinique rivers.

This chapter provides the results of monitoring of this contamination in Martinique rivers from 2008 to 2012 (Bargier and Desrosiers 2009; Labeille and Bargier 2011; Labeille and Vergès 2011; Labeille and Vergès 2012). This survey covered the whole island and complemented the monitoring conducted by the DIREN (now DEAL, the regional environmental authority) from 1999 to 2006 and by the Martinique water office (ODE) since 2007. This chapter also presents the adaptive and iterative approach implemented by the local authorities, the technical and field survey teams, and the laboratory in order to establish methodological frameworks for aquatic environment quality monitoring implementation. This sampling program occurred in a context of crisis management situation, rather than a scientific one, which should bring answers on how the processes involved. Our aim was to gain an overview of the distribution and intensity of the contamination in order to support future environmental and public health policies, and to identify further prior lines of research actions in this field that need financing. The surveying itself was conducted by Asconit Consultants under the supervision of ODE and DEAL, and was financed by these institutions with the help of the French national water office (ONEMA).

To understand the peculiarity of chlordecone contamination, a synthesis on the contamination of rivers by all pesticides was conducted from traditional monitoring networks in Martinique and was discussed in the chapter.

6.2 METHOD

In this section, we describe the adaptive nature of our approach used by discussing the technical surveying criteria regarding (1) the localization of stations, (2) the makeup of batches for the analysis of fauna, (3) the choice of species depending on specific field conditions, (4) the laboratory's own needs, and finally (5) the quantification efforts upon which the precision of our results depended.

6.2.1 Definition of the Sampling Program

The aim of the sampling program was to assess the geographical extent and degree of the chlordecone and 5b-hydro-chlordecone contamination in aquatic habitats. This assessment was articulated around four phases, each composed of one to three campaigns carried out between 2008 and 2012. A campaign was the period during which a series of stations were monitored. Here, different campaigns were selected according to hydrologic periods: wet or dry season. A station was a specific sampling point of the river.

The first phase was conducted for a first overview of the situation with three sampling campaigns conducted between October 2008 and May 2009. The sampling plan included about 90 stations. Water analyses were performed in all of them, and sediment and fauna analyses in about 50 of them, not necessarily the same. These stations were located (1) in areas with known contamination or in areas potentially contaminated due to the presence of former banana production sites, (2) in rather large watershed areas greater than 5 km^2, and (3) in areas of special health concerns about water abstractions such as drinking water supply areas, aquaculture installations, or agriculture irrigation (river abstraction higher than 150 m^3 h^{-1}) (Bocaly 2008). The results showed a broad geographical extent of contamination in aquatic habitats, and a particularly intense contamination in aquatic fauna. These were considered so alarming that local authorities decided to ban fishing in all Martinique rivers and to continue with the assessment of nearly all Martinique rivers.

In light of the alarming initial results, three other phases were set up. They were conducted between June 2010 and February 2012, each made up of one or two campaigns. The same three environmental compartments, water-sediments-aquatic fauna, were analyzed for approximately 30 stations. The purpose of this additional sampling was (1) to investigate upstream areas of rivers where downstream contamination was already established, (2) to finalize an analysis of aquatic fauna levels of contamination and the assessment of certain sites where only water and sediments were studied, and (3) to assess areas far from areas *a priori* removed from all sources of contamination and popular with fishing enthusiasts (these were located with the help of FDAAPPMA, the Martinican fishermen's federation). These last results alleviated some concerns regarding the spatial extent of aquatic habitat contamination and the higher levels of fauna contamination. The general diagnosis, however, remained unchanged.

To conclude, after 5 years of study, we got a quasi-exhaustive sampling of the 70 main rivers of Martinique. The very small rivers and inaccessible zones, often upstream, were excluded. The rivers, often those with a small length, might be subject to a single sampling, in the integrative downstream zone, near the outlet. When the downstream zone was brackish, the electric fishing protocol could not be used and the intermediate zone was then assessed. The main rivers were most often assessed on several zones along their course that allowed assessing whether or not there were differences between upstream and downstream contamination of fauna and waters.

6.2.2 Sampling Methods

Sampling was carried out according to sampling protocols described in NF EN ISO 5667 (AFNOR 2003), a French standard for water that was also included in the technical sampling guide published by the Agence de l'Eau Loire-Bretagne (Gay Environment 2006) on water and sediment sampling. For fauna, the sampling mode and methods were governed by the French standard NF EN 14011. Fish and crustacean samples were collected by electrofishing technique at the water and sediment sampling point.

Water and sediments were frequently sampled in Martinique rivers. It was not the case for fauna. A review was needed for the selection of sentinel species to be sampled and on the minimum sample weight needed to ensure good conditions both for the analytical process and the quality of results. It was decided that the

TABLE 6.1
Fish and Crustacean Sentinel Species Priority

Priority	Fish	Crustacean
1	*Sicydium* sp. (sentinel species)	*Atya scabra* and/or *A. innocuous*
2	*Eleotris perniger*	*Macrobrachium crenulatum* and *M. heterochirus*
3	*Anguilla rostrata*	*Macrobrachium carcinus*

stations would be assessed based on analyses carried out on batches of different species of fauna. Each batch of one species comprised at least three whole subjects, if possible homogeneously sized. Egg-carrying female of crustaceans were released. Each station provided between one and three batches, including at least, where possible, one crustacean and one fish species. Priority was given to the sampling of *Macrobrachium faustinum*/*Atya* sp. for shrimps and *Agonostomus monticola* for fish during the first phase. Collecting those species was, however, not always possible because of the scarcity and low biomass of fauna. Other species had to be included. The samples collected during the first phase allowed us to define the most frequently caught species and to draw a prioritized list of sentinel species for the subsequent phases (see Table 6.1). When these primary taxons were absent, second or third species were to be collected when they could satisfy the recommendations cited earlier (number, size, etc.).

The laboratory initially requested that each batch weigh a minimum of 200 g. This stipulation was very inconvenient when applied to less frequent species. The size and weight of the macrofauna, particularly of fish, in the French West Indies were usually reduced compared with the specimens in European rivers (Fiévet 1995). According to this specification, the laboratory teams adapted the threshold value that was reduced to 100 g for low-biomass species. In case of real difficulty and after 1 h 30 of fishing, a minimum of 60 g was deemed acceptable, which was the very limit to analyze chlordecone.

6.2.3 Analysis Methods

Analysis on all matrixes was performed in the Laboratoire Départemental de la Drôme, LDA 26. The laboratory is accredited by Cofrac, the French Accreditation Committee, for pesticide analysis on all matrixes. Moreover, the laboratory uses methods recommended by ANSES, the National Food, Environmental and Work Safety Agency, that guarantees their technical capacity and reliability, as well as good management practices.

The analyzed molecules were chlordecone, and its co-product, 5b-hydrochlordecone.

Results were provided with the same quantification threshold of 5 $\mu g\ kg^{-1}$ of fresh weight (FW) for fauna throughout the study. It was not the case for "water" and "sediment" compartments. LDA 26 lowered the quantification thresholds on these two compartments during the second phase of the study. The quantification threshold was reduced from 0.01 to 0.003 μg L^{-1} in water and from 10 to 0.15 μg kg^{-1} dry weight (DW) for sediments. In the third and fourth phases, ODE and DEAL decided to continue working with the reduced threshold of 0.003 μg L^{-1} in water. But for sediment analysis, the cost–benefit balance was found to be insufficient and so the former limit of 10 μg kg^{-1} DW was reinstated.

6.2.4 Choices Regarding Data Use

At the start of the study in 2008, based on a consensus obtained with all stakeholders concerned, we chose to compare the results with the existing regulatory standards for water and fauna:

- 0.1 μg L^{-1} for water, which was the upper limit for pesticides according to the French Watercourse Water Quality Assessment System, SEQ-Eau v2 (MEDD & Agences de l'Eau Loire-Bretagne 2003) and for drinking water supply (decree of January 11, 2007).
- 20 μg kg^{-1} FW of fauna, which was the upper limit for residues as established by ANSES (decree of June 30, 2008), beyond which threshold the sale of fish, crustaceans, and so on for freshwater fishing was no longer permitted.

The four quality levels set up to depict the results of the study and perform situational analysis were based on these regulatory standards for water and fauna (see Figure 6.1). As no environmental quality standards existed for sediments, three groups were drawn based on quantification thresholds.

We were able to study some stations more than once. In such cases, only the highest result was kept for analysis and shown subsequently. Results for aquatic fauna were also given depending on the number of batches tested.

Chlordecone and 5b-hydrochlordecone data were analyzed separately, according to the same regulation thresholds.

6.3 RESULTS

The results for chlordecone and its co-product, 5b-hydrochlordecone, are presented in the following paragraphs. The values obtained are compared with the regulatory

Contamination	Absent	Low	Medium	High
Water & fauna	Below QT	Below RS	Less than 10 times over RS	More than 10 times over RS
Sediments	Below QT	Less than 10 times over QT		More than 10 times over QT

FIGURE 6.1 **(See color insert.)** Quality classes for water, fauna and sediments (colors are the same as used on the maps fig. 2 and 4; QT=quantification threshold, RS=regulatory standards).

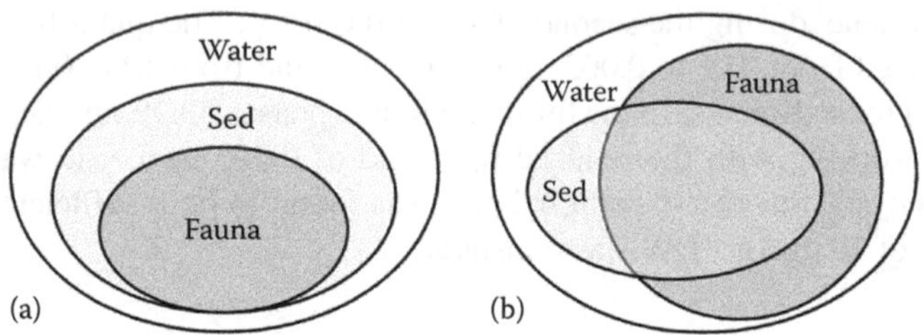

FIGURE 6.2 Relationship between the different sampling types for watersheds (a) and stations (b).

standards in place and show that contamination is not uniformly spread throughout the island.

From October 2008 to February 2012, the following environmental compartments were analyzed:

- "Water" in 44 watersheds and 116 stations
- "Sediments" in 41 watersheds and 73 stations
- "Aquatic fauna" in 38 watersheds and 82 stations.

Sediment and aquatic fauna have always been collected from a station where the water was analyzed. But sediment and aquatic fauna have not always been taken from the same stations, as shown in Figure 6.2.

6.3.1 Water

On the 116 stations, Figure 6.3 shows:

- An absence of contamination in 35% of stations, thus two-thirds of the sampling points were polluted, with the majority above the regulatory standard. The noncontaminated stations are mainly located in the North Caribbean and South Caribbean areas and upstream some rivers in other areas (see Figure 6.4).
- A low-level contamination, below the regulatory standard, in 23% of stations. These are located in the south, in the upstream zone in the North Caribbean area, and to a lesser extent in the Fort-de-France urban area.
- A medium to high level of contamination in 42% of stations, with results exceeding the regulatory standard. It should be noted that close to one-fifth of sampling points are highly contaminated and present over 1 $\mu g\ L^{-1}$, which is 10 times above the regulatory standard. These stations are located exclusively in the Central and North Atlantic areas, and in almost all of the prospected watersheds. On the North Atlantic side, mostly downstream stations are affected, whereas in the Central area, the whole system is severely impacted. Surprisingly, some upstream points were polluted by a medium

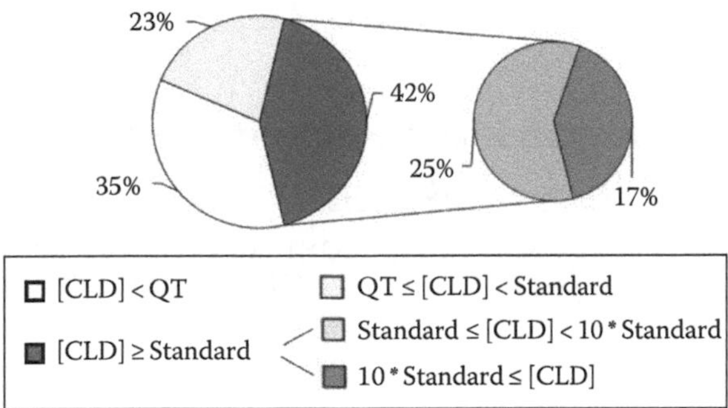

FIGURE 6.3 **(See color insert.)** Distribution of results for water by station (qualified on their worst results) according to the quality classes defined for chlordecone (in % of the number of stations—CLD, chlordecone; QT, quantification threshold; Standard, regulatory standard).

to high level of contamination. Maximum values of 6.10 and 4.28 μg L^{-1} are recorded downstream of the Sainte-Marie River and its tributary, the Bambou River, in the North Atlantic area.

Whenever two or more analyses were performed on the same station, the values obtained were by and large in the same order of magnitude. This suggested a chronic pollution of waters.

Presence of 5b-hydrochlordecone was detected in 46% of the studied stations. Only six stations showed a concentration above or equal to the 0.1 μg L^{-1} standard. They were all located in the North Atlantic area. The maximum value obtained was 0.26 μg L^{-1}, which was far less than the maximum levels observed with chlordecone.

6.3.2 Sediments

Sixty-four percent of stations studied were not contaminated by chlordecone; thus more than one-third of the stations were contaminated. This was a surprising result since two-third of the sampling site for water was contaminated. One explanation was that the residence time of sediment is short probably due to the torrential flow of rivers in Martinique, with a potential renewing with a non-contaminated material originating from non-contaminated areas of the watershed. The majority of concentrations ranged from 10 to 50 μg kg^{-1} DW. The five stations showing concentrations above 100 μg kg^{-1} DW, which was 10-times over the quantification threshold, were located on the North Atlantic coast and in the Central area. The results presented in Figure 6.4 underline the fact that the contamination of sediments largely followed the distribution of water contamination, albeit with a lower intensity.

In the Central and North Atlantic areas, we found only two stations where 5b-hydrochlordecone was detected, at 26 and 14 μg kg^{-1} DM relatively low levels.

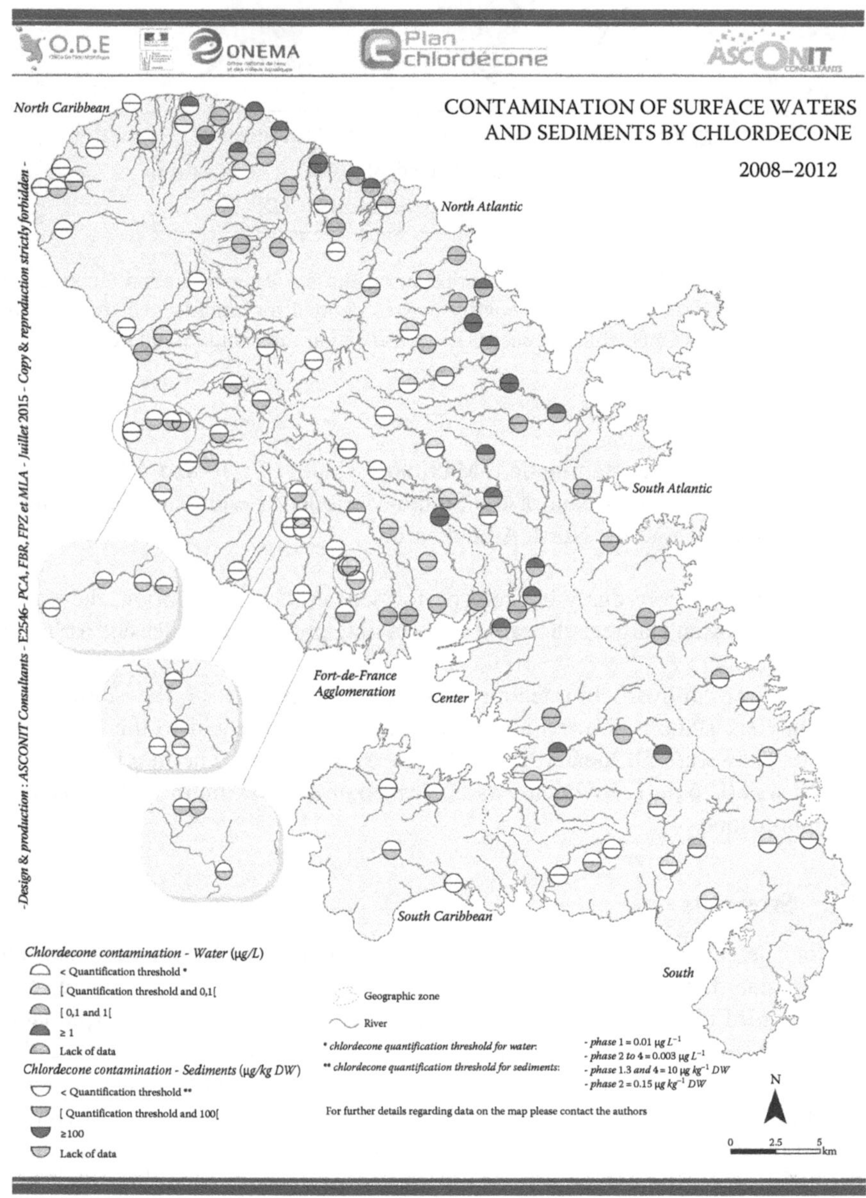

FIGURE 6.4 **(See color insert.)** Map of chlordecone contamination in Martinique surface waters and sediments. (Data from O.D.E, Deal Martinique, ASCONIT 2012, BD CARTHAGE®.)

6.3.3 Aquatic Fauna

The results for aquatic fauna from the 82 stations and 466 batches showed (Figure 6.5) the following:

- An absence of contamination in 16% of stations and 21% of batches. It meant that more than four-fifths of stations and batches were contaminated. Figure 6.6 shows that these pristine stations were mainly located in the North Caribbean area. Noncontaminated batches and stations were marginally found in the all of the zones except in North Atlantic.
- A low-level contamination, below the standard of 20 μg kg^{-1} FW, in 12% of stations and 14% of batches. These stations were most of the time situated in the North Caribbean area.
- A medium to high level of contamination with results exceeding the regulatory standard in 72% of stations and 65% of batches. This was a relatively severe contamination since almost half of all stations and batches show individual stations with chlordecone concentration were at least 10 times above regulatory standard of 20 μg kg^{-1} FW. These maximum values were found in all watersheds of the North Atlantic area and in most of the watersheds in the South Atlantic, Southern, and Northern areas.

In the Fort-de-France urban area, contamination levels showed wide variations. On the North Atlantic side, the whole stretch of rivers was affected by very high intensities of contamination. It was indeed in this area that six stations exceeded the 10,000 μg kg^{-1} FW concentration, thus 500 times the regulatory standard value. Twelve batches (2.6% of batches) were involved. The maximum values were slightly above 30,000 μg kg^{-1} FW in the Sainte Marie and Grande Anse Rivers.

Those 12 batches included species from all ecological guilds: grazing herbivores (1 of *Sicydium* sp.), detritivores (3 of *Atya* sp.), omnivores (4 of *Eleotris perniger* and 2 of *Macrobrachium acanthurus*), and carnivores (2 of *Anguilla rostrata*). Also in

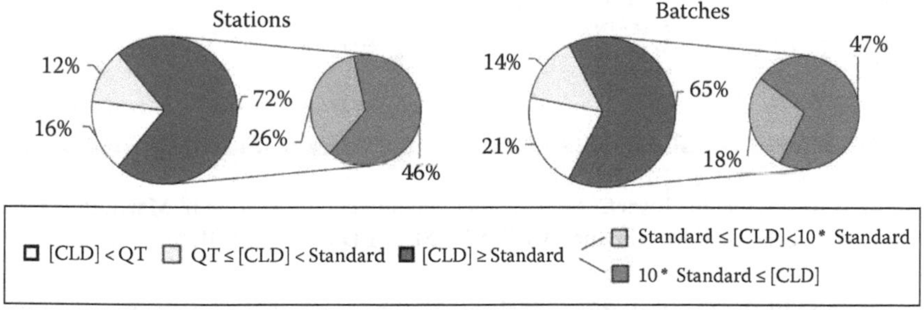

FIGURE 6.5 (See color insert.) Distribution of results for aquatic fauna by station and batch (qualified on their worst results) according to quality classes defined for chlordecone (in % of stations and % of batches—CLD, chlordecone; QT, quantification threshold; Standard, regulatory standard).

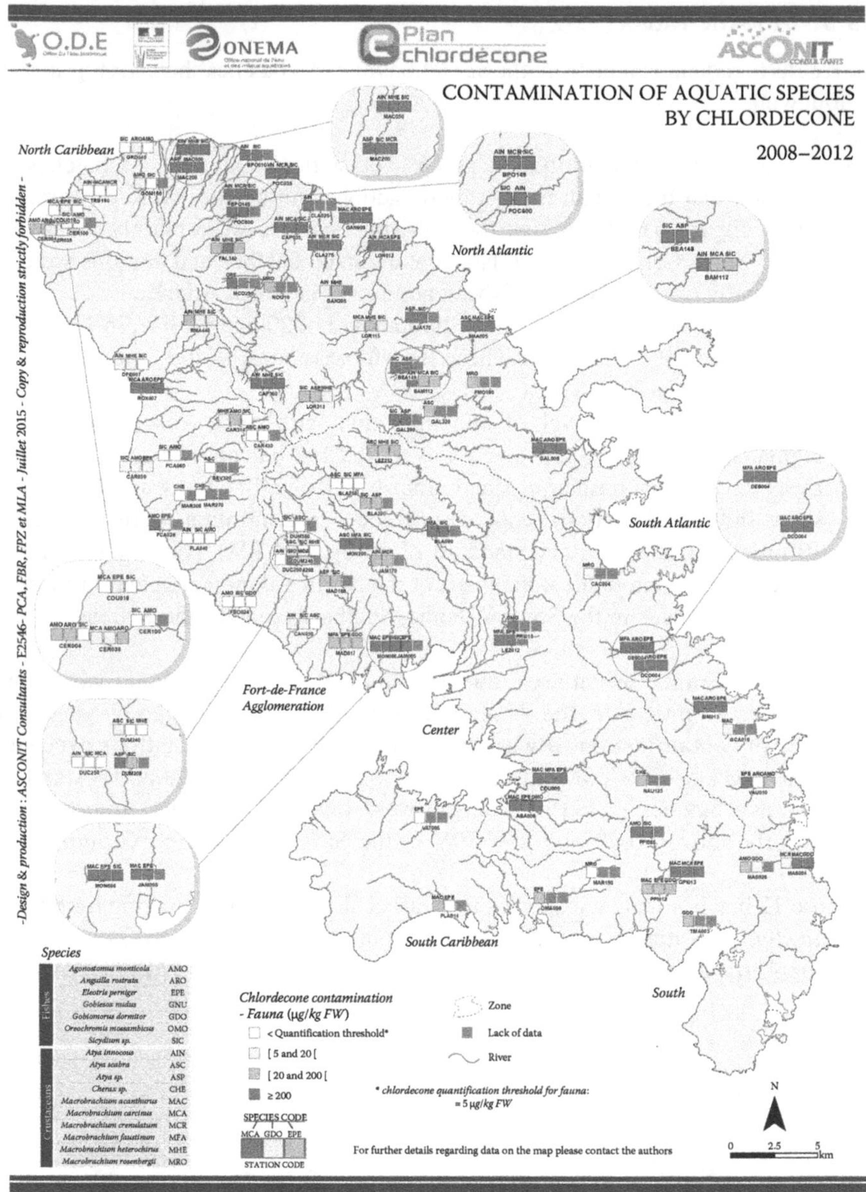

FIGURE 6.6 **(See color insert.)** Map of aquatic fauna contamination in Martinique rivers. (Data from O.D.E, Deal Martinique, ASCONIT 2012, BD CARTHAGE®.)

this area and in the Central area, we observed that contamination grew upstream to downstream. These were the most severely contaminated zones, specifically where river stretches were the longest.

In the stations where more than one batch was studied, the order of magnitude of the contamination remained the same, but values varied from one species to another.

Surprisingly, it was important to note that the analysis of aquatic fauna revealed many areas of contamination, whereas the water and sediments were uncontaminated.

To conclude, Figure 6.6 confirms that the North Caribbean area was characterized by lower levels of contamination than other areas even if most of its watersheds were being contaminated.

Fifty-seven percent of stations and 55% of batches sampled and tested for 5b-hydrochlordecone were free from contamination. The 5b-hydrochlordecone concentrations varied from 10 to 754 µg kg^{-1} FW. Only one-third of stations showed levels of contamination above 20 µg kg^{-1} FW the regulatory standard. Five stations with concentrations above 200 µg kg^{-1} FW, thus 10 times the regulatory standard, were located on the North Atlantic side, mainly in downstream areas. On the 458 batches, one-third showed concentrations of 5b-hydrochlordecone above the regulatory standard and 3% reached concentrations exceeding 200 µg kg^{-1} FW.

6.4 DISCUSSION

6.4.1 Limits and Bias

This first diagnosis of a large-scale aquatic habitat contamination for almost all of Martinique's permanent watercourses strongly increased the volume of available data and knowledge. It is, however, necessary to throw some light on the difficulties and limits that we encountered in the course of undertaking this assessment.

The iterative process this study operated upon proved very beneficial, as it allowed for various improvements and methodological reframing over the five years of the study that are still in use today. It included the definition of sentinel species and improvements in laboratory techniques for lowering quantification thresholds and the fresh matter weight requirements. These advances, however, created difficulties regarding the exploitation and presentation of data: Various species sampled, increasing the precision of results for lower values related to the change of quantification threshold (but it did not matter because the low values were largely below the regulatory standard).

The answers to our initial inquiry on the geographical extent and levels of contamination in aquatic habitats had indeed been reached. But our initial goals and the nature of the sampling plan made it difficult to provide any statistical analysis or other exploitation of the data: drawing links between the concentrations found in water-sediments-fauna, assessment of bioaccumulation in the food web, and definition of sentinel species on criteria other than mere abundance. Analyses indeed focused on different species, at different times and in a wide range of stations.

ODE requested ANSES and its National Laboratory of Reference to validate its dataset from analyses performed on aquatic fauna. The main point to consider was that the quantification of very highly contaminated samples, with concentrations above 1500 µg kg^{-1} FW, could not be validated as sufficiently reliable. Nonetheless, the orders of magnitude of the values found beyond this limit remained acceptable (Hommet 2010). ODE had deemed it sufficient for the purposes of environmental assessment.

One of the more important insights obtained from this diagnosis was about the presence of mildly to severely contaminated fauna in uncontaminated water and sediments. The probable explanations for this were the bioaccumulation phenomena and/or the mobility of living organisms because of their amphidromic behaviors—a trait common to most species present in Martinique rivers (Lim et al 2002)—possibly causing population shifts from one watershed to another. Thus, more studies would be necessary on the movements of the species according to their life cycle, and sampling would have to be adapted to them. The secondary explanation could be the limited technical capacities of the laboratory or the molecules' hydrophobic properties.

6.4.2 Decisions Taken by Authorities

These results led the aquatic habitat decision makers and local French authorities to make several decisions. The samples collected during the study were not intended to provide a representative view to assess the contamination of food products available for human consumption, but to perform an environmental assessment of freshwater habitats. However, fishing and commercial exploitation of freshwater fish and crustacean species were prohibited by prefectoral decree as early as 2009 in Martinique (N° 09-03540 of September 25, 2009). This decree has since then been extended annually.

The first available results allowed confirming the importance of the issue. Indeed in 2010, the French authorities decided to include chlordecone in the list of compounds to be monitored according to the European Water Framework Directive. Therefore, the regulation now officially compels French West Indies authorities to set up monitoring systems to qualify the ecological status of river basins. The environmental quality standards for water and fauna cited and used earlier were therefore adopted and confirmed by a ministerial decree (January 25, 2010 modified July 8, 2010).

6.4.3 Comparison of Chlordecone Pollution with Other Pesticide Pollution in Martinique

In Martinique, local authorities (Martinique water office) started conducting monthly surveys of chemical contaminants of surface waters from 2007. The regulatory Water Framework Directive list had been complemented with pesticides currently used in agricultural and local practices. 191 pesticides were assessed in water compartment, in 28 sampling sites. The data used in this part of this chapter concern 6 years of monitoring (from 2008 to 2014) extracted from the Martinique water office database.

For detection frequency, 102 molecules were detected once. Figure 6.7 presents the 40 pesticides that were detected at a frequency more than 1%, which meant that per 100 measures we found them at least once in surface waters for 100 performed analyses. Fourteen pesticides were found with detection rates higher than 10%. Interestingly, among those, seven of these compounds are now forbidden in European Member States. Not surprisingly, the organochlorine residues chlordecone

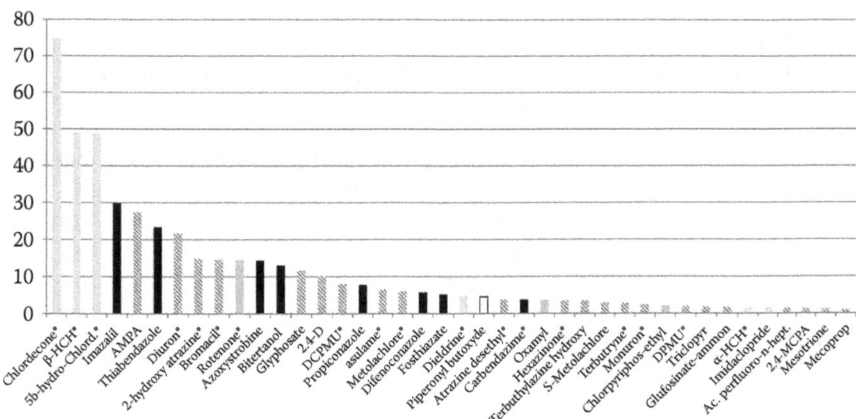

FIGURE 6.7 Percentage of detection (>1%) of pesticide residues in surface freshwaters in Martinique between 2008 and 2014 (gray, insecticide; black, fungicide; hatched, herbicide; white, synergist; *, molecules unauthorized now and their metabolites). (Adapted from Rateau 2013).

(75%), β-HCH (49%), and 5b-hydro-chlordecone (49%) were the most frequently detected (Figure 6.7), leading to chronic contamination of surface waters. These three insecticides, forbidden today, were all used in banana production. Afterward, in a different degree came post-harvest treatment fungicides, imazalil and thiabendazole, and then AMPA, the metabolite of the herbicide glyphosate. All of them are allowed and in use now.

For concentrations, means were calculated per sampling sites using analysis performed between 2008 and 2014. Only 12 pesticide residues were quantified at an average concentration by site higher than 0.1 μg L^{-1}, the regulatory standard (Table 6.2).

Thus, contamination of rivers in Martinique could be described as the result of co-occurrence of residues related to past and current uses of insecticides, fungicides, and herbicides.

6.5 CONCLUSION

A large array of monitoring systems had been set up during the 2008–2012 period, which allowed us to supplement the information provided by frequent quality assessment monitoring on the Water Framework Directive and pesticides, thus aiming to deliver a more precise assessment of the data available on contamination of aquatic ecosystems. As for the freshwater ichthyofauna and carcinofauna, this was the first large-scale diagnostic effort.

The three environmental compartments provided a clearer picture of chlordecone contamination in aquatic habitats. Analyzing aquatic fauna, which was far less common before this study, proved very helpful in supplementing the results of water testing in rivers that appeared not to be contaminated, or only mildly so.

TABLE 6.2
Pesticides Detected at Least on a Site with an Average Concentration (2008–2014) above the Regulatory Standard

Molecule	Use	Cropping Systems	Percentage Sites with [Avg] > 0.1 (μg L^{-1})	[Avg] Max by Site (μg L^{-1})
2,4-D	Herbicide	SC	7	0.13
Glyphosate	Herbicide	SC+B+P+U	32	0.2
AMPA	Herbicide	SC+B+P+U	50	0.45
Asulam*	Herbicide	SC	21	0.31
Bromacil*	Herbicide	P+O	7	0.81
Chlordecone*	Insecticide	B	71	1.97
5b-Hydrochlordecone*	Insecticide	B	7	0.16
β-HCH*	Insecticide	B	14	0.55
Imazalil	Fungicide	B	25	1.84
Thiabendazole	Fungicide	B	11	1.16
Azoxystrobin	Fungicide	B+O	18	0.48
Bitertanol*	Fungicide	B	14	0.21

Note: B, banana; SC, sugarcane; P, pineapple; O, other local crops; U, urban; Avg, average concentration.
* Molecules unauthorized now and their metabolites.

Ultimately, very few rivers were found to be free of contamination (water, sediments, fauna). These were located only in the South Caribbean and North Caribbean areas of Martinique, where bananas were not grown 30 years ago. The highest values of contamination were found in the Atlantic area, mainly in the North. Of the 15 species sampled, all were contaminated by chlordecone. A majority of rivers showed contamination levels exceeding the European environmental regulatory threshold and sanitary limits for water and/or fauna. The map (Figure 6.4) showing river water contamination is largely consistent with that of potentially contaminated soils (Desprats 2004). This was not the case with fauna contamination, which tends to spread even in areas with an absence, or very low risk, of contamination. Contamination was, however, much less important with 5b-hydrochlordecone and other pesticides, both in extent and intensity. It did not mean that their cumulated impacts on aquatic habitats were not to be dismissed. Vigilance has to be set on the high fungicide concentration and the extent of herbicides, all authorized and largely used now.

The present study highlights the lack of specific tools for the French islands' tropical habitats available to decision makers and technical teams, as well as the strong need for research on this topic and on the ecology and ethology of aquatic species. The results underline the relevance of the "fauna" compartment, which, through integration of all micro-pollutant contaminations, provides a deepened assessment, particularly in areas that were only mildly affected.

At last, the French authorities, within the European Water Framework Directive (decree of July 27, 2015), strongly lowered the environmental regulatory standards:

5×10^{-6} μg L^{-1} for water and 3 μg kg^{-1} for aquatic fauna. This is going to pose new challenges to the laboratory teams, the scientists, and the decision makers.

ACKNOWLEDGMENTS

The authors would like to especially thank J.G. Lacas, M. Bocaly, C. Figueras from DIREN-DEAL (Martinique regional environmental authority); L. Mangeot and F. Rateau from Martinique water office (ODE); N. Bargier, C. Desrosiers, F. Pezzato, and E. Lefrançois from Asconit Consultants; F. Massat and B. Planel from LDA 26; and everyone who came out and participated in the various field programs and other program steps.

REFERENCES

AFNOR, 2003. NF EN 14001, Qualité de l'eau—Échantillonnage des poissons à l'électricité. AFNOR, La Plaine Saint-Denis, France.

Agence de l'Eau Loire Bretagne, MEDD, 2006. Le prélèvement d'échantillons en rivière—Techniques d'échantillonnage en vue d'analyses physico-chimiques—Guide technique, Gay environnement, 134pp. http://www.eau-loire-bretagne.fr/espace_documentaire/documents_en_ligne/guides_milieux_aquatiques/Guide_prelevement.pdf. Accessed July 2015.

Arrêté 09-3540 du 25 septembre 2009 portant interdiction de la pêche et de la commercialisation des poissons et des crustacés pêchés dans les rivières situées sur le territoire de la Martinique. http://www.martinique.developpement-durable.gouv.fr/IMG/pdf/PECHE_AP_121120_INTERDICTIONPECHERIVIERES_cle615439.pdf. Accessed July 2015.

Arrêté du 08 juillet 2010 modifiant l'arrêté du 25 janvier 2010 relatif aux méthodes et critères d'évaluation de l'état écologique, de l'état chimique et du potentiel écologique des eaux de surface pris en application des articles R.212-10, R.212-11 et R.212-18 du code de l'environnement. http://www.legifrance.gouv.fr/affichTexte.do?cidTexte=LEGITEXT000022733577&dateTexte=20151021. Accessed July 2015.

Arrêté du 11 janvier 2007 relatif aux limites et références de qualité des eaux brutes et des eaux destinées à la consommation humaine mentionnées aux articles R. 1321-2, R. 1321-3, R. 1321-7 et R. 1321-38 du code de la santé publique. http://www.legifrance.gouv.fr/affichTexte.do?cidTexte=JORFTEXT000000465574. Accessed July 2015.

Arrêté du 27 juillet 2015 modifiant l'arrêté du 25 janvier 2010 relatif aux méthodes et critères d'évaluation de l'état écologique, de l'état chimique et du potentiel écologique des eaux de surface pris en application des articles R. 212-10, R. 212-11 et R.212-18 du code de l'environnement. http://www.legifrance.gouv.fr/eli/arrete/2015/7/27/DEVL1513989A/jo. Accessed July 2015.

Arrêté du 30 juin 2008 relatif aux limites maximales applicables aux résidus de chlordécone que ne doivent pas dépasser certaines denrées alimentaires d'origine végétale et animale pour être reconnues propres à la consommation humaine. http://www.legifrance.gouv.fr/affichTexte.do?cidTexte=JORFTEXT000019117823. Accessed July 2015.

Bargier N., Desrosiers C., 2009. Détermination de la contamination des milieux aquatiques par la chlordécone et les organochlorés. Asconit Consultants, Ducos, Martinique, France; Project owner and ed.: Office de l'Eau Martinique, 89pp. http://www.eaumartinique.fr/IMG/pdf_E0948_RF4.pdf. Accessed July 2015.

Bocaly M., 2008. Définition d'un programme d'échantillonnage améliorant la connaissance de la contamination des cours d'eau de la Martinique par le chlordécone. Mémoire de Master, Université Victor Segalen, Bordeaux, France. DIREN Martinique, Fort-de-France, Martinique, France, 71pp.

Desprats, J., Comte, J., Chabrier, C., 2004. Cartographie du risque de pollution des sols de Martinique par les organochlorés, Rapport phase 3: synthèse, s.l.: BRGM/RP-53262-FR.

Fiévet E., 1995. Comparaison de deux méthodes d'échantillonnage quantitatif et étude de la co-structure faune-milieu de quatre cours d'eau de Guadeloupe. Mémoire de DEA Université Claude-Bernard—Lyon 1, Villeurbanne, France. URA CNRS N 1974, Parc National de Guadeloupe, Guadeloupe, France, 35pp.

Hommet F., 2010, Résultats des analyses du contrôle qualité portant sur les analyses de chlordécone dans des échantillons d'animaux d'eau douce. Unité POP, ANSES, Maison Alfort, France.

Labeille M., Bargier N., 2011. Détermination de la contamination des milieux aquatiques par la chlordécone—Volet 2: Investigations complémentaires—Nouvelles zones et réévaluation. Asconit Consultants, Ducos, Martinique, France; Project owner and ed.: Office de l'Eau Martinique, 37pp. http://www.eaumartinique.fr/IMG/pdf_E1717-rapport-vf2_BR.pdf. Accessed July 2015.

Labeille M., Vergès C., 2012. Détermination de la contamination des milieux aquatiques par la chlordécone—Volet 4: Investigations complémentaires—Renforcement du maillage géographique sur les cours d'eau d'intérêt piscicole. Asconit Consultants, Ducos, Martinique, France; Project owner and ed.: Office de l'Eau Martinique, 48pp. http://www.eaumartinique.fr/IMG/pdf_E2546-PNA-Chlordecone-Volet4_vf.pdf. Accessed July 2015.

Labeille M., Vergès C., 2011. Détermination de la contamination des milieux aquatiques par la chlordécone—Volet 3: Investigations complémentaires—Têtes de bassins versants et Sud Martinique. Asconit Consultants, Ducos, Martinique, France; Project owner and ed.: Office de l'Eau Martinique; 68pp. http://www.eaumartinique.fr/IMG/pdf_E1978-PlanchlordeconeVolet3_BR.pdf. Accessed July 2015.

Lim P., Meunier F.J., Keith P., Noël P.Y., 2002. *Atlas des poisons et des crustacés d'eau douce de la Martinique*, Muséum National d'Histoire Naturelle, Paris, France. *Patrimoines Naturels*, 51, 120pp.

MEDD & Agences de l'eau, 21 mars 2003. Grilles d'évaluation version 2—Système d'évaluation de la qualité de l'eau des cours d'eau (SEQ-Eau). MEDD, Paris, France, 40pp. http://rhin-meuse.eaufrance.fr/IMG/pdf/grilles-seq-eau-v2.pdf. Accessed July 2015.

Ministry of Public Health, The Chlordecone National Action Plan, 2008–2010. General Direction of Public Health, Paris, France, 16p. http://www.developpement-durable.gouv.fr/IMG/pdf/plan_chlordecone_-2.pdf. Accessed June 2016.

Rateau F., 2013. *Les produits phytosanitaires dans les cours d'eau de Martinique 2008–2012*. Office de l'Eau Martinique, Fort-de-France, Martinique, France, 48pp. http://www.eaumartinique.fr/IMG/pdf_Atlas_pesticides.pdf. Accessed July 2015.

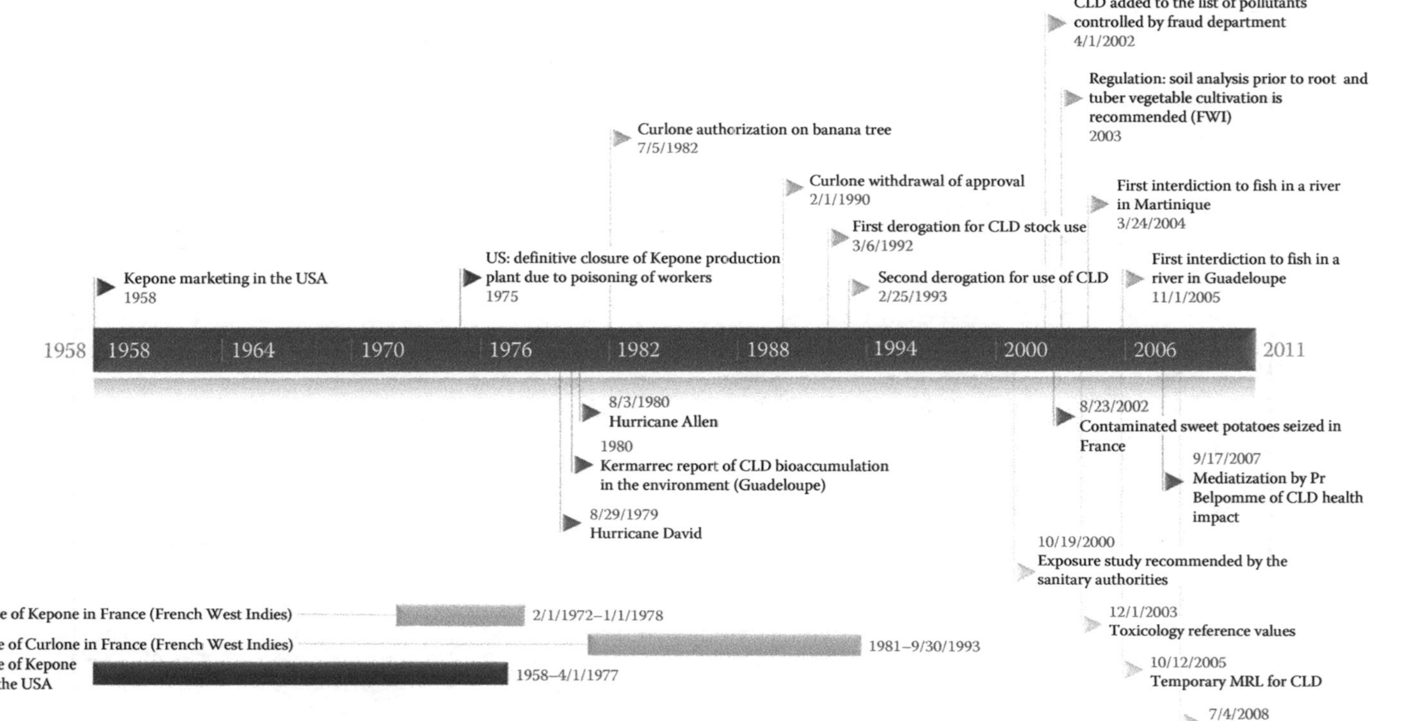

FIGURE 1.4 Timeline of chlordecone (CLD) use and pollution management.

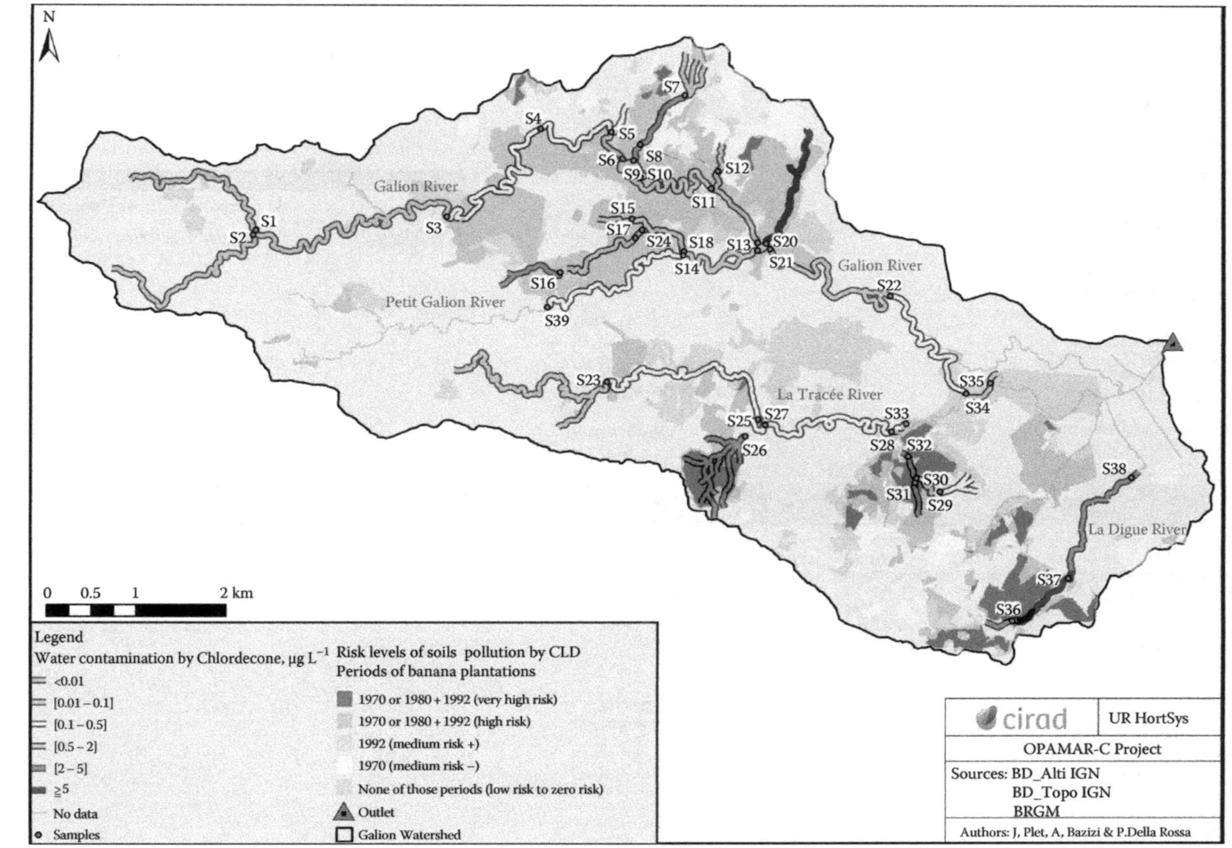

FIGURE 4.1 Map of soil pollution risk and of polluted river segments for the Galion watershed, Martinique.

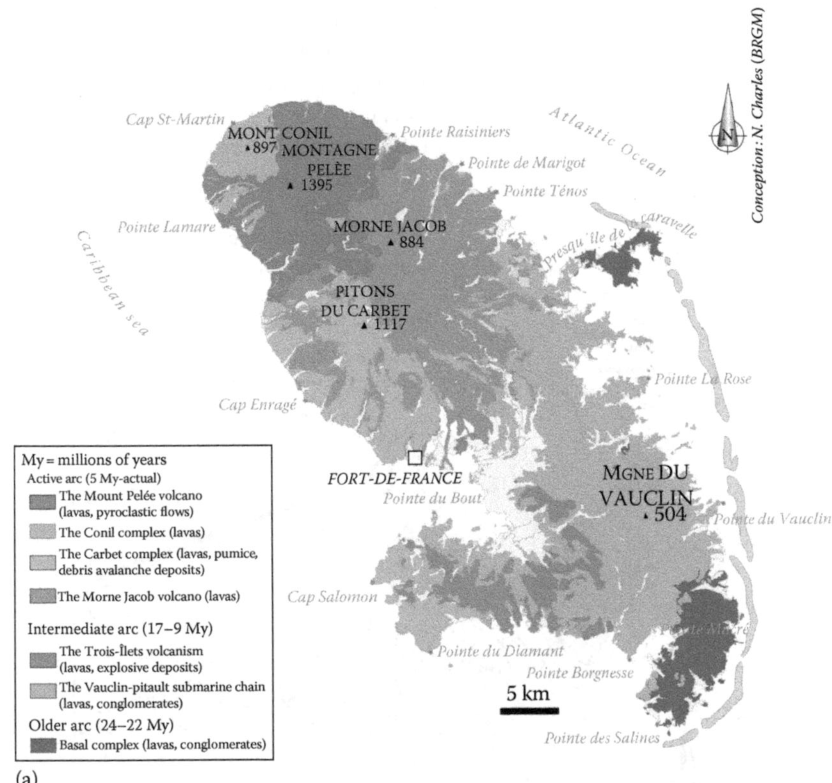

FIGURE 5.1 (a, b) Geology and structure of Martinique and Guadeloupe. *(Continued)*

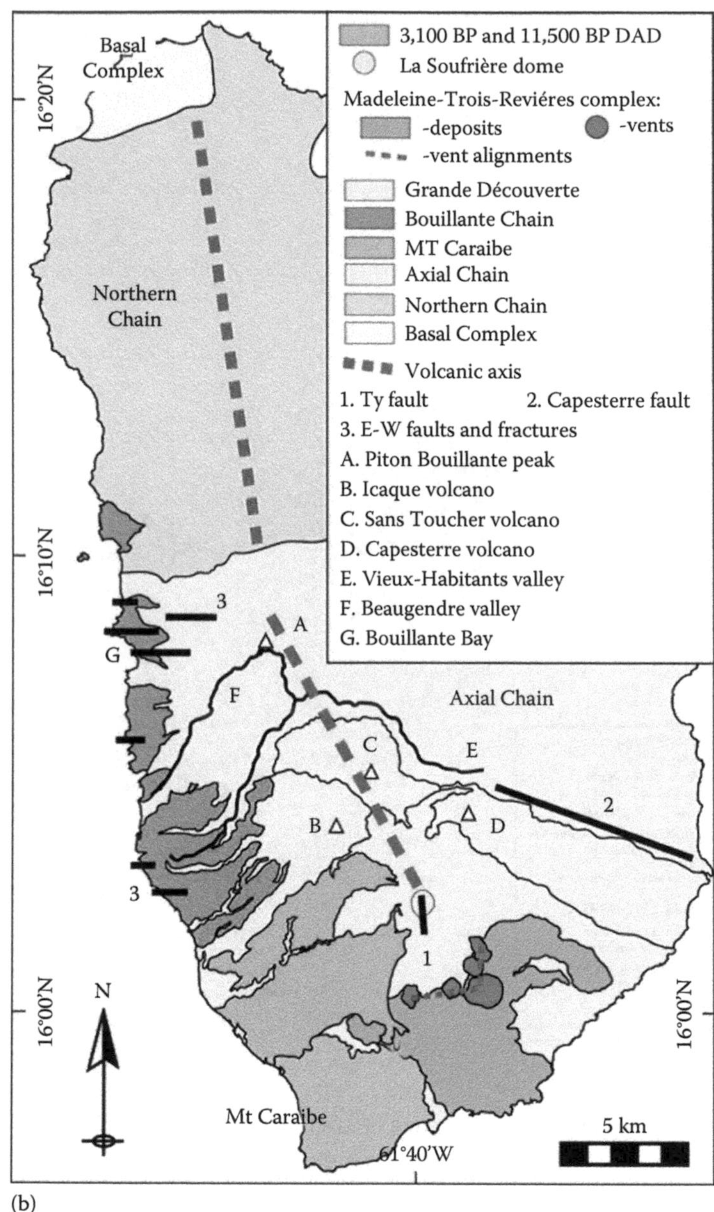

FIGURE 5.1 ***(Continued)*** (a, b) Geology and structure of Martinique and Guadeloupe. ([a]: From Audru et al., 2015; Modified from Westercamp, D. et al., 1989 Carte géologique de la France (1/50,000), feuille Martinique. BRGM. Notice explicative par W.D., Andreieff P., 1990; [b]: Adapted from Mathieu, L. et al., *Bull. Volcanol.* 75:700, 2013; Modified from Dagain 1981; Boudon et al. 1988; Gadalia et al. 1988; Samper et al., 2007.)

Contamination	Absent	Low	Medium	High
Water & fauna	Below QT	Below RS	Less than 10 times over RS	More than 10 times over RS
Sediments	Below QT	Less than 10 times over QT		More than 10 times over QT

FIGURE 6.1 Quality classes for water, fauna and sediments (colors are the same as used on the maps fig. 2 and 4; QT = quantification threshold, RS = regulatory standards).

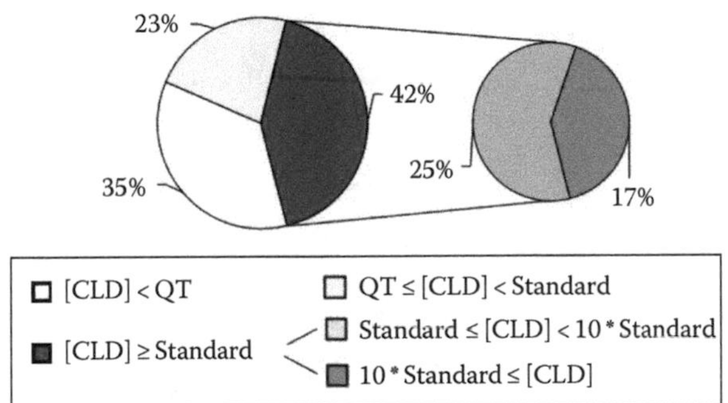

FIGURE 6.3 Distribution of results for water by station (qualified on their worst results) according to the quality classes defined for chlordecone (in % of the number of stations—CLD, chlordecone; QT, quantification threshold; Standard, regulatory standard).

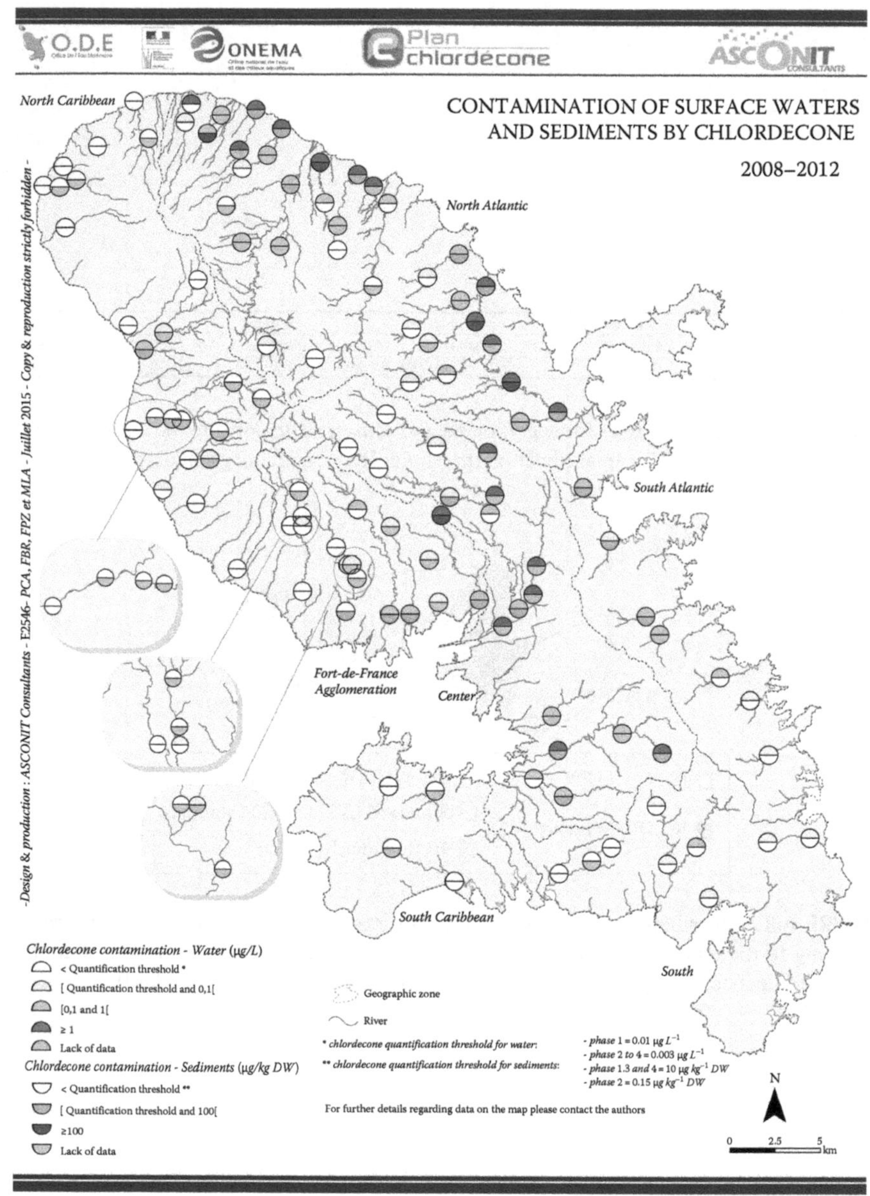

FIGURE 6.4 Map of chlordecone contamination in Martinique surface waters and sediments. (Data from O.D.E, Deal Martinique, ASCONIT 2012, BD CARTHAGE®.)

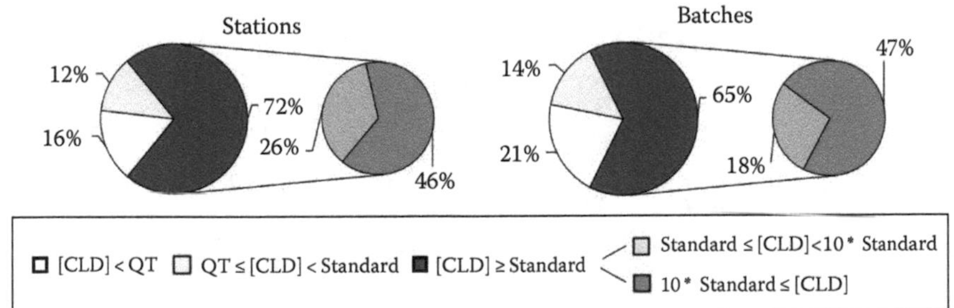

FIGURE 6.5 Distribution of results for aquatic fauna by station and batch (qualified on their worst results) according to quality classes defined for chlordecone (in % of stations and % of batches—CLD, chlordecone; QT, quantification threshold; Standard, regulatory standard).

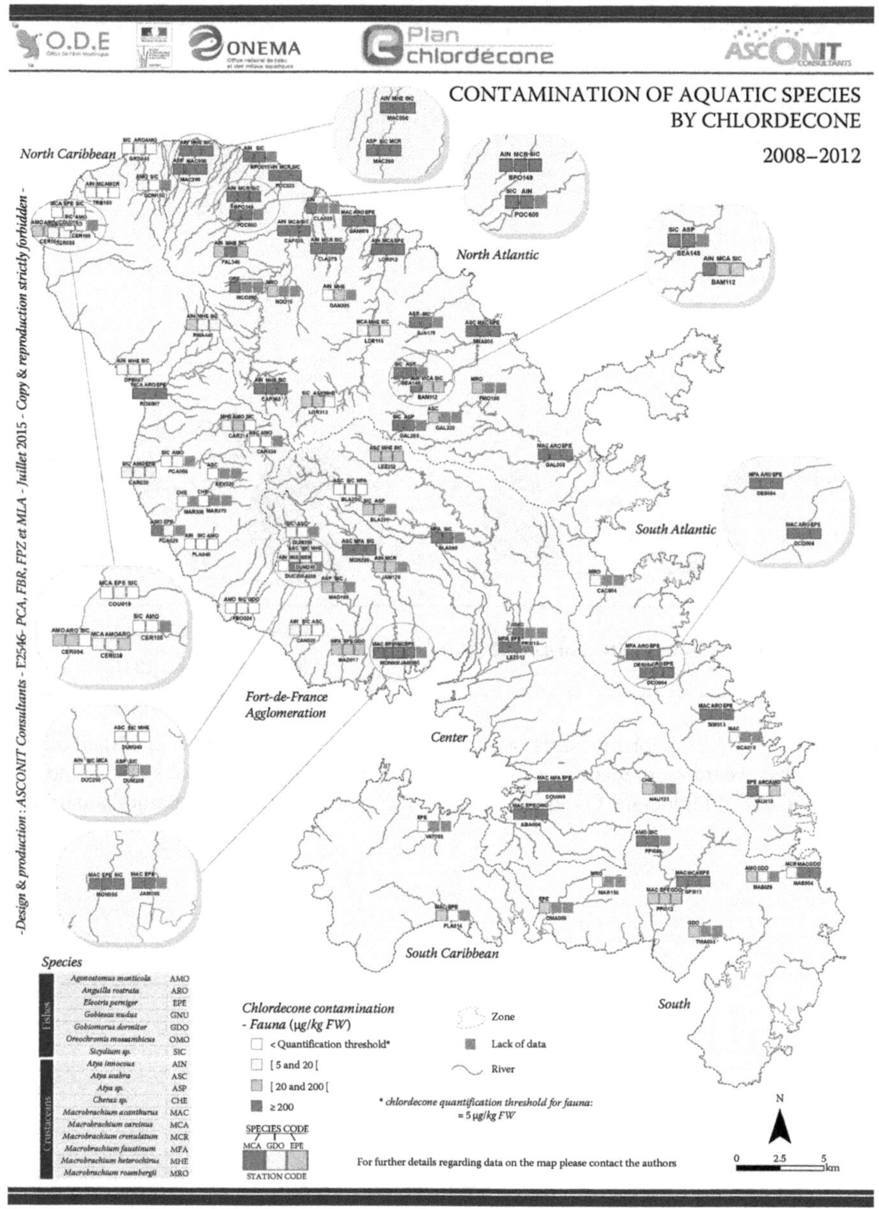

FIGURE 6.6 Map of aquatic fauna contamination in Martinique rivers. (Data from O.D.E, Deal Martinique, ASCONIT 2012, BD CARTHAGE®.)

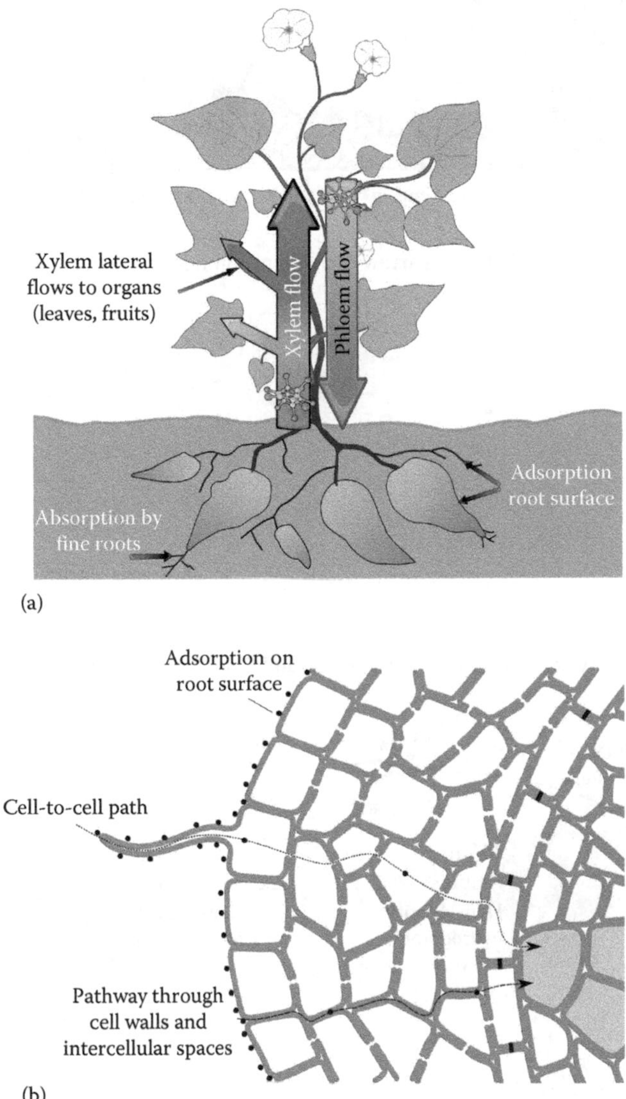

FIGURE 10.2 (a) Mechanisms of contamination by chlordecone at plant scale (Adapted from Clostre 2015a, CC BY-NC-ND) and (b) focus at root scale on pathways to reach xylem vessels (Adapted from Clostre 2015b, CC BY-NC-ND). Chlordecone, dark round points.

No- or low-uptakers
Mid-uptakers
High-uptakers
1 mg kg⁻¹ dry soil
0.1 mg kg⁻¹ dry soil

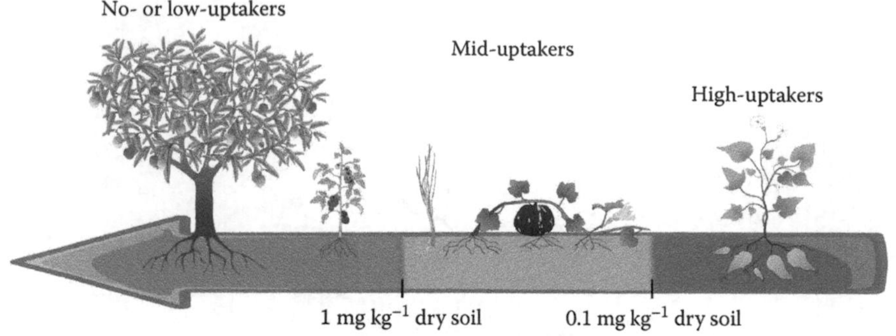

FIGURE 17.1 Soil thresholds and arrow of the noncompliance risk according to crop uptake potential.

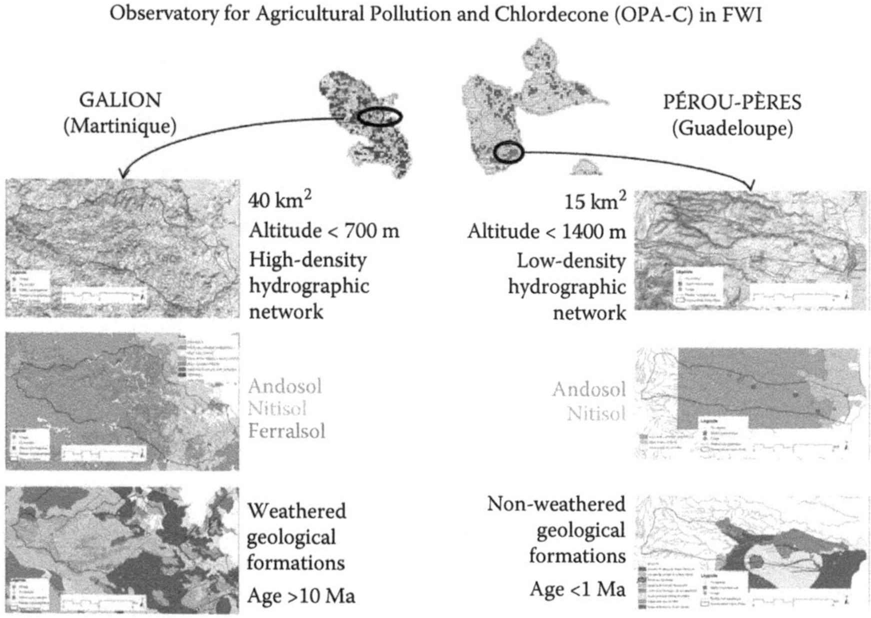

Observatory for Agricultural Pollution and Chlordecone (OPA-C) in FWI
GALION
(Martinique)
PÉROU-PÈRES
(Guadeloupe)
40 km2
Altitude < 700 m
High-density
hydrographic
network
15 km2
Altitude < 1400 m
Low-density
hydrographic
network
Andosol
Nitisol
Ferralsol
Andosol
Nitisol
Weathered
geological
formations
Age >10 Ma
Non-weathered
geological
formations
Age <1 Ma

FIGURE 19.4 Illustration of the main environmental characteristics of the Observatory of Pollution from Agriculture and Chlordecone (OPA-C).

7 Contamination of Freshwater Fauna

Dominique Monti, Philippe Rey, and Jean-Pierre Thomé

CONTENTS

7.1 THE CONTEXT

The small island of Guadeloupe imports thousands of tons of various pesticides every year (Bonan and Prime 2001). Most of the pesticides used in agriculture and urban areas are leached and get into the aquatic environment. In the early 2000s, a significant contamination of some drinking water resources was detected, at doses up to 100 times the "drinking water" standards. The three incriminated molecules were chlordecone (CLD) (the most worrying), the beta isomer of hexachlorocyclohexane, and dieldrin (Monti 2005, 2008, Coat et al. 2006). After the publication of these first alarming results, the Direction Régionale de l'Environnement of Guadeloupe quickly sought to know the degree of contamination of species living in the rivers. The objective was to assess the level of contamination found in fish and crustaceans, which were highly consumed, to be able to quickly take the necessary administrative measures. The first study evaluating the contamination in the rivers of the south of Basse-Terre revealed an extreme concentration of CLD in the muscle and viscera of the sampled species (Monti 2005). This inventory gradually continued, to the north of the windward coast (2006), first, and then to the Leeward coast of Basse-Terre (2007). A general assessment all around the volcanic island was thus conducted and has been coupled, more recently, with an analysis of species living in the three major freshwater reservoirs retaining the water before irrigation.

Beyond these simple but urgent health-related questions, scientific hypotheses were tested to identify (1) the most contaminated species in the rivers, (2) possible selective accumulation in tissues of organisms (muscle, viscera, or carapace), and (3) general mechanisms explaining, at best, the bioaccumulation of this molecule in freshwater species, which are almost all migratory in this island.

7.2 MATERIAL AND METHODS

7.2.1 Sampling

The collection of individuals was performed by using an electrofishing portable device (DEKA 3000, DEKA Gerätebau, Marsberg, Germany), and realized between 2008 and 2014. For fish, preference was given, when possible, to carnivorous or benthic fish, and for crustaceans, the genus *Macrobrachium* sp. was retained primarily because it is very popular and heavily consumed by the population. When these species were not present, the retained species were the most representative of the place. Preference was given to (1) the larger sizes of individuals, most attractive for fishing or consumption, and (2) the juvenile stages of the same species. The samples were composed of 10–25 individuals of the same species, of neighboring standard length, to provide the minimum mass required for analysis of pollutants. Batches of individual samples were measured, wrapped each in a resistant foil of aluminum, and placed in two water-impermeable plastic bags. The samples were stored on the field in a freezer until their delivery to the laboratory, and then grouped before shipment.

7.2.2 Chemical Analyses

Two laboratories, using their own methods, had realized the determination of CLD content in fish and crustaceans. The samples that led to a regulation text were analyzed at the Laboratoire Départemental d'Analyse de la Drôme (Valence, France), accredited by the French Accreditation Committee (COFRAC). Lipid fraction, containing organochlorine (OC) pesticides, was extracted from homogenized animal tissues in 1:1 acetone–dichloromethane using an Accelerated Solvent Extractor (ASE) 200® apparatus (Dionex 200, Thermo Scientific, Sunnyvale, California, USA). The solvent was removed by evaporation, while purification was achieved by acid hydrolysis. Concentrations of OC were measured by gas chromatography with electron capture detector (GC/ECD: Varian, Les Ulis, France) and confirmed by gas chromatography coupled with mass spectrometry (GC/MS: ion trap Varian, Les Ulis, France). Pure CLD was used as reference compound (calibration) and it was introduced every 10 samples in the analysis process (quality control). Five points determined the concentration range, and linearity was situated between 3 and 100 $\mu g\ L^{-1}$. The calibration curve allowed for determination of pesticide concentration, according to the area of the peak obtained. PCB 101 and HBB-TPP (hexabromobenzene triphenylphosphate) with known concentrations were used as injection and extraction tracers, respectively.

For the specific analysis of juvenile stages and the carapace content of CLD, that is, very small masses of tissue, analyses were performed by the CART-LEAE Laboratory (Université de Liège, Belgique) according to a method derived from Debier et al. (2003). Briefly, biological materials were freeze-dried for 20 hours with a Benchtop 3 L Sentry Lyophilisator (VirTis, Gardiner, New York, USA). The lyophilized samples were weighed in order to determine water content. The extraction of CLD was performed with a solvent mixture of *n*-hexane: dichloromethane (90:10; v:v; Biosolve-Chimie, Dieuze, France) using an ASE 200 at 80°C and under a pressure of 1500 psi. Before the extraction, 100 μL of a hexanic solution of PCB congener 112 (Dr. Ehrenstorfer, Augsburg, Germany) was added to the samples as a surrogate internal standard to obtain a final concentration of 50 pg μL^{-1}. The solvent, containing the extracted fat and/or CLD, was collected in pre-weighed vials and evaporated at 40°C under a gentle nitrogen flow using a TurboVap LV (Zymark, Hopkinton, Massachusetts, USA) until a constant weight of residues was obtained. Then, the lipid content was determined gravimetrically. The residues containing lipids and/or CLD were resuspended in 2 mL of *n*-hexane and transferred to a test tube. An acid cleanup was performed on the extract by adding 2 mL of 98% sulfuric acid (Merck, Darmstadt, Germany) to remove organic matter (e.g., lipids, lipoproteins, carbohydrates) as described in Debier et al. (2003). Nonane 5 μL was added as a keeper. This extract was evaporated under a gentle nitrogen stream using a Visidry evaporator (Supelco®, Bellefonte, Pennsylvania, USA) before being resuspended with 45 μL of *n*-hexane and 50 μL of a solution of PCB 209 (100 pg μL^{-1} in *n*-hexane) as an injection volume internal standard (Dr. Ehrenstorfer, Augsburg, Germany). The samples were analyzed for CLD by high-resolution gas chromatography using a ThermoQuest Trace 2000 gas chromatograph equipped with an Ni^{63} ECD detector (Thermo Scientific, Sunnyvale, California, USA) and an autosampler Thermo Quest AS 2000. The analytical parameters are described in Multigner et al. 2010. A procedural blank and quality control (QC) were carried out for all 10 samples. The QC was a CLD-free biological matrix (freeze-dried muscle of *Penaeus monodon*) spiked with an acetonic solution of CLD in order to obtain a final concentration of 2.5 ng g^{-1} wet weight.

The recovery efficiency based on the CLD recovery in QC and on the recovery of the surrogate internal standard (PCB112) was (mean ± standard deviation) 88 ± 4% and 115 ± 5%, respectively; it was thus within the limits recommended by SANCO (i.e., ranged from 60% to 140%—Document No. SANCO/12495/2014 European Union 2014). The limit of detection was fixed at three times the background noise of the chromatogram (i.e., 0.02 ng g^{-1} wet weight). The limit of quantification was determined with freeze-dried prawn muscle spiked with various concentrations of CLD and was established at 0.06 ng g^{-1} wet weight.

Selective accumulation of CLD between muscle and viscera was sought by calculating the selective accumulation factor: [muscle]/([muscle] + [viscera]) (Erdogrul et al. 2005).

7.2.3 Demographic Analyses

Total lengths of shrimps were measured to the nearest millimeter. Size distribution histograms (size-class interval 2 mm) were plotted through time. Due to a strong overlap between age groups, cohort extraction analyses were performed by using the package "Mixdist" (Macdonald and Du 2010) running under software R (R Development Core Team 2012) to be able to visualize overlapping cohorts and obtain the mean lengths and proportions of each age group.

7.2.4 Modeling

The consistency in the general shape and the different characteristics of experimental curves of CLD contamination of aquatic organisms led to the development of a general model describing the accumulation of CLD by living aquatic organisms in natural ecosystems over time (Bahner and Oglesby 1979). This generalized equation of accumulation of CLD by aquatic animals was originally calibrated on a large set of data contamination (natural logarithms of the pesticide residues) and is of the general form of Equation 7.1.

$$Y = \left[A + 10^{(-C\times days)}\right]^{-1} - \left[A + 10\ e^{(-D\times(days-E))}\right]^{-1}, \qquad (7.1)$$

where

A is the maximum concentration at equilibrium
C is the slope or bioaccumulation rate
D is the slope of the generalized equation of depuration
E is the origin, when there is a change of environment

This nonlinear equation requires an iterative method to estimate its parameters, which was formerly built with a Statistical Analysis System (SAS) routine. A script in R environment was conceived for an update. The optimization of the parameters was done using the minpack.lm package, under R environment.

7.3 RESULTS

7.3.1 The Extent and Magnitude of Biocontamination

The average contamination of organisms living in 24 water bodies spread over the entire island and the altitude of the sampling stations are shown in Table 7.1.

These results show numerous values that are among the global contamination records for this molecule. Moreover, these values far exceed, for the vast majority of them, the French maximum residue limits for animal consumption (20 μg kg^{-1} FW). Running waters and reservoirs are both affected. Two of these reservoirs (Letaye and Gaschet), located far away from banana plantations in the island of Grande-Terre, show significant contamination and validate the hypothesis of a long-distance transfer of CLD through irrigation pipes. In running waters, it appears there is no clear determinism as a function of altitude for the molecule of CLD: The small spatial scale variability is more important than the influence of some altitudinal gradient. When located in the same environment and when they are adults, fish are generally more contaminated than crustaceans.

TABLE 7.1
Fish and Crustacean Contamination by CLD in the Rivers

River	Latitude	Longitude	Altitude (m)	CLD μg kg^{-1} FW	
				Fish	Crustaceans
Etang Gommier	16.982852	−61.644879	170	**12971**	**3392**
Grande Rivière à Goyaves	16.221102	−61.660999	59	**93**	**35**
Grande Rivière à Goyaves	16.189541	−61.658943	128	—	**373**
Grande Rivière de Capesterre	16.058880	−61.567996	28	**9026**	**704**
Grande Rivière de Vieux Habitants	16.062508	−61.762777	13	**51**	**39**
Grande Rivière de Vieux Habitants	16.082074	−61.728037	219	**73**	**72**
Petite Rivière à Goyaves	16.126379	−61.584302	16	**1450**	**490**
Ravine Bleue	16.224780	−61.781200	8	<10	<10
Ravine des Coudes	16.323755	−61.418820	11	**21**	<10
Rivière aux Herbes	15.994868	−61.729205	8	**1782**	**950**
Rivière aux Herbes	16.001943	−61.719957	94	**1642**	**1490**
Rivière Bourceau	16.154554	−61.775557	4	<10	<10
Rivière Bras David	16.191385	−61.676088	134	—	**200**
Rivière Corossol	16.172221	−61.688492	251	10	<10
Rivière Deshaies	16.303653	−61.793710	8	11	—
Rivière du Grand Carbet	16.022476	−61.575086	11	**3565**	**2725**
Rivière du Grand Carbet	16.045115	−61.630801	435	**2922**	**1298**
Rivière du Plessis	16.037945	−61.742048	105	**745**	**918**
Rivière Grande Anse	15.962750	−61.659446	7	**11733**	**3281**
Rivière Grande Anse	15.973077	−61.663381	30	**23180**	**3120**
Rivière La Rose	16.147503	−61.585006	10	**247**	10
Rivière Lézarde	16.201738	−61.609292	6	**150**	**171**
Rivière Lostau	16.162384	−61.772102	7	<10	<10
Rivière Moreau	16.126931	−61.591609	37	**540**	**636**
Rivière Moustique Petit Bourg	16.181988	−61.598450	14	**120**	**1480**
Rivière Petite Plaine	16.223297	−61.781007	7	12	<10
Réservoir Gaschet	16.413997	−61.490345	5	—	**344**
Réservoir Letaye	16.304708	−61.320433	22	—	**71**
Réservoir Dumanoir	**16.041471**	**−61.607194**	**212**	—	**1334**

Note: Bold, the values above the maximum residue limits for human consumption (20 μg kg^{-1} FW).

7.3.2 Selective Accumulation

The values obtained in different tissues but same individuals collected in various contaminated areas helped assess the possible selective concentration of CLD in muscle, viscera, and shrimp carapaces. Figure 7.1 shows the relative contamination of muscles and viscera in crustaceans and fish.

The values greater than 0.50 reveal preferential accumulation in muscles (part consumed). These results reveal preferential accumulation in viscera for fish,

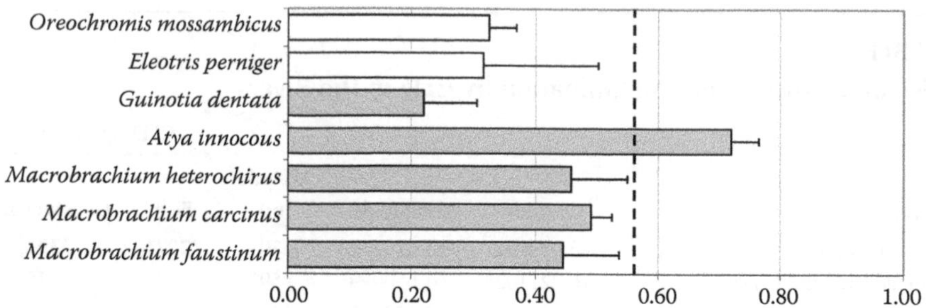

FIGURE 7.1 Selective accumulation of CLD in the tissues of the sampled organisms (pools of individuals). White, fish; grey, crustaceans.

represented here by *Oreochromis mossambicus* (Peters 1852), tilapia, detritivore or herbivore, and *Eleotris perniger* (Cope 1871), small sleeper, carnivorous. The most prized crustaceans, *Macrobrachium* spp., show homogeneous contamination between the tails (muscle) and the cephalothorax (viscera). Special mention must be made of a crustacean species belonging to the Atyidae (*Atya innocous* [Herbst 1792], filter feeder), in which there is always more pollutant in its muscle than in its viscera.

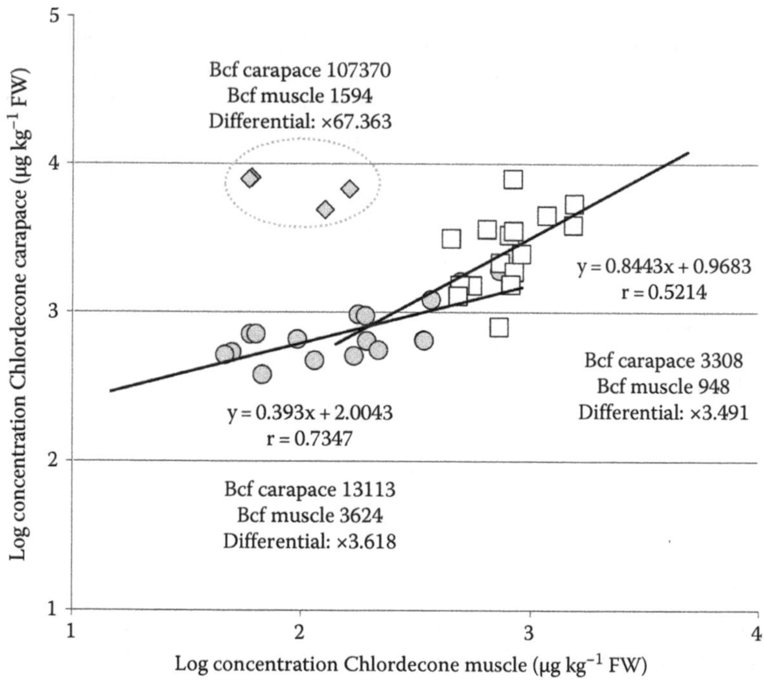

FIGURE 7.2 Selective accumulation of CLD in the tissues of crustaceans (pools of individuals). Squares—important contamination; circles—moderate contamination; diamonds—crustaceans freshly molted.

The examination of the CLD content in the carapace versus muscle in crustaceans (Figure 7.2) always shows more concentration of the pollutant in the carapace (bioconcentration factor: $Bcf_{carapace}$) than in the muscle (Bcf_{muscle}), with a multiplicative factor of 3.49–3.61. Irrespective of the water pollution level, a significant proportional relationship is found. Analysis of freshly moulted crustaceans shows an extreme concentration of the pollutant in the new carapace in formation, with a bioconcentration factor 67 times greater than in muscles.

7.3.3 Evolution of Contamination after Reentering the Rivers

It was possible to apply the underlying ideas of Bahner and Oglesby's model (1979) (see Section 7.2.4) to data on diadromous fauna available in Guadeloupe. The most interesting part of this approach is the evolution of the contamination of the juveniles, reentering the river. To be able to do that, a specific work on demographic data was first conducted so as (1) to extract the different age groups from a mixture of sizes in the population and (2) to mobilize the knowledge of the chosen species (obtained from previous studies) to be able to attribute an age to an observed average size and, thus, a residence time in the environment. The candidate species for such a work of population dynamics are rare in the West Indian rivers because they have to generate a large number of individuals. *Macrobrachium faustinum* (De Saussure 1857) develops populations with all these qualities and is a species for which the relationship between size and age has been extensively established by adjusting a growth model on several rivers and conditions. Six cohorts were observed in *M. faustinum* population sampled in the Grande Anse river (Guadeloupe) ranging from 14 to 49 mm (Figure 7.3) and

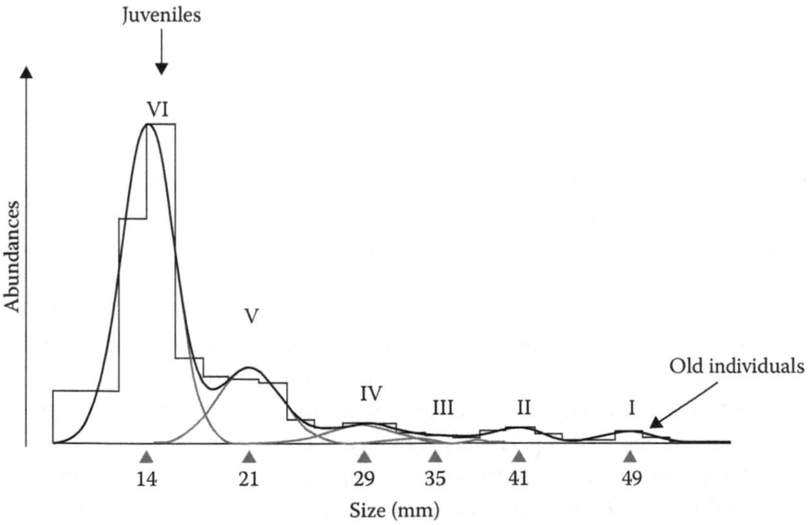

FIGURE 7.3 Cohorts of *M. faustinum* population in the Grande Anse River, extracted with mixdist package. Roman figures identify the different cohorts.

TABLE 7.2
Length to Age Conversion for *M. faustinum*: Grande Anse River

Age (Days)	Size (mm)
30	5.5
60	8.8
90	12.0
120	14.9
150	17.7
180	20.3
210	22.7
240	25.1
270	27.2
300	29.3
330	31.2
360	33.0
390	34.8
420	36.4
450	37.9
480	39.3
510	40.7
540	41.9

Source: Monti, D., *Etude du niveau de contamination des organismes aquatiques d'eau douce par les pesticides, en Guadeloupe*, Convention Direction régionale de l'Environnement Guadeloupe, 2005.

it needs to be noted that the growth at this place is known to be particularly slow for this species.

With the help of previous studies conducted on growth models at the same place (Monti 2005), it was then possible to establish a conversion between sizes and age of individuals (Table 7.2), and then link this age to the observed mean values of contamination.

When the juveniles reenter the river with an average size of 14–15 mm (i.e., an average age of 110 days), they undergo a change of conditions related to the transition between seawater and freshwater. This disruption of the ecological conditions can be explored with the parametrization of the global model with, here, the origin of the environmental change E being 110 days (Figure 7.4).

Using the mean water concentration in this part of river for the concentration at time zero and the CLD concentration in mature eggs of *M. faustinum* for the second early point of biocontamination, the model was successfully adjusted through the entire set of cohorts (Figure 7.5). The following parameters were determined for Equation 7.1: A = 0.15228 ± 0.00993, Pr(>|t|) = 4.86e−06***; C = 0.04459 ± 0.01014,

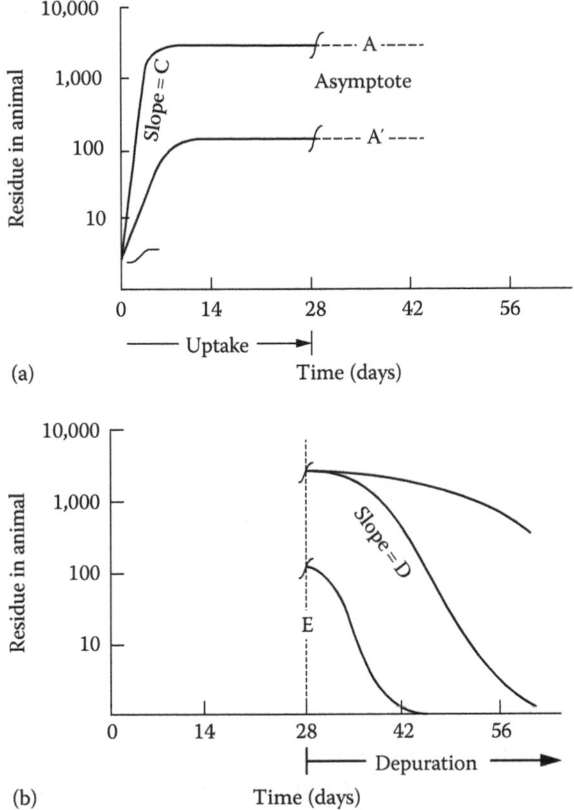

FIGURE 7.4 Generalized equation of accumulation of CLD by aquatic animals in the wild: (a) uptake and (b) depuration. E—origin of the environmental changes; D—slope of an eventual depuration. (Data from Bahner, L.H. and Oglesby, J.L., Test of a model for predicting Kepone accumulation in selected estuarine species, in *Aquatic Toxicology*, ASTM STP 667, Marking L.L. and Kimerle R.A., eds., American Society for Testing and Materials, Philadelphia, PA, 1979, pp. 221–231.)

Pr(>|t|) = 0.00459**, and D = 0.02130 ± 0.08326. Looking at these results (see Figure 7.5), it is interesting to see that (1) the juveniles are more contaminated than the adults, (2) the contamination of the age groups of *M. faustinum* in the Grande Anse River shows a minimal decline over time during the 2 months after the juveniles have returned to the river, and (3) this decline, here, is temporary and the global decrease is small (D = 0.021).

7.4 DISCUSSION

The extreme contamination of consumable freshwater fauna by CLD led the French government to prohibit fishing and to issue consumption warnings in 12 of the 15 districts of Basse-Terre. Studies conducted over the past 10 years in Guadeloupe

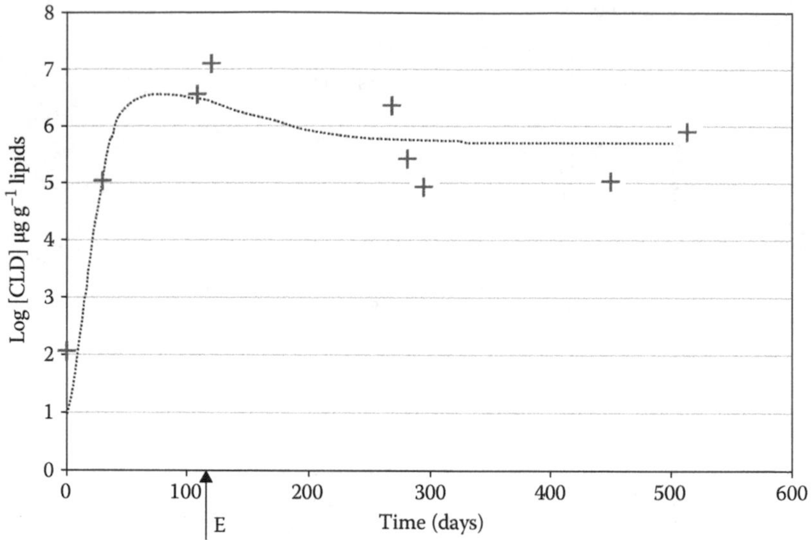

FIGURE 7.5 Generalized model of accumulation of CLD by *M. faustinum* in the Grande Anse River.

have shown that all aquatic food web components are polluted. Concentrations measured in the freshwater species are among the highest values detected worldwide and exceed the highest levels monitored in the James River biota (Virginia, USA) after disasters involving Kepone, causing millions of dollars of damage (Bahner et al. 1977, Nichols 1990). In the French West Indies, CLD presents a major hazard for ecosystems (Coat et al. 2009) because of its stability and the tropical climate, which continually comes in contact with fauna and contaminants originating from soils or underwater resources, without any chance of perennial sequestration. The simulations on the persistence of CLD in the soils of Antilles forecast a very long remanence (Cabidoche et al. 2009). The effects of this molecule that is known to cause growth reductions in invertebrate or vertebrate populations and affect ecosystem productivities through alteration of egg productions or species fertility (Schimmel et al. 1979, Sanders et al. 1981, Goodman et al. 1982) are still completely to be assessed. Decades of research will be needed to determine the long-term effects on, and damages suffered by, the aquatic ecosystems of the French Antilles.

The adjustment of the bioaccumulation model for CLD to the age groups of wild *Macrobrachium faustinum* population in the Grande Anse River showed a greater concentration of CLD in juveniles. This information, validated by numerous analysis of the CLD content in aquatic organisms in the French Antilles, raises the questions of favorable habitats for biocontamination and mechanisms determining the entry of the molecule in aquatic living organisms. The particular life cycle of diadromous organisms helps us answer the first point: The young larvae of *Macrobrachium* spp. leave freshwaters very early (usually after 2 or 3 days) and then spend 3 months in mid-saline waters before reentering the rivers. Thus, in these very particular intermediate areas (mangroves and estuaries) the organisms receive their massive load of pollutants. The large amount of suspended solids in these biotopes probably reinforces the entry of pollutants (Garnas et al. 1978, Huggett et al. 1980, Nichols 1980, Adams 1987), but the mechanisms are not yet precisely known. These results identify natural areas for which it would be necessary to intensify research on the transfer of CLD to living organisms.

The slight decrease of pollution through age in the freshwaters, visible on the model adjustment (D = 0.02130), is, in part, probably due to a "dilution caused by growth" (ratio volume/surface less favorable in larger individuals) or molting processes, but the strong persistence of the contamination through time seems a characteristic of the crustaceans. This value (0.02130) is of the same order of magnitude of those obtained for the modeling of the evolution of the contamination in the crustacean *Palaemonetes pugio* (Holthuis 1949) put in depuration process after previous CLD contamination (Bahner and Oglesby 1979), which was 0.098 for 0.023 μg L^{-1} of CLD in the contaminating water and 0.099 for 0.04 μg L^{-1}. This little difference in depuration is probably due in part to the high discharge of CLD into the Grande Anse River, which influences the pollutant equilibrium after the juveniles had reentered the river. These elements also confirm that crustacean species poorly eliminate this contaminant; therefore, it should be addressed with priority during contamination assessments as CLD bioindicator and special warning must be issued to the population.

The partition of the pollutant in organisms shows a distribution in all tissues analyzed, but with a prominent accumulation in viscera for fish and carapace for crustaceans. Considering (1) the wide distribution of this pollutant in all aquatic ecosystems of the French Antilles, (Guadeloupe and Martinique, sea and rivers) and (2) the very high consumption of fish and specially crustaceans by local and tourist populations, it would be probably useful to disseminate such information to limit the impregnation of the populations by this pesticide.

ACKNOWLEDGMENTS

We sincerely thank the Direction de l'Aménagement et du Logement (DEAL) of Guadeloupe and the Office de l'Eau de la Guadeloupe, the Office de l'Eau de la Martinique, and the Conseil Général de la Guadeloupe for their financial support and for giving us permission to use environmental data. The experiments comply with the current laws of the country in which they were performed.

REFERENCES

Adams W.J. 1987. Bioavailability of neutral lipophilic organic chemicals contained in sediments. In *Fate and Effects of Sediment-Bound Chemicals in Aquatic Systems*, Dickson K.L., Maki A.W., and Brungs W.A. (eds.). Pergamon Press, New York, pp. 219–244.

Bahner L.H. and Oglesby J.L. 1979. Test of a model for predicting Kepone accumulation in selected estuarine species. In *Aquatic Toxicology*, ASTM STP 667, MarkingL.L. and Kimerle R.A. (eds.). American Society for Testing and Materials, Philadelphia, PA, pp. 221–231.

Bahner L.H., Wilson A.J., Jr., Sheppard J.M., Patrick J.M., Jr., Goodman L.R., and Walsh G.E. 1977. Kepone bioconcentration, accumulation, loss, and transfer through estuarine food chains. *Chesapeake Science*, 18: 299–308.

Bonan H. and Prime J.-L. 2001. Rapport sur la présence de pesticides dans les eaux de consommation humaine en Guadeloupe. Ministère de l'Emploi et de la solidarité (Rapport IGAS n°2001-070), Ministère de l'Aménagement de territoire et de l'environnement (IGE n°01/007), Paris, France, 86pp.

Cabidoche Y.-M., Achard R., Cattan P., Clermont-Dauphin C., Massat F., and Sansoulet J. 2009. Long-term pollution by chlordecone of tropical volcanic soils in the French West Indies: A simple leaching model accounts for current residue. *Environmental Pollution*, 157: 1697–1705.

Coat S., Bocquené G., and Godard E. 2006. Contamination of some aquatic species with the organochlorine pesticide chlordecone in Martinique. *Aquatic Living Resources*, 19: 181–187.

Coat S., Monti D., Bouchon C., and Lepoint G. 2009. Trophic relationships in a tropical stream food web assessed by stable isotope analysis. *Freshwater Biology*, 54(5): 1028–1041.

Debier C., Pomeroy P.P., Dupont C., Joiris C., Comblin V., Le Boulenge E., Larondelle Y., and Thome J.P. 2003. Quantitative dynamics of PCB transfer from mother to pup during lactation in UK grey seals *Halichoerus grypus*. *Marine Ecology Progress Series*, 247: 237–248.

Erdogrul O., Covaci A., and Schepens P. 2005. Levels of organochlorine pesticides, polychlorinated biphenyls, and polybrominated diphenyl ethers in fish species from Kahramanmaras, Turkey. *Environment International*, 31: 703–711.

Garnas R.L., Richard L., Bourquin A.W., and Pritchard P.H. 1978. The fate of 14-C Kepone in estuarine microcosms. Presented at *175th National Meeting of the American Chemical Society*, Anaheim, CA, Paper 59.

Goodman L.R., Hansen D.J., Manning C.S., and Faas L.F. 1982. Effects of Kepone on the sheepshead minnow in an entire life-cycle toxicity test. *Archives of Environmental Contamination and Toxicology*, 11(3): 335–342.

Huggett R.J., Nichols M.N., and Bender M.E. 1980. Kepone contamination of the James River estuary. In *Proceedings of Symposium on Contaminants in Sediments*, *Ann Arbor Science*, Ann Arbor, MI, vol. I, Baker R.A. (ed.), pp. 33–52.

Macdonald P. and Du J. 2010. Mixdist: Finite mixture distribution models. R package version 0.5-3. http://CRAN.R-project.org/package=mixdist. Accessed March 21, 2016.

Monti D. 2005. *Etude du niveau de contamination des organismes aquatiques d'eau douce par les pesticides, en Guadeloupe.* Convention Direction régionale de l'Environnement, Guadeloupe, France, 35pp. + annexes, juin 2005.

Monti D. 2008. *Evaluation de la biocontamination en Chlordecone de crustacés et poissons de rivières du Nord-Ouest de la Basse-Terre, et synthèse à l'échelle de la Guadeloupe.* Convention UAG/Direction Régionale de l'Environnement, Guadeloupe, France, 31pp. + annexes.

Multigner L., Ndong J.R., Giusti A., Romana M., Delacroix-Maillard H., Cordier S., Jégou B., Thome J.P., and Blanchet P. 2010. Chlordecone exposure and risk of prostate cancer. *Journal of Clinical Oncology*, 28: 3457–3462.

Nichols M., 1990. Sedimentologic fate and cycling of Kepone in an estuarine system: Example from the James River estuary. *Science of the Total Environment*, 97/98: 407–440.

R Development Core Team. 2012. *R: A Language and Environment for Statistical Computing.* R Foundation for Statistical Computing, Vienna, Austria. ISBN 3-900051-07-0. http://www.R-project.org. Accessed April 2016.

SANCO. 2014. Guidance document on analytical quality control and validation procedures for pesticide residues analysis in food and feed. SANCO/12571/2013 Supersedes SANCO/12495/2011. Implemented by 01/01/2014, EUROPEAN COMMISSION, HEALTH & CONSUMER PROTECTION DIRECTORATE-GENERAL (ed), pp. 48.

Sanders H.O., Huckins J., Johnson B.T., and Skaar D. 1981. Biological effects of kepone and mirex in freshwater invertebrates. *Archives of Environmental Contamination and Toxicology*, 10(5): 531–539.

Schimmel S.C., Patrick J.M., Faas L.F., Oglesby J.L., and Wilson A.J. Jr. 1979. Kepone: Toxicity and bioaccumulation in blue crabs. *Estuaries*, 2(1): 9–15.

8 Assessment of Chlordecone Content in the Marine Fish Fauna around the French West Indies Related to Fishery Management Concerns

Jacques A. Bertrand, Alain Abarnou, Xavier Bodiguel, Olivier Guyader, Lionel Reynal, and Serge Robert

CONTENTS

8.1 INTRODUCTION

Chlordecone ($C_{10}Cl_{10}O$) has a very low aqueous solubility and conversely a high hydrophobicity (UNEP 2007). With bioconcentration factor values of up to 6,000 in algae, of up to 21,600 in invertebrates, and of up to 60,200 in fish and because of its lipophilic nature and its persistence, chlordecone has a high potential for bioaccumulation and biomagnification. Based on its physicochemical properties and modeling data, chlordecone can be transported long distances (UNEP 2007). Chlordecone is toxic to aquatic organisms; the invertebrates (crustaceans) are the most sensitive, which is not surprising for a substance with insecticidal properties (UNEP 2007).

Historically, the first major accidental pollution by chlordecone in aquatic systems happened in the James River from a plant located in Hopewell, Virginia (USA), which produced the molecule from 1966 to 1975 when the production was halted after health problems were detected in workers (UNEP 2007). The contamination of sediments and shellfish in the river downstream the plant has been observed since 1967, very soon after the beginning of chlordecone production (Huggett and Bender 1980). As regards human health, the high level of contamination was of concern to close fishing activity in the river for 13 years. It took about 30 years for the river to recover after the source of contamination was removed (Luellen et al. 2006).

In the French West Indies conditions, after rainy episodes, rainwater washed out chlordecone from soil, to the surface waters (Cattan et al. 2008) and then to the marine coastal waters. The transport of the substance from soil plantations to the sea, either dissolved or bound to soil particles, may be important and rapid because of the importance and the violence of precipitations, because of the mountainous relief and the relative narrowness of the territory of the islands. It was hypothesized that the eroded contaminated material was transported by rivers to the marine environment and settled close to the seashore, particularly in sheltered bays, where it became a source of contamination for marine organisms, and a potential hazard to marine life and finally to human health through seafood consumption.

At the beginning of the research around the French West Indies, the main example of chlordecone dissemination in the aquatic environment was the one in the James River, a tributary of the Chesapeake Bay. But the situation was probably not fully comparable to the one in the French West Indies as discharges from the manufacture stopped with the production, and due to freshwater of the river (Shen and Lin 2006). In this area, two major sinks of chlordecone-contaminated bed sediments were found, at the original source of contamination in Hopewell, on the one hand, and downstream in the high turbidity at the interface between fresh and saline water, on the other (Luellen et al. 2006). In biota, the highest chlordecone residues were found in zooplankton. Connolly and Tonelli (1985) predicted that diet was responsible for about 90% of the observed contamination in three coastline fish species. Furthermore, they identified a positive relationship with concentration in vegetal detritus. Chlordecone concentrations showed a high variability between fish species, areas, and periods, illustrating their sensitivity related to the species' biology and their environment, including a positive relationship with turbidity and contaminated sediments.

Diet is one of the most intricate interactions between fish and their environment. It induces variable flow with ontogeny and space, among species and ecosystems. Most of the marine species have a composite diet, changing according to life cycle stages with opportunistic adaptation with local conditions (Sierra et al. 2001).

This chapter reports on the work done to improve knowledge on the contamination of the marine fishery species between 2000 and 2011. It gives an overview of the extent of contamination in the marine fish fauna around the islands and also about the results of some approaches carried out to contribute to its characterization.

8.2 THE FIRST SIGNS OF MARINE FISH CONTAMINATION (2000–2007)

In the French West Indies, the presence of chlordecone had been detected in rivers and wild fauna since the end of the 1970s (Snegaroff 1977; Kermarrec 1980). The first systematic investigations on the coastal marine fauna were carried out around Guadeloupe in the early 2000s (Bouchon and Lemoine 2003, 2007). In Martinique, the first studies focused on water and sediments in river plumes along the coastline (Bocquené and Franco 2005) and on fishery species from samples collected in harbors (Coat et al. 2006). These studies demonstrated contamination of the coastal marine system. The measurements in the marine species revealed a much lower contamination in the marine environment than in the freshwater ecosystems; nevertheless, these data did not throw light on the possible extent of this contamination toward the open sea. The concentrations encountered in the marine species were generally under the maximum residue limit (MRL) recommended at that time for marine products by the French food safety agency (200 µg kg^{-1} wet weight, Afssa 2005).

At that time, two main features contributed to increased research at sea. One was that these first observations might be considered as a warning about a possible expansion of the contamination in the marine system, but it did not throw light on the extent and factors acting on the contamination processes. The second was the decision of the French food safety agency that drastically lowered the MRL to 20 µg kg^{-1} (wet weight) in marine products for human consumption (Anon 2008). As this pollution became a great concern in the islands, a governmental Action Plan was launched to better assess the pollution and also to ensure consumer protection. In accordance with this plan, various studies have been carried out since 2008 on the marine fish fauna contamination around the islands.

8.3 LARGE-SCALE CHARACTERIZATION OF THE CONTAMINATION (2008–2011)

A large-scale characterization of the contamination of the fish fauna around the French West Indies was carried out between 2008 and 2011. Sampling surveys were planned by veterinary services and research institutes using common protocols and also by sharing theirs means and results (Bertrand et al. 2009). The general objective of this task was to describe the contamination of the fish fauna considering the high diversity of the commercially harvested fish species and their distribution in various habitats. Taking into account the diversity of the system and the limited number of samples, the sampling schemes were established according to three assumptions related to (1) the input of the contaminant in the marine environment, (2) the transmission routes toward the fish fauna, and (3) the biology of the species.

At that time, the transport pathways from the spreading areas to the marine environment were not well established. Nevertheless, the two main routes for pesticides identified by Cattan et al. (2008), through surface runoff and groundwater percolation, induced a likely relationship between the contamination of the catchment basins and that of the marine environment. The maps of the soil contamination in

Guadeloupe (Tillieut 2007) and in Martinique (Desprat et al. 2004) associated with the knowledge of coastal hydrodynamics gave first hypotheses about the spreading of contamination in the marine areas.

To manage difficulties to capture wild specimens of a wide range of often scarce and scattered species, the sampling plan was designed to integrate the following criteria: (1) site (sampling position); in both islands, the spatial sampling design was based on the water masses set for the implementation of the European Water Framework Directive (WFD, Diren 2005; Pareto et al. 2007; Figure 8.1) defined by morphology of the coast, sediment, hydrology, and hydrodynamic conditions; and (2) selection of species according to their biology (lifestyle, feeding habits, etc.). The species were split into four trophic types defined by their trophic level and their diet: detritivorous, herbivorous, or two carnivorous levels.

To have a comprehensive idea of the contamination of the marine fauna, the first sets of samples were equally distributed within the different areas and types of species. The fish specimens were collected by commercial fishermen using appropriate technical guidelines. The general results showed in this chapter are based on 2680 samples collected between 2008 and 2010 in Guadeloupe (1319 samples) and from 2008 to 2011 in Martinique (1361 samples), concerning 131 species including fish (100), crustaceans (22), mollusks (8), and echinoderm (1). The samples were prepared and analyzed for chlordecone according to specific recommendations of the French food safety agency (Afssa 2007).

The 2008–2011 investigations confirmed the presence of chlordecone with a large variability in marine organisms, from below the quantification limit to more than 1000 μg kg^{-1} wet weight. In 27.7% of the samples, concentrations exceeded the MRL in fish set at 20 μg kg^{-1} (Anon 2008). This contamination of the marine fauna was characterized by two features.

1. Most of the highly contaminated samples were caught in areas downstream the contaminated watersheds (Figure 8.1). In Guadeloupe, these areas are mainly located along the southern coast of the island, and in the head of the northern bay. In Martinique, they were found along the east coast of the island and in the sheltered western bay receiving waters from the main river of the island running through banana plantations (Lézarde River). Generally, the specimens living in sheltered bays were more exposed to chlordecone than those living outside where terrigenous fluxes are more dispersed. The results showed a relationship between the spatial distribution of contamination and the zonation established for the WFD.
2. In the most exposed areas in both islands, the highest concentrations were measured in species living very close to the mouth of the rivers, like Mugilidae, bottom and open-water crustaceans like crabs *Callinectes* sp. or the shrimp *Farfantepenaeus subtilis*, and coastal inshore small pelagic fish (Table 8.1). The more contaminated species, detritic feeders and carnivorous, were related to a strong relation with substratum or to a high trophic level, in which the highest values were found in crustaceans, including the highest value in *Callinectes* sp. (15,200 μg kg^{-1} wet weight), followed by fish and mollusks. For some species, the signal was very different between the two islands.

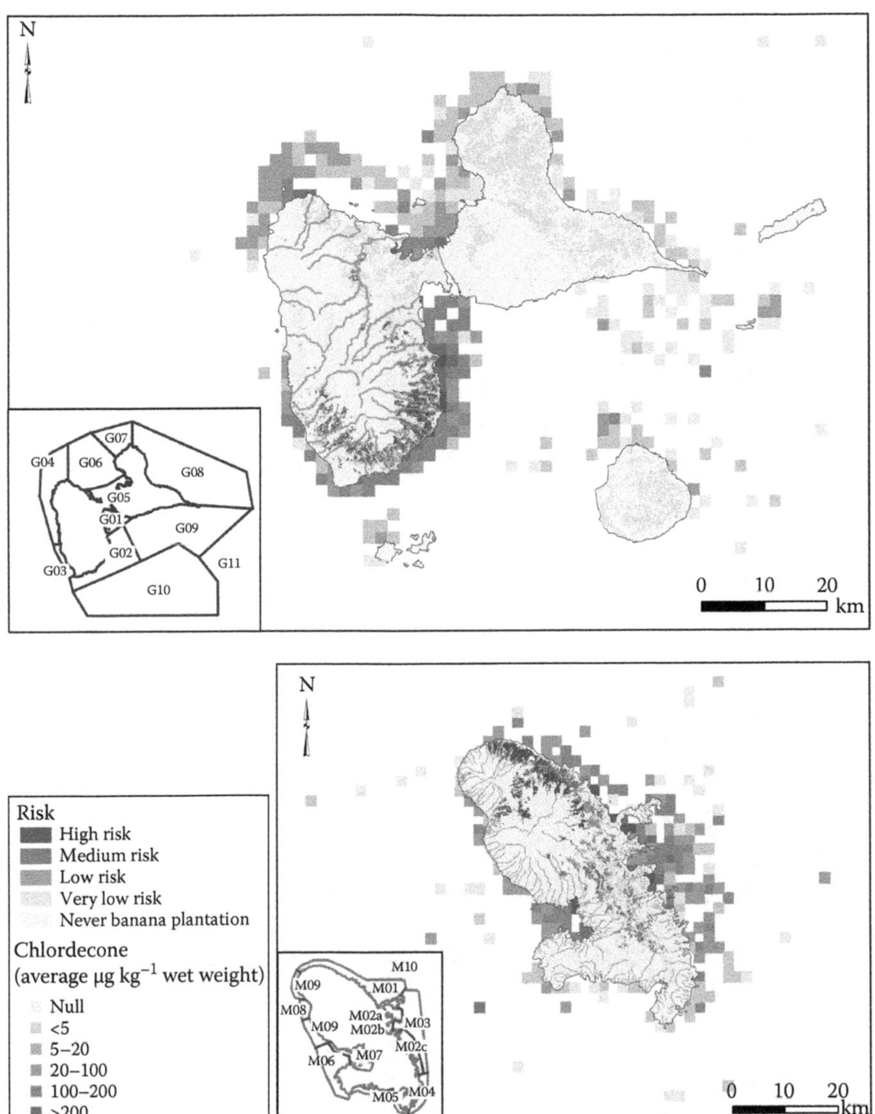

FIGURE 8.1 Chlordecone contamination of the marine macrofauna in Guadeloupe and Martinique from the data collected between 2008 and 2011. Soil contamination risks from Tillieut (2007, Guadeloupe) and Desprat et al. (2004, Martinique).

TABLE 8.1

Mean Chlordecone Concentration (µg kg^{-1} Wet Weight) by Species and by Sub Area in Martinique (Left) and Guadeloupe (Right) for the Species and Areas the Most Impacted

	Species	M01	M02	M07	M05	M03		Species	G02	G01	G04	G03	G05
Fish	*Mugil cephalus*		705				Fish	*Mugil curema*	659	261			45
	Scomberomorus sp.	23	614					*Megalops atlanticus*	499	85			45
	Centropomus undecimalis	534						*Oreochromis mossambicus*	342				64
	Lutjanus analis			443	<5	<5		*Centropomus undecimalis*	16	38	337		
	Oreochromis mossambicus	156		399				*Engraulidae* and *Atherinidae*	177				
	Scomberomorus cavalla	362	98	93		24		*Gymnothorax funebris*	146	23	15	7	
	Diapterus rhombeus		44	330				*Caranx latus*	136	88	12	9	<5
	Haemulon bonariense		268	21				*Harengula humeralis*	121				
	Caranx hippos	258				18		*Cephalopholis fulva*	90		4	49	<5
	Selene vomer	214						*Holocentrus adscensionis*	72	53	4	33	8
	Caranx crysos	14	156	13	10	22		*Caranx crysos*	22	71	7		
	Chloroscombrus chrysurus	145	156			67		*Holocentrus rufus*	70			48	20
	Sphyraena guachancho	155						*Epinephelus guttatus*	68	<5	<5	<5	<5
	Engraulidae and *Atherinidae*	136		55				*Lutjanus synagris*	63		<5	15	6
	Mugil curema		9	130	43			*Haemulon plumieri*	25	50	6	13	5

(Continued)

TABLE 8.1 (*Continued*)
Mean Chlordecone Concentration (µg kg^{-1} Wet Weight) by Species and by Sub Area in Martinique (Left) and Guadeloupe (Right) for the Species and Areas the Most Impacted

	Species	M01	M02	M07	M05	M03		Species	G02	G01	G04	G03	G05
Crustacean	*Callinectes* spp.	7631	14	175			Crustacean	*Callinectes* spp.	357		92		88
	Callinectes sapidus		986		1025			*Panulirus guttatus*	171	82		87	10
	Callinectes danae	136	256	922	137			*Caranx ruber*	0	79			<5
	Farfantepenaeus subtilis		445					*Mithrax spinosissimus*	44	<5	<5	<5	
	Xiphopenaeus kroyeri		342					*Panulirus argus*	35	5	6	39	22
	Callinectes larvatus		317			207							
	Callinectes bocourti		<5		287								
	Mysidacea		226										
	Callinectes ornatus	134											
	Panulirus guttatus	105	92			33							
	Panulirus argus	22	84	37	7	79							

8.4 COMPLEMENTARY APPROACHES TO THE CONTAMINATION ROUTES AND THE IMPACTS ON FISHERIES

From the general picture drawn in 2008–2009, complementary approaches were investigated to explore the possible routes of contamination from the catchment basins to the megafauna: the biomagnification through the food web, the contamination of superficial sediments, the contamination trends of a sensitive species in the open sea, and the extent of the contamination in the fisheries.

8.4.1 BIOMAGNIFICATION IN THE FOOD WEB

Chlordecone contamination was measured in the food web downstream to a contaminated catchment basin in Martinique, from the primary producers (phanerogam seagrass) to the second-level carnivores (Bodiguel et al. 2011). Sampling was carried out along a transect in the bay from the shoreline to its entrance covering communities of phanerogam seagrass (*Thalassia testudinum*) as well as coral reef. The contaminant analyses were carried out on samples from superficial sediment (22 samples), phanerogam (*Thalassia testudinum*, 26 samples), and macrofauna matrix (260 samples); the trophic levels were determined through stable isotope analyses (^{15}N and ^{13}C).

The signal given by statistical multivariate analyses was very low. Concentrations were below the detection limit in all the sediment samples (less than 0.5 μg kg^{-1} dry weight) and in *T. testudinum* (with or without its epiphytes; less than 2 μg kg^{-1} wet weight). Conversely, only 10 animal samples were below the detection limit (0.5 μg kg^{-1} wet weight for this matrix). The highest mean values were found in the detritivorous crustacean *Callinectes danae* and in a second-level carnivorous fish species, *Centropomus undecimalis*, with respectively 178 and 159 μg kg^{-1} wet weight.

A hierarchical classification defined from the mean value of ^{15}N and ^{13}C enrichment ($\delta^{15}N$ and $\delta^{13}C$) allowed to identify four groups, one for the vegetal (*Syringodium filiforme*, *Thalassia testudinum*, and epiphytes of *T. testudinum*), and three for the animals. The relationship between the trophic level illustrated by the $\delta^{15}N$ and the contamination level allowed determining a bioamplification factor between 1.4 and 1.9 from the primary producers to the second-level carnivores.

A positive relationship between terrigenous inputs, characterized by $\delta^{13}C$ and the contamination of the fish fauna, was detected. The species that were able to integrate organic material from land-based sources had higher chlordecone concentrations than other species with similar trophic level. For instance *Larimus breviceps* and *Selene vomer* presented significant contamination (more than 100 μg kg^{-1} wet weight) compared with other second-level carnivorous like *Archosargus rhomboidalis* or *Lutjanus apodus* having a higher $\delta^{13}C$ signature. Nevertheless, the signal may be merged with feeding ground, individuals from a same species being more contaminated when living close to river runoff (e.g., *Callinectes danae* or *Sparisoma viridis*).

Furthermore, for some highly contaminated species, like *Chloroscombrus chrysurus* and *Callinectes danae*, signs of an increased bioaccumulation with size have been observed. On the contrary, a significant decrease of the chlordecone concentration has been measured in the biggest individuals (total length > 23 cm) of the spiny lobster *Panulirus argus*, from a site outside the bay and particularly in the case of mature females. Finally, this study has confirmed a combined effect of their individual habitat and diet on the macrofauna contamination.

8.4.2 Marine Sediment Contamination

The hypothesis of transport of contaminant in relation with suspended particulate matter prompted investigation of the sediment contamination from the alluvial cones to assess its possible extension from the estuaries, and the potential use of the sediment deposits as a chronological marker of long-term chlordecone flux toward the marine system.

A study was carried out in two adjacent bays of Martinique downstream from contaminated basins (Bay of Robert and Bay of Galion). It was based on a description of the surface sediments (within the first 40 cm) and chlordecone analysis from 32 sediment cores distributed in 22 positions all over the bays (Robert 2012). The sediment samples were analyzed to determine their origin, sedimentation rate, adsorption capacity, and compaction.

The results allowed identification of some muddy and muddy/sandy facies in both the bays. They showed a clear distinction between thin and cohesive terrigenous sediment, besides marine sediment, mainly coralligenous. Terrigenous material predominates in the areas close to the mouth of the rivers. In each bay, their morphological and sedimentary characteristics differ according to the soil type of their catchment basin.

In the Bay of Robert, which presents a low sedimentation rate (0.5–1 cm year^{-1}), the chlordecone concentrations were very low in all the locations (close or below the quantification threshold: 6 µg kg^{-1} dry weight). In the Bay of Galion, highest values were recorded in two locations close to the mouth of the Galion River, generally between 15 and 30 µg kg^{-1} dry weight, with a maximum to 68 µg kg^{-1} dry weight. In the cores, these values covered the previous 3–4 years, without any trend between the depth levels; they were lower than one measure in the estuary of the river Cacao (Bay of Robert) in 2008 (23 µg kg^{-1} dry weight), and far from the maximum value of 552 µg kg^{-1} measured in the estuary of another river in Martinique (Lézarde River) the same year (Bertrand et al. 2009).

Finally, the surface sediments in the marine system did not appear to be an important sink for chlordecone. In both bays, current chlordecone concentrations are generally low (most often below the quantification threshold). Significant values have been recorded only in the thin terrigenous sediments very close to the mouth of the single important river in the area (Galion River), suggesting a strong decrease in the sediment contamination between the freshwater and the marine systems.

8.4.3 Spreading of the Contamination in the Spiny Lobsters in Martinique

Spiny lobsters are an important component of the coastal fisheries around the Martinique island, about 25% of the professional annual landings in 2009 (in value), with 20% accounting for *Panulirus argus* and the remaining for *P. guttatus* (Reynal et al. 2013). The importance of this fishery and the previous information on the contamination of these species contributed to the focus on the presence of chlordecone in these two species. The investigations took place in the main fishery area all along the eastern coast of Martinique. In order to help further fishery management measures, the objectives were to precisely find the distribution of the spiny lobster contamination in the area, to explore possible relationships with biological parameters such as individual size, gender, and to compare the contamination in the muscle to that in the head.

The two species have different biology that justifies considering them separately. *P. argus* is a gregarious and migratory species. Young specimens live mainly in the coastal zones, whereas the adults move down to about 90 m depth (Carpenter 2002). This distribution has been confirmed in the study area where the total mean size of *P. argus* increases from 20 cm in the very nearshore waters up to 25 cm when they live out of the coral. On the contrary, *P. guttatus* spends all its life in shallow waters.

Analyses were done of 125 samples of spiny lobsters collected from 2008 to 2010 all around the island, and from a new specific set of 167 samples collected in 2011 in the study area (tail muscle). The whole Martinique coastal area has been divided in three main areas, the no-take regulatory zone defined in 2010 (Anon 2010), a complementary zone defined by the line of 30 m depth named as sensitive areas, and all other places (named as external areas).

As a general trend, the results demonstrated a contamination of the young *P. argus* when living in shallow waters down to contaminated catchment basins. This contamination may reach open zones out of the bays, all over the coral shoals. Then, it tends to decrease when animals move down far from those areas as they get older.

We concluded that an increase of the minimum size of species from 22 cm (the current minimum legal size) to about 25 cm could be explored as an alternative or complementary approach to manage this fishery, combining both human health considerations and productive scopes, anticipating that an increase of the minimum legal size could also improve the yield of the species.

For *P. guttatus*, a difference between the sensitive and the external areas was also identified, with mean concentration of 30 and 12 μg kg^{-1} wet weight respectively. Nevertheless, the weak sign on a potential relation with size may be linked with the absence of migration for this species spending all its lifespan in the coastal areas.

To anticipate potential management measures on minimal legal size, a relationship between chlordecone concentration and individual size of the *P. argus* was tested by simulating a shift of the minimum size of the kept fraction from 22 cm (the current minimum legal size) to 26 cm, with a step of 1 cm. In the external areas, the average chlordecone concentration remained at a low level irrespective of the size threshold. The results from the sensitive zone (Figure 8.2) highlighted a decreasing mean concentration from about 50 μg kg^{-1} wet weight to the reference value (20 μg kg^{-1} wet weight) for the kept fraction, with a length size threshold moving from 22 to 25 cm. The difference was significant between the two fractions for the thresholds between 24 and 26 cm.

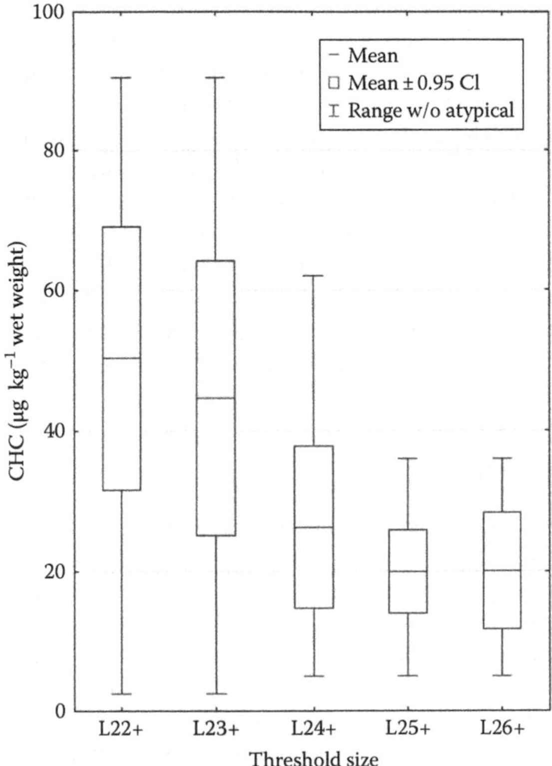

FIGURE 8.2 Chlordecone concentrations in *P. argus* around Martinique for the kept fraction in the sensitive area with different minimum size thresholds (2008–2011 data).

In freshwater crustaceans, the tail muscle appeared less contaminated than the carapace soft flesh (Monti and Thomé, in Anon 2011). A test to compare chlordecone concentration between the tail muscle (TM) and the cephalothorax soft flesh (CSF) of 20 samples of *P. argus* and 25 samples of *P. guttatus* gave a wide concentration range between the two matric, with a mean ratio of 2.5 (CSF/TM).

8.4.4 Relationship between the Marine Macrofauna Contamination and Fisheries

To assess the scope of contamination in the fisheries, the results of the contamination in fish were coupled with the distribution of the fishing activity all around the West mainland of Guadeloupe. The fishing activity in Guadeloupe has been described through extensive surveys carried out every year since 2008. The information was collected through a geographical segmentation of 10 minute side rectangles and according to a grouping of the species by faunal categories. The detailed data on fish contamination from 2008 to 2010 described earlier were drawn through this grid.

From the fishery data collected in 2010, almost one-third of the fishing effort of the Guadeloupe fishing fleet was deployed in areas more or less concerned by the contamination (Bertrand et al. 2013). The total production in the selected areas was estimated at about 690 tons among approximately 3270 tons for all the main Guadeloupe islands. These global results showed wide variations between métiers and species. In these areas, 50% of the landings are made with *Scaridae* (parrotfishes) and *Lutjanidae* (snappers). These species were mainly caught with traps, gill nets, and seines. But for these gears, only a part of the fishing effort is involved in targeting inshore species more or less associated with river entrances such as *Mugilidae*, *Megalops* sp., *Centropomidae*, *Callinectes* sp., and *Clupeidae*. All these species constitute a small part of the landings of the benthic and demersal fisheries in Guadeloupe (about 2%). Nevertheless, they may constitute more than 40% of the landing for gears targeting inshore small pelagic species.

8.5 CONCLUSION

The investigations in the marine macrofauna around the French West Indies in 2008–2011 have given a first global picture of the chlordecone contamination. They have allowed pointing out the strong relationship with the contaminated catchment basins, in connection with the freshwater discharges. Most of marine species may accumulate chlordecone in their tissues. The contamination decreases when going toward the open sea, with modalities linked to the coastal configuration. Moreover, the contamination in fish depends on biological factors, particularly their lifestyle and feeding habits. Some groups evidenced a strongest signal, like the crustaceans that presented the highest concentrations. As strongly observed in case of fish, the species living inshore close to the river entrances were the most sensitive, whereas, at the opposite, the mollusks were less contaminated.

Signs of bioamplification through the food web have been detected. Conversely, decontamination might occur for individuals able to move far from the source of contamination during their lifespan; this was particularly observed among adults of spiny lobster *P. argus* once they left the coastal areas. This suggests a possibility of combining human health and fishery management objectives by increasing the minimum legal size.

Besides this assessment of the contamination of the marine fauna, the investigations on the role of sediment did not allow precisely find any relationship between polluted soil particles and the contamination of living species. Indeed, the concentrations in the marine sediments were generally very low, apart from those close to the mouths of the rivers.

These general trends could be improved by deeper investigations on the distribution of the contamination of the marine species, and on the pesticide transfer from banana plantation grounds to the contaminated basins toward the fish fauna. More particularly, the role of microfauna and dead materials might be considered at the interface between the freshwater and the marine system. Furthermore, the question on the effect of chlordecone on the contaminated macrofauna is still open.

Finally, considering the concentrations still reported in the marine system 15 years after the ban on the use of chlordecone, without any trend during the survey period, and also considering the projections of agro-environmentalists on soil decontamination, the problem of the contamination of the French West Indies marine systems by chlordecone may persist for a long time.

REFERENCES

Afssa, 2005. Avis de l'Agence française de sécurité sanitaire des aliments concernant deux projets d'arrêtés relatifs à la teneur maximale en chlordécone que doivent présenter certaines denrées d'origine végétale et d'origine animale pour être reconnues propres à la consommation humaine. Afssa, Maisons-Alfort, France. Saisine n° 2005-SA-0279, 2pp. http://www.afssa.fr/Documents/RCCP2005sa0279.pdf. Accessed 24, May 2015.

Afssa, 2007. Avis de l'Agence française de sécurité sanitaire des aliments relatif à l'actualisation des données scientifiques sur la toxicité du chlordécone en vue d'une éventuelle révision des limites tolérables d'exposition proposées par l'Afssa en 2003. Afssa, Maisons-Alfort, France. Saisine N°2007-SA-0305, 6pp. http://www.afssa.fr/Documents/RCCP2007sa0305.pdf. Accessed 24, May 2015.

Anon, 2008. Arrêté du 30 juin 2008 relatif aux limites maximales applicables aux résidus de chlordécone que ne doivent pas dépasser certaines denrées alimentaires d'origine végétale et animale pour être reconnues propres à la consommation humaine. JORF, Paris, France. 4 juillet 2008. NOR: AGRG0816067A, 10pp.

Anon, 2010. Arrêté préfectoral N° 10-3275 du 7 octobre 2010 réglementant la pêche et la mise sur le marché des espèces de la faune marine dans certaines zones maritimes de la Martinique en lien avec les bassins versants contaminés par le chlordécone. Préfecture, Fort-de-France, Martinique, France, 4pp.

Anon, 2011. En Action N° 12. Le mensuel du plan Chlordécone en Martinique et en Guadeloupe. Préfecture, Fort-de-France, Martinique. N° 12, 2pp.

Bertrand J. A., A. Abarnou, G. Bocquené, J. F. Chiffoleau, and L. Reynal, 2009. Diagnostic de la contamination chimique de la faune halieutique des littoraux des Antilles françaises. Campagnes 2008 en Martinique et en Guadeloupe. Ifremer, Martinique, France, 136pp. http://archimer.ifremer.fr/doc/00000/6896/. Accessed 24, May 2015.

Bertrand J. A., O. Guyader, and L. Reynal, 2013. Caractérisation de la contamination de la faune halieutique par la chlordécone autour de la Guadeloupe. Résultats des campagnes de 2008 à 2011 (projet CarGual). Ifremer, Brest, France, 39pp. http://archimer.ifremer.fr/doc/00136/24762/. Accessed 24, May 2015.

Bocquené G. and A. Franco, 2005. Pesticide contamination of the coastline of Martinique. *Marine Pollution Bulletin* **51**: 612–619.

Bodiguel X., J. A. Bertrand, and J. Frémery, 2011. Devenir de la chlordécone dans les réseaux trophiques des espèces marines consommées aux Antilles (Chloretro). Ifremer, Martinique, France, 46pp. http://archimer.ifremer.fr/doc/00036/14684/. Accessed 24, May 2015.

Bouchon C. and S. Lemoine, 2003. Niveau de contamination par les pesticides des chaînes trophiques des milieux marins côtiers de la Guadeloupe et recherche de biomarqueurs de génotoxicité. UAG-DIREN, Pointe-à-Pitre, Guadeloupe, France, 71pp.

Bouchon C. and S. Lemoine, 2007. Contamination par les pesticides des organismes marins de la baie du Grand Cul-de-Sac Marin (île de la Guadeloupe). UAG-Dynecar, Pointe-à-Pitre, Guadeloupe, France, 148pp.

Carpenter K. E. (ed.), 2002. *The Living Marine Resources of the Western Central Atlantic.* FAO species identification guide for fisheries purposes. FAO, Rome, Italy, 2127pp. http://www.fao.org/docrep/009/y4160e/y4160e00.htm.

Cattan P., E. Barriuso, Y.-M. Cabidoche, J. B. Charlier, and M. Voltz, 2008. Quelques éléments clés sur l'origine et le mode de pollution des eaux par les produits phytosanitaires utilisés en agriculture. *Les cahiers du PRAM* **7**: 12–19.

Coat S., G. Bocquené, and E. Godard, 2006. Contamination of some aquatic species with the organochlorine pesticide chlordecone in Martinique. *Aquatic Living Resources* **19**: 181–187.

Connolly J. P. and R. Tonelli, 1985. Modeling Kepone in the striped bass food chain in the James River Estuary. *Estuarine, Coastal and Shelf Science* **20**: 349–366.

Desprat J. F., J. P. Comte, and C. Chabrier, 2004. Cartographie du risque de pollution des sols de Martinique par les organochlorés. Rapport phase III: synthèse. BRGM, Orléans, France, RP/53262-FR, 25pp.

Diren, 2005. Etat des lieux 2005 du district hydrographique de la Martinique. DIREN, Martinique, France, 369pp. http://www.martinique.ecologie.gouv.fr/download/etatdes-lieux%20district%20martinique%202005.pdf. Accessed 24, May 2015.

Huggett R. J. and M. E. Bender, 1980. Kepone in the James River. *Environmental Science and Technology* **14**: 918–923.

Kermarrec A., 1980. *Niveau actuel de contamination des chaînes biologiques en Guadeloupe: pesticides et métaux lourds 1979–1980.* INRA, Petit-Bourg, Guadeloupe, France, 155pp.

Luellen D. R., G. G. Vadas, and M. A. Unger, 2006. Kepone in James River fish: 1976–2002. *Science of the Total Environment* **358**: 286–297.

Pareto, Impact-Mer, Arvam, Asconit, and R.N. St-Martin, 2007. Directive cadre sur l'eau. Définition de l'état de référence et du réseau de surveillance pour les masses d'eau littorales de la Guadeloupe. Période 2007–2009. Phase 1. Définition des sites de référence et de surveillance. DDE Guadeloupe, Basse-Terre, Guadeloupe. P.07.138, 58pp.

Reynal L., S. Demanèche, O. Guyader, J. Bertrand, P. Berthou, C. Dromer, E. Maros et al., 2013. Premières données sur la pêche en Martinique (2009–2010). Projet pilote du Système d'informations halieutiques (SIH) Martinique (2007–2010). Ifremer, Martinique, France, 67pp. http://archimer.ifremer.fr/doc/00156/26762/. Accessed 24, May 2015.

Robert S., 2012. Historique de la contamination des sédiments littoraux des Antilles françaises par la chlordécone (Chlosed). Rapport final. Ifremer, L'Houmeau, France, 93pp. http://archimer.ifremer.fr/doc/00071/18247/. Accessed 24, May 2015.

Shen J. and J. Lin, 2006. Modeling study of the influences of tide and stratification on age of water in the tidal James River. *Estuarine, Coastal and Shelf Science* **68**: 101–112.

Sierra L. M., R. Claro, and O. A. Popova, 2001. Trophic biology of the marine fishes of Cuba. In *Ecology of the Marine fishes of Cuba*, Claro R., Lindeman K.C., and Parenti L.R. (eds.). Smithsonian Institution Press, Washington, DC, pp. 115–148.

Snegaroff J., 1977. Les résidus d'insecticides organochlorés dans les sols et les rivières de la région bananière de Guadeloupe. *Phytiatrie-Phytopharmacie* **26**: 251–268.

Tillieut O., 2007. Etat de la cartographie de la pollution. In *Le chlordécone en Guadeloupe. Environnement, santé, société*, Verdol P. (ed.). Jasor, Pointe-à-Pitre, Guadeloupe, France, pp. 28–37.

UNEP, 2007. Report of the Persistent Organic Pollutants Review Committee on the work of its third meeting. Revised risk profile on chlordecone. Nairobi, Kenya. UNEP/POPS/POPRC.3/20/Add.10, 24pp. http://www.pops.int/documents/meetings/poprc/POPRC3/POPRC3_Report_e/POPRC3_Report_add10_e.pdf. Accessed 24, May 2015.

Section IV

Pollutant Transfers

9 From Fields to Rivers *Chlordecone Transfer in Water*

Charles Mottes, Jean-Baptiste Charlier, Nicolas Rocle, Julie Gresser, Magalie Lesueur Jannoyer, and Philippe Cattan

CONTENTS

9.1 INTRODUCTION

Repeated applications of chlordecone on banana fields for 20 years while not taking into account the long-term stability of the molecule have resulted in the accumulation of the molecule in soils (Chapter 3). In the case of nonplowed fields, the major fraction of chlordecone in soil stayed in the upper soil layers, while tillage and deep tillage (until 1 m depth with backhoe) integrated chlordecone throughout the depth of tillage (Clostre et al. 2014a). Nowadays, soils are reservoirs of chlordecone that act as starting points of its environmental fate. Given its high stability, the only significant fate of chlordecone is consequently to transfer to other environmental compartments and to dilute in the environment.

This chapter aims at tackling the following questions according to current knowledge:

- What are the pathways of chlordecone transfer to rivers and their implications in terms of pollutions?

- How long will chlordecone transfer toward rivers?
- What is the potential for managing pollution of rivers by chlordecone?
- What have we learnt for the actual use of pesticides?

In this chapter, we propose an overview of the different pathways of chlordecone transfers toward rivers that were questioned. Also, we compare the incidence of these different pathways on the characteristics of pollution they generate.

9.2 PATHWAYS OF PESTICIDE TRANSFERS TO RIVERS

There are three major pathways for pesticide transfers from fields to rivers: groundwater transfers, surface runoff transfers, and atmospheric transfers. In the specific case of chlordecone, because of the molecular structure of the molecule, some dissipation processes are almost inhibited: notably volatilization and degradation. Figure 9.1 synthesizes the significant processes of chlordecone fate according to the current knowledge. It notably highlights the possible transfer of chlordecone bounded to soil particles due to the high affinity of chlordecone for soil organic carbon as well as the slow leaching of chlordecone into groundwater. The following section describes and characterizes the different transfer pathways of chlordecone to rivers.

According to the different research works, we are able to conclude that the transfer toward leaching and aquifers is the one that contributes to the chronic pollution of water. Nevertheless, studying other pathways shows that particulate transfers during runoff events might transfer more chlordecone on an annual basis than the amount of chlordecone transferred in the dissolved fraction.

9.2.1 Chlordecone Potential Transfers with Water in Soil and with Eroded Soil

In order to assess the ability of chlordecone to transfer with water and with soil, we can use the soil organic matter—water coefficient: K_{oc}. At a given soil organic carbon content, the highest the K_{oc}, the more adsorbed the molecule to the soil. The K_{oc} of chlordecone was found to be extremely variable in the literature: a value calculated from chemical characteristics was 2500 L kg^{-1} (Kenaga 1980), while values obtained from laboratory experiments ranged from 1218 to 2547 L kg^{-1} depending on soil type (Fernandez-Bayo et al. 2013). Modeling studies based on *in situ* observation gave K_{oc} ranging from 2,500 to 20,100 L kg^{-1} (Cabidoche et al. 2009, Mottes et al. 2015). Even for the same soil type, the values vary depending on the methods used to determine the K_{oc}. This effect was reported by Delle Site (2001). Indeed, the method used to determine K_{oc} in laboratory experiments considers only a part of the processes involved in the molecule retention. Notably, batch experiments destroy soil structure, which is an important factor behind chlordecone retention in ash soils (Woignier et al. 2012, 2013). Although these methods are highly reproducible, they do not give a clear indication of the repartition of the molecule during the solid and liquid phases of a structured soil during water leaching for instance (Mottes et al. 2015).

Nevertheless, in spite of the high variability of K_{oc} observed among the different studies, its value is higher than 1218 L kg^{-1}, which means that chlordecone is highly

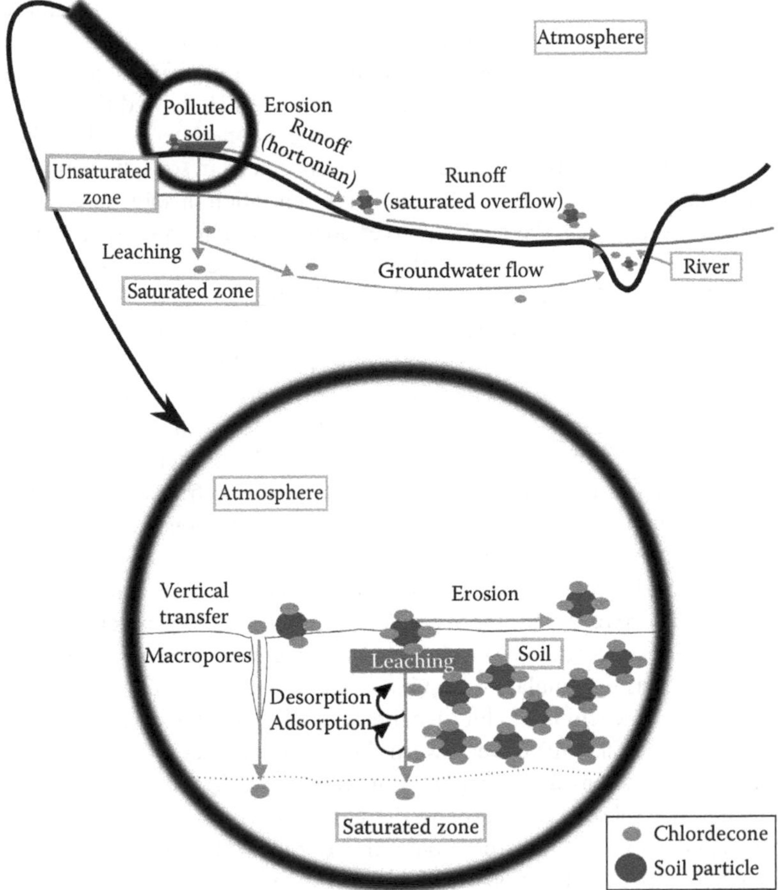

FIGURE 9.1 Major processes and transfers pathways involved in the chlordecone environmental fate.

adsorbed to soil. It has two main consequences: (1) Chlordecone is slowly released into the soil water that leaches and (2) The transfer of polluted soil particles might transfer large amounts of chlordecone adsorbed to it. Both mechanisms are illustrated in the following sections.

9.2.2 Slow Leaching of Chlordecone Has Led to a Chronic Pollution of Rivers

In the French West Indies, most rivers located in agricultural areas are polluted by chlordecone in a chronic manner (Rateau 2013). Two modeling studies based on field observations were performed to assess chlordecone transfers at both the field and the watershed scales. Both studies focused on the dissolved fraction of chlordecone and showed that slow leaching of chlordecone was responsible for chronic pollution of rivers.

At the field scale, the WISORCH model (Cabidoche et al. 2009) simulates the annual budget of chlordecone leaching from contaminated fields. The approach made it possible to estimate the K_{oc} of chlordecone for the different soils along the altitudinal gradient. According to the authors, WISORCH showed that taking into account the effects of tillage and leaching on chlordecone fate in soil made it possible to account for the current chlordecone concentrations in soils. It means leaching is the main pathway to decrease chlordecone concentration in soils.

At the watershed scale, the study of river contamination was needed to account for groundwater contribution from highly contaminated aquifers as illustrated by Arnaud et al. (2013), Charlier et al. (2015), and Chapter 5.

For that, the WATPPASS model (Mottes et al. 2015) was designed to integrate both field scale processes and watershed scale transfers. In tropical volcanic context, given the high amount of rainfalls on permeable soil and multilayered geological formations, it appears that there is a high connectivity between the different aquifers and surface waters (Charlier et al. 2008, 2011, 2015, Mottes et al. 2015). As a result, WATPPASS accounts for soil macropore flow and delayed transfers of chlordecone from several groundwater compartments linking river pollution and soil chlordecone contamination (Mottes et al. 2015).

WATPPASS has been assessed against weekly chlordecone concentrations and loads, and was able to reproduce the average concentration as well as temporal dynamics of chlordecone concentrations and loads at the watershed outlet (Figure 9.2b and c).

Finally, comparing simulation results to observations is consistent with the hypothesis that the leaching of chlordecone is indeed the major pathways that generate chronic pollution at the outlet toward the discharge of aquifers into the river.

Finally, using leaching as the major chlordecone transfers in the dissolved phase, modeling and field observations explain the current chlordecone concentration residues in soil and give a correct estimation of the average concentration of chlordecone at a watershed outlet and the global shape of the weekly temporal dynamic of chlordecone concentrations in the river.

Based on these results, we can argue that the leaching of dissolved chlordecone toward aquifers drained by the rivers is the major transfer pathway involved in the chronic pollution of rivers by chlordecone in the French West Indies.

9.2.3 Soil Erosion Leads to High Amounts of Chlordecone Transferred during Flood Events

Due to the affinity of chlordecone for the soil organic carbon, chlordecone could be carried bounded with soil particles into rivers in the case of erosion from contaminated fields. Chlordecone transfer with soil particles in surface runoff has been investigated on the Robert Bay site of Martinique at the subcatchment scale. This study site is characterized by soil rich in montmorillonite (swelling clay) with a topsoil (0–30 cm) organic matter content of 2.2 ± 0.16%. Such soils are known to be highly prone to erosion (Pinte 2006). In their study, Rocle et al. (2009) showed that on a very small watershed where the average topsoil concentration of banana fields was 0.42 mg kg^{-1}, the concentration of chlordecone on suspended solids was 0.3 mg kg^{-1}. Thus, the contaminated fields, which are prone to erosion, seem to be

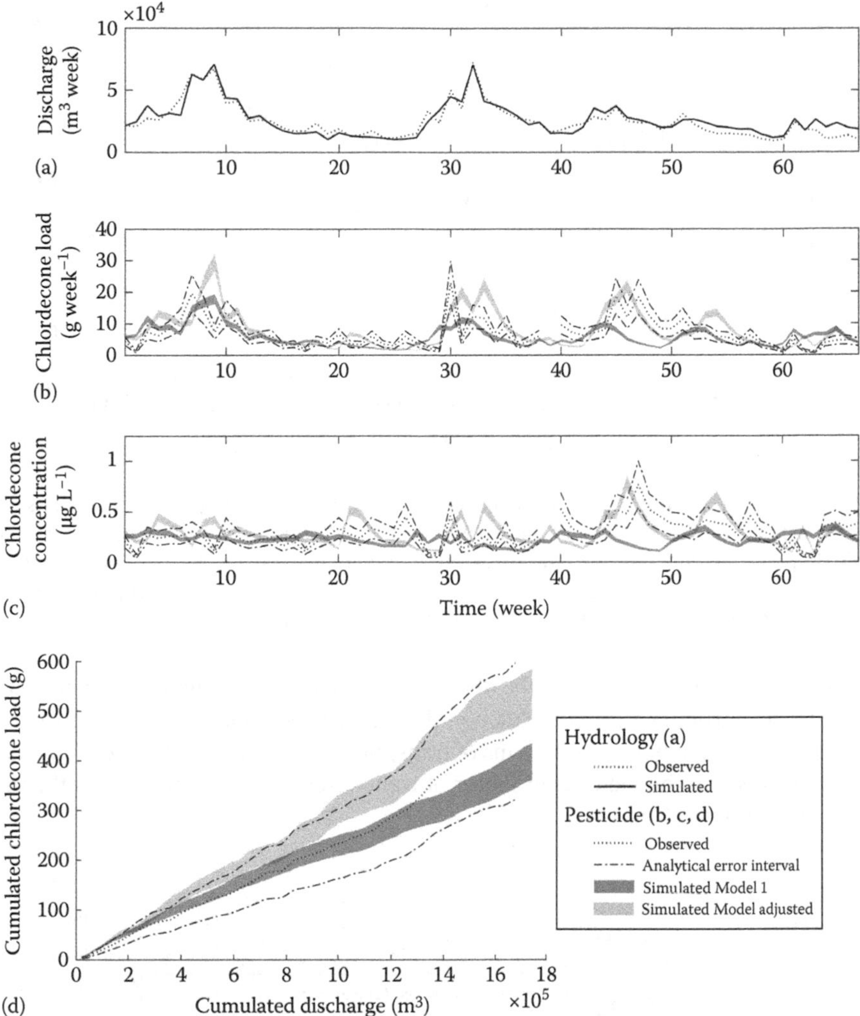

FIGURE 9.2 Simulation results for the hydrologic and pesticide modules of WATPPASS at weekly time step. (a) Discharge (m^3 week^{-1}), (b) chlordecone loads, (c) chlordecone concentrations, (d) cumulated chlordecone load. Upper limit of pesticide simulation area: K_{oc} = 14,400 L kg^{-1}, lower limit: K_{oc} = 17,500 L kg^{-1}. (Reprinted from *J. Hydrol.*, 529, Mottes, C., Lesueur-Jannoyer, M., Charlier, J.-B., Carles, C., Guéné, M., Le Bail, M., and Malézieux, E., Hydrological and pesticide transfer modeling in a tropical volcanic watershed with the WATPPASS model, 909–927, Copyright (2015), with permission from Elsevier.)

the major contributors to the total erosion coming from fields and channels. At the flood scale, Figure 9.3 shows that the concentration of the dissolved fraction plus the particle fraction under 1 µm (hereafter named dissolved fraction) is very low during high flow, while the total chlordecone concentration (including particles over 1 µm) is at its highest concentration during high flow. During floods, when discharge

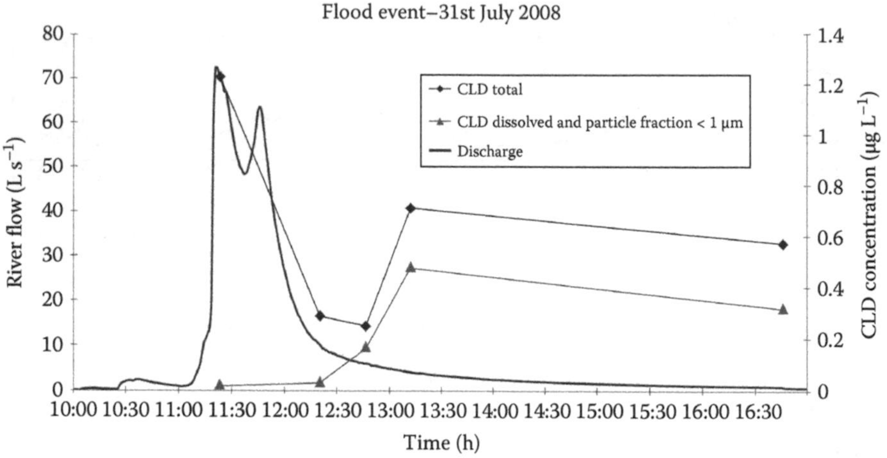

FIGURE 9.3 Evolution of chlordecone concentrations during a flood event (Mansarde watershed). CLD, chlordecone. (After Rocle, N. et al., Gestion agro-environnementale intégrée du risque de contamination de la ressource halieutique par les produits phytopharmaceutiques. Cas du transfert de la chlordécone dans la baie du Robert, Martinique, Rapport Cemagref-Ifremer, Ministère de l'Outre-mer, Paris, France, 2009, 140pp.)

decreases, the concentrations of the dissolved fraction increase while the concentrations of the particulate (≥1 μm) fraction decrease. When analyzing the impacts of such concentration dynamics in terms of transported mass of chlordecone, Figure 9.4 clearly shows that the major contributor of transferred chlordecone mass is the particulate fractions (≥1 μm).

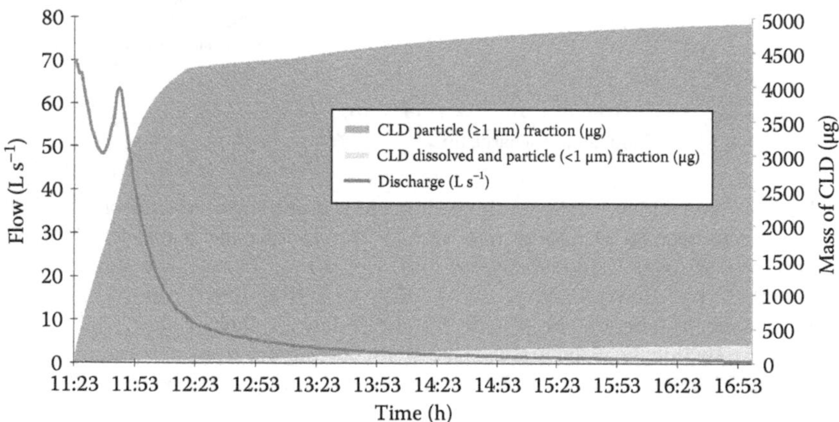

FIGURE 9.4 Cumulated chlordecone loads in dissolved and particulate phases during a flood event (Mansarde watershed). CLD, chlordecone. (After Rocle, N. et al., Gestion agro-environnementale intégrée du risque de contamination de la ressource halieutique par les produits phytopharmaceutiques. Cas du transfert de la chlordécone dans la baie du Robert, Martinique, Rapport Cemagref-Ifremer, Ministère de l'Outre-mer, Paris, France, 2009, 140pp.)

Rocle et al. (2009) also calculated the annual budget of chlordecone transported with eroded sediments and in the dissolved phase. They found that the amount of chlordecone transported by sediments (6.7 g year^{-1}) was almost one order of magnitude higher than in the dissolved fraction (0.75 g year^{-1}). These results can be related to the intensive erosion in this watershed. These results highlight that in terms of the transferred mass of chlordecone on an annual basis, erosion of contaminated soil particles could be a major process. According to locations, the relative amounts of chlordecone transferred via sediments and dissolved phase can vary according to the erosion potential of contaminated soils and to the relative contribution of a contaminated aquifer.

9.2.4 Chlordecone Transfers from Soil to Rivers toward the Atmosphere

The transfer of chlordecone from the soil to the atmosphere is very unlikely, first because the molecule has a very low Henry constant: 2.53×10^{-3} Pa m^3 mol^{-1} and a low vapor pressure 3.5×10^{-5} (Footprint 2013), indicating an extremely low potential of volatilization. Second, Donon and Boullanger (2013) measured pesticides in the air of Martinique (integrative sampling 1 m^3 h^{-1} of air sampled for 1 week) and they did not show chlordecone over the detection threshold (0.2 µg sample^{-1}) during the monitored weeks. In the light of these results and the molecular characteristics of the molecule, it is very unlikely that the atmosphere is a significant pathway for the contamination of water by chlordecone.

9.3 HOW LONG WILL CHLORDECONE TRANSFER TOWARD RIVERS?

Cabidoche et al. (2009) state that soil "pollution is bound to last for several decades for nitisol, centuries for ferralsol and half a millennium for andosol" according to their modeling experiments where leaching is the only mean of chlordecone fate. This result has to be complemented with the ones from Mottes et al. (2015), who modeled chlordecone transfers delayed by several months in shallow aquifers. Alternatively to this modeling approach, the geochemical approach of Gourcy et al. (2009) showed that the pesticide's concentrations are highly variable in aquifers that drain cultivated areas. The groundwater residence time of several years is found to be an explanation of this variability. It means that the aquifers can act as buffers that release chlordecone in rivers for several years to several decades even if the soil reservoirs do not hold significant concentrations of chlordecone anymore. So, even if a remediation method were to be found, the kinetics of the transfers of chlordecone toward aquifers would probably not result in a fast decrease of the chronic pollution of rivers or in deep aquifers. Nevertheless, it would induce the beginning of the dilution without reload of the remaining fraction in the groundwater compartment.

9.4 HOW TO MANAGE CHLORDECONE TRANSFERS TOWARD WATER?

In terms of management options, reducing soil erosion, either on field by covering the soils with cover crops or by installing grass strip to favor sediment deposition, would in the end result in the slow leaching of the chlordecone in the immobilized or intercepted soil particles. It would result in a replacement of the transfer pathway. As a result, reducing chlordecone transport by runoff may favor infiltration and then spread transfers to rivers over time.

While research is looking for a remediation potential (Chapter 14), one option to reduce transfers toward water is to reduce chlordecone transfers from soil to other compartments by favoring its sequestration in soil (Clostre et al. 2014b). Here, the option to add organic matter into soil could make it possible to reduce the leaching potential of chlordecone. Nevertheless, to limit transfers with soil particles, that kind of practice requires an association with practices that limit erosion (such as soil mulching and cover crops). Such option is not a long-term solution, but in the meanwhile, it is a solution that could limit transfers of chlordecone to other environmental compartments while waiting for an effective and applicable remediation option to be found (Chapter 15).

Another solution to limit transfer would be to limit leaching by limiting the amount of percolated water. In the tropical context, the tropical storm season concentrates a large amount of rainfall during a specific period. As a result, the benefit from increasing evapotranspiration over the whole annual cycle would not be significant to reduce chlordecone infiltration. Finally, due to frequent storm events over the year, flow occurs in saturated conditions when rainfall amount is higher than the evapotranspiration. In fact, the only long-term solution would be to fully mineralize chlordecone, but such a solution has not been found yet.

Dynamic models such as WATPPASS could be used to identify periods of increase or decrease of the average chlordecone concentration in water. Such model could be mobilized to help with the management of drinking water pumping during the less polluted periods by anticipating the potential variation of chlordecone concentration in water. It may also help manage the renewal of activated carbon used in filtration devices by assessing carbon saturation time from the computation of chlordecone flows.

9.5 CONCLUSION: WHAT HAVE WE LEARNT FROM THE ACTUAL USE OF PESTICIDES?

Chlordecone is a specific molecule that was applied in the past and that keeps polluting soils as well as the surrounding environment because of its movements with water. Various other pesticides are still in use in agricultural fields. The knowledge of chlordecone transfer pathways from fields to rivers gives insights on the dispersion pathways for other molecules: high leaching rate under abundant rainfalls on permeable soils and high erosion rate from contaminated fields.

In particular, we are aware of the importance of underground transfers that contaminate aquifers and may delay the arrival of pollutants to rivers for years. Additionally, this knowledge can be implemented in models for assessing the potential of mid- to long-term contamination of water resources by pesticides. For instance, as stated earlier, WATPPASS can be used to identify the potential evolution of pesticide concentrations in rivers and also to detect and tackle priority situations. For instance, it suggests that AMPA, a degradation product of glyphosate having a high virtual potential to accumulate in aquifers, has a high risk of groundwater contamination. As a result, weed management alternatives to herbicides should be prioritized (Mottes 2013). However, such modeling results require additional inquiries to further enhance them.

To conclude, due to the lack of solutions to which chlordecone confronted us, pesticides should be thoroughly tested in local conditions before they are used.

REFERENCES

Arnaud, L., N. Baran, and L. Gourcy (2013). Etude du transfert de la chlordécone vers les eaux souterraines en Martinique. Rapport final. Rapport BRGM/RP-61767-FR, Fort-de-France, Martinique 73pp.

Cabidoche, Y. M., R. Achard, P. Cattan, C. Clermont-Dauphin, F. Massat, and J. Sansoulet (2009). Long-term pollution by chlordecone of tropical volcanic soils in the French West Indies: A simple leaching model accounts for current residue. *Environmental Pollution* 157(5): 1697–1705. doi: 10.1016/j.envpol.2008.12.015.

Charlier, J.-B., L. Arnaud, L. Ducreux, B. Ladouche, and B. Dewandel (2015). CHLOR-EAU-SOL—Volet EAU—Caractérisation de la contamination par la chlordécone des eaux et des sols des bassins versant polites guadeloupéen et martiniquais. Rapport final. Rapport BRGM/RP-64142-FR, 160pp.

Charlier, J.-B., P. Cattan, R. Moussa, and M. Voltz (2008). Hydrological behaviour and modelling of a volcanic tropical cultivated catchment. *Hydrological Processes* 22(22): 4355–4370. doi: 10.1002/hyp.7040.

Charlier, J.-B., P. Lachassagne, B. Ladouche, P. Cattan, R. Moussa, and M. Voltz (2011). Structure and hydrogeological functioning of an insular tropical humid andesitic volcanic watershed: A multi-disciplinary experimental approach. *Journal of Hydrology* 398(3–4): 155–170. doi: 10.1016/j.jhydrol.2010.10.006.

Clostre, F., M. Lesueur-Jannoyer, R. Achard, P. Letourmy, Y.-M. Cabidoche, and P. Cattan (2014a). Decision support tool for soil sampling of heterogeneous pesticide (chlordecone) pollution. *Environmental Science and Pollution Research International* 21(3): 1980–1992. doi: 10.1007/s11356-013-2095-x.

Clostre, F., T. Woignier, L. Rangon, P. Fernandes, A. Soler, and M. Lesueur-Jannoyer (2014b). Field validation of chlordecone soil sequestration by organic matter addition. *Journal of Soils and Sediments* 14(1): 23–33. doi: 10.1007/s11368-013-0790-3.

Delle Site, A. (2001). Factors affecting sorption of organic compounds in natural sorbent/water systems and sorption coefficients for selected pollutants. A review. *Journal of Physical and Chemical Reference Data* 30(1): 187–439. doi: 10.1063/1.1347984.

Donon, E. and C. Boullanger (2013). Evaluation des concentrations des produits phytosanitaires dans l'air ambiant. Martinique 2012–2013, MadiniAir. Rapport 06/13/PESTICIDES2012-13, Fort-de-France, Martinique, 66pp.

Fernandez-Bayo, J. D., C. Saison, C. Geniez, M. Voltz, H. Vereecken, and A. E. Berns (2013). Sorption characteristics of chlordecone and cadusafos in tropical agricultural soils. *Current Organic Chemistry* 17(24): 2976–2984.

Footprint (2013). The Pesticide Properties Database (PPDB) developed by the Agriculture & Environment Research Unit (AERU), University of Hertfordshire, funded by UK national sources and the EU-funded FOOTPRINT project (FP6-SSP-022704). http://sitem.herts.ac.uk/aeru/footprint/fr/index.htm.

Gourcy, L., N. Baran, and B. Vittecoq (2009). Improving the knowledge of pesticide and nitrate transfer processes using age-dating tools (CFC, SF6, 3H) in a volcanic island (Martinique, French West Indies). *Journal of Contaminant Hydrology* 108(3–4): 107–117. doi: 10.1016/j.jconhyd.2009.06.004.

Kenaga, E. E. (1980). Predicted bioconcentration factors and soil sorption coefficients of pesticides and other chemicals. *Ecotoxicology and Environmental Safety* 4(1): 26–38. doi: 10.1016/0147-6513(80)90005-6.

Mottes, C. (2013). Evaluation des effets des systèmes de culture sur l'exposition aux pesticides des eaux à l'exutoire d'un bassin versant. Proposition d'une méthodologie d'analyse appliquée au cas de l'horticulture en Martinique. PhD thesis, AgroParisTech—ABIES, Paris, France.

Mottes, C., M. Lesueur-Jannoyer, J.-B. Charlier, C. Carles, M. Guéné, M. Le Bail, and E. Malézieux (2015). Hydrological and pesticide transfer modeling in a tropical volcanic watershed with the WATPPASS model. *Journal of Hydrology* 529(3): 909–927. doi: 10.1016/j.jhydrol.2015.09.007.

Pinte, K. (2006). Diagnostic de l'érosion sur le bassin versant de la baie du Robert en Martinique. Master II, INA-PG, Paris, France, 184pp.

Rateau, F. (2013). Les produits phytosanitaires dans les cours d'eau de Martinique—Atlas des pesticides. Office de l'eau Martinique, Fort-de-France, Martinique, 48pp.

Rocle, N. (coord.), A. Abarnou, R. Achard, A. Arimone, J.-A. Bertrand, G. Bocquené, A. Catlow et al. (2009). Gestion agro-environnementale intégrée du risque de contamination de la ressource halieutique par les produits phytopharmaceutiques. Cas du transfert de la chlordécone dans la baie du Robert, Martinique. Rapport Cemagref-Ifremer, Ministère de l'Outre-mer, Paris, France, 140pp.

Woignier, T., F. Clostre, H. Macarie, and M. Jannoyer (2012). Chlordecone retention in the fractal structure of volcanic clay. *Journal of Hazardous Materials* 241–242: 224–230. doi: 10.1016/j.jhazmat.2012.09.034.

Woignier, T., P. Fernandes, A. Soler, F. Clostre, C. Carles, L. Rangon, and M. Lesueur-Jannoyer (2013). Soil microstructure and organic matter: Keys for chlordecone sequestration. *Journal of Hazardous Materials* 262: 357–364. doi: 10.1016/j.jhazmat.2013.08.070.

10 From Soil to Plants *Crop Contamination by Chlordecone*

Florence Clostre, Magalie Lesueur Jannoyer, Jean-Marie Gaude, Céline Carles, Philippe Cattan, and Philippe Letourmy

CONTENTS

10.1 INTRODUCTION

Chlordecone is a very stable molecule characterized by high hydrophobicity, high affinity for organic matter, and low solubility in water (Dawson et al. 1979, Kenaga 1980, U.S. Environmental Protection Agency 2012). These properties drive its environmental fate and partly explain its persistence in soils, which in turn can pollute crops, thus raising public health concerns (Multigner et al. 2016).

Plants are exposed to soil pollutants via different routes: Roots can be contaminated by direct contact and uptake, while above-ground shoots can be contaminated by contact with soil airborne suspension particles and through pollutant volatilization. Bioavailability of pollutant in soil is a key parameter in root contamination, and the subsequent diffusion inside plants depends on the pollutant properties, and species and variety of crops (Topp et al. 1986, Mattina et al. 2000, Collins et al. 2006).

We studied crops as they are the main contributors to dietary intake of chlordecone in the French West Indies (Dubuisson et al. 2007), mainly cucurbits and root vegetables that can be polluted at levels above regulatory thresholds (Clostre et al. 2014b, 2015b). We cultivated these crops in different soil types representative of cultivated soils in the French West Indies (andosol, nitisol, and ferralsol) and under different growing conditions: field and greenhouse. The aim of our experiments was to have a better insight in the contamination processes and to assess the variables influencing the uptake of chlordecone from the soil by the plants.

10.2 CONTAMINATION PATHWAYS FOR PLANTS AND ROOT UPTAKE

Due to its bishomocubane chemical structure and high steric hindrance, degradation (metabolization) of chlordecone in plants is unlikely (Dawson et al. 1979, Dolfing et al. 2012, Clostre et al. 2015a).

Contamination pathways for organic compounds, plant uptake, and diffusion inside plants are driven by physicochemical properties of pollutants (solubility, hydrophobicity, polarity, etc.), soil properties (organic matter, clays, etc.), growing conditions, and plant characteristics (Calvelo Pereira et al. 2006, White et al. 2006a, Gent et al. 2007, Donnarumma et al. 2009).

10.2.1 Soil to Plant Route Is the Major Contamination Pathway

Results from our chlordecone uptake experiments support the hypothesis that soil to plant route is the major contamination pathway (Figure 10.1).

We observed a gradient of chlordecone content from roots to top of the plant (roots > stems > leaves and fruits); chlordecone contents in fine roots were of the same order of magnitude as in soil, and higher contamination was observed in fruits of trained cucumber plants in greenhouse compared to those of nontrained cucumber plants in field (Clostre et al. 2014b).

Moreover, the hypothesis of soil to plant route as the major contamination pathway is consistent with uptake models for hydrophobic pollutants (Brudenell et al. 1995, Fujisawa et al. 2002, Li et al. 2005, Dettenmaier et al. 2009). Likewise, soil to plant route is the dominant pathway for crop contamination by other persistent organochlorine pesticides such as lindane, DDE, and heptachlor (Mattina et al. 2000, Collins et al. 2006, Gent et al. 2007, Murano et al. 2009). Additionally, chlordecone has a low volatility and partitions poorly to air; contamination through molecule volatilization is thus unlikely (Dawson et al. 1979, Cabidoche et al. 2009). Contaminated airborne soil particles can contribute to above-ground plant contamination. Nevertheless, according to measurements close to a chlordecone production site in Virginia, USA, chlordecone content in airborne particles was reduced from 40% to none after factory closure despite soil-persistent pollution (Dawson et al. 1979).

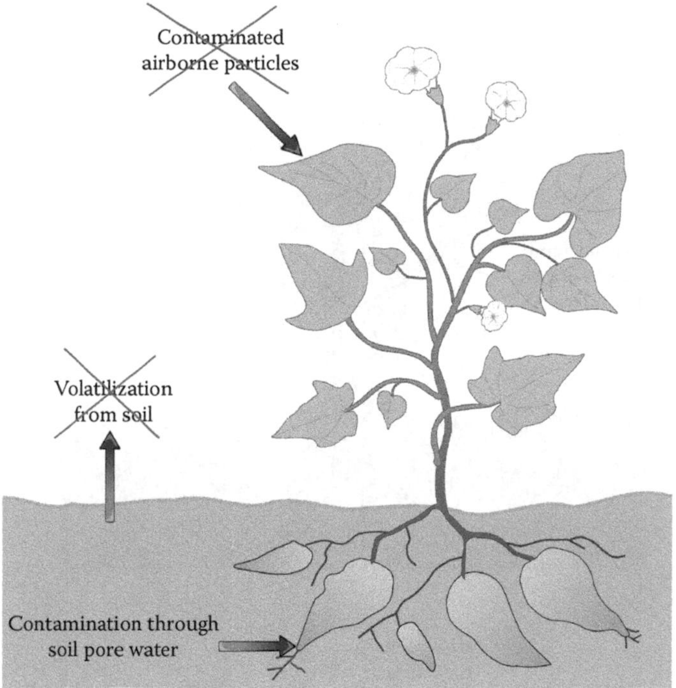

FIGURE 10.1 Exposure pathways of plants to chlordecone (Adapted from Clostre 2015a, CC BY-NC-ND).

10.2.2 Contamination by Adsorption to Root Surface

sIn general, adsorption to root surface (Figure 10.2a and b) depends upon interaction between soil characteristics (organic matter, soil type, etc.), physicochemical properties of pollutants, and root composition (nature and content in lipid compounds). With a log K_{ow} (octanol/water partition coefficient) over 4, chlordecone is considered to be very hydrophobic (U.S. Environmental Protection Agency 2012). Different authors demonstrated that the more lipophilic the pollutant, the higher the partition into plant root surface (Briggs et al. 1982, Fismes et al. 2002, Collins et al. 2006).

Adsorption of lipophilic pollutants to root epidermis is followed by very slow diffusion in root tissues, from epidermis to the inner tissues (Fujisawa et al. 2002, Trapp 2002). This was experimentally confirmed for chlordecone by Létondor et al. (2015) on radish.

In the case of chlordecone, the relative contribution of the different pathways to contamination of roots and aerial tissues is still unknown, even if absorption through transpiration stream does not seem to be the major pathway (Létondor et al. 2015).

10.2.3 Contamination by Absorption with Transpiration Fluxes

Pollutants can also be absorbed by plant root hairs (Figure 10.2a and b) via the soil pore water along with transpiration fluxes (Fujisawa et al. 2002, Juraske et al. 2011). During

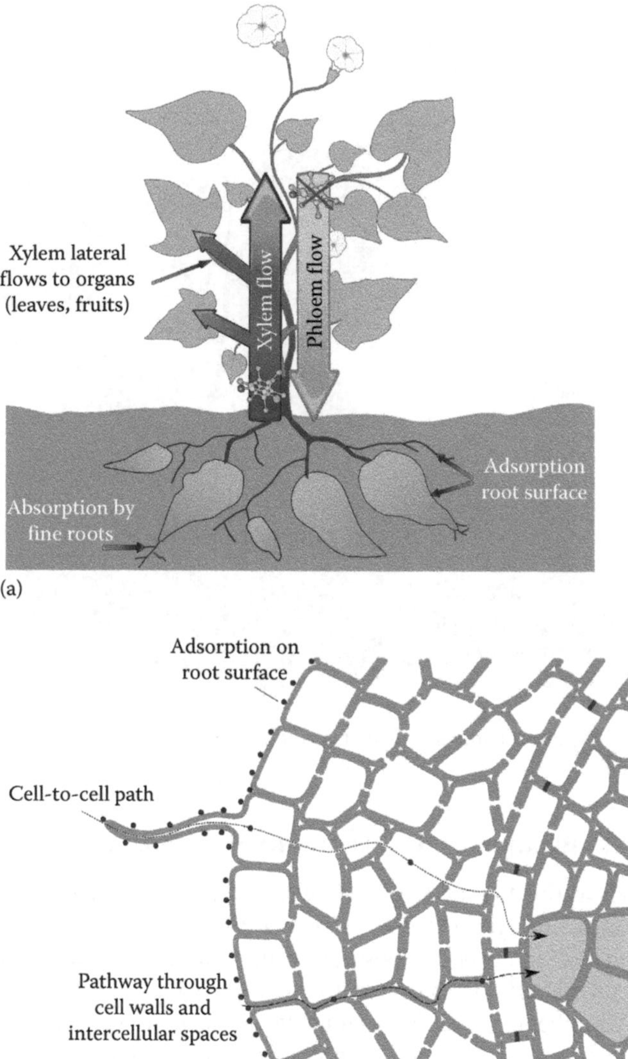

FIGURE 10.2 **(See color insert.)** (a) Mechanisms of contamination by chlordecone at plant scale (Adapted from Clostre 2015a, CC BY-NC-ND) and (b) focus at root scale on pathways to reach xylem vessels (Adapted from Clostre 2015b, CC BY-NC-ND). Chlordecone, dark round points.

plant transpiration, pollutants are taken up in the rhizosphere soil through two mechanisms: diffusion by gradient difference in pollutant content and water movements due to water uptake by plants. Pollutant content in soil pore water is influenced not only by the pollutant's molecular properties but also by soil characteristics (organic matter, soil moisture) and by plant physiology (root exudates); thus, it may vary with growing conditions. As chlordecone has a poor solubility and a high affinity for soil organic matter and root surface, its uptake through the transpiration stream should be limited (Trapp 2002, Khan et al. 2008, U.S. Environmental Protection Agency 2012).

Absorption is followed by a transport of the pollutant to above-ground shoots. This is a passive process through xylemic pathway along the transpiration stream.

10.2.4 Contamination Depends on Soil Type and Growing Conditions

Soil type is known to affect pesticide bioavailability and diffusion into the environment (Alexander 2000, Kumar and Philip 2006). In the case of chlordecone, andosols are more contaminated but release less chlordecone to water than do nitisol and ferralsol (Cabidoche et al. 2009). We showed that crops cultivated in andosols had lower chlordecone uptake ratios than those cultivated on ferralsols and nitisols. Plant uptake was about three times lower in andosols than in nitisols and/or ferralsols for lettuce, yam, and dasheen (Woignier et al. 2012). For cucumber, transfers from soil to fruits were about twice higher in nitisols and ferralsols than in andosols both under field and greenhouse conditions (Clostre et al. 2014b). Likewise for root vegetables, soil type (organic matter and clay microstructure) had a highly significant effect on chlordecone uptake (Clostre et al. 2015b). These differences can be explained by the higher organic matter content in andosols compared to nitisols and ferralsols, and to the peculiar fractal microstructure of allophane clays in andosols (Woignier et al. 2012, Clostre et al. 2015b). Given the aforementioned contamination processes, soils are not equivalent in their ability to transfer chlordecone to plants.

Our cucurbit experiment showed that greenhouse conditions favored plant uptake both in andosols and in nitisols/ferralsols, with transfers 40%–60% higher compared to field conditions (Clostre et al. 2014b). Other authors showed that growing conditions could affect the contamination of plants (Ficko et al. 2011). Higher soil moisture led to higher contamination of pumpkin roots by DDE, and planting density had an effect on zucchini and cucumber contamination by DDE (Wang et al. 2004, Kelsey et al. 2006).

10.3 THE CHLORDECONE FLUXES IN THE PLANT: FROM ROOTS TO OTHER PLANT PARTS

Contamination levels depend on the crop and, at the plant scale, on the plant organ. Chlordecone distribution inside the plant is driven by contamination mechanisms and sap flows.

10.3.1 Contamination of Root Vegetables

In an exposure study, Dubuisson et al. (2007) reported higher chlordecone contents in root vegetables than in other crop products, which is consistent with the fact that they are directly exposed to soil contamination. Similarly, higher contents were observed in root vegetables for various organochlorine pesticides (Mattina et al. 2000, Poulsen and Andersen 2003, Zohair et al. 2006).

In our field experiments, transfers of chlordecone from soil to root vegetables (ratio of content in crop to content in soil) ranged between 0.02 and 0.1 in andosols and 0.07 and 0.6 in nitisols/ferralsols for three root vegetables (dasheen, sweet potato, and yam) (Clostre et al. 2015b). Even if these ratios are higher than transfers to cucurbit fruits or lettuce leaves (Woignier et al. 2012, Clostre et al. 2014b), they remained below 1.

Two contamination mechanisms account for contamination levels found in underground organs: adsorption to root surface by direct contact, and absorption with transpiration stream followed by xylemic diffusion inside the roots. For underground storage organs, a third mechanism is involved: growth dilution. During filling processes, storage organs are loaded via the phloem, which induces a dilution of the contamination, as downward translocation of organochlorine pesticides via the phloem assimilation stream is negligible (Trapp 2002, Paraíba and Kataguiri 2008).

10.3.2 Translocation to Above-Ground Shoots: Stem, Leaf, and Fruit Contamination

In our experiments, cucurbit roots were more contaminated than stems, and stems more contaminated than leaves (Clostre et al. 2014b). This is in accordance with the results we obtained in sugarcane and tomato plants (Figure 10.3a and b).

Moreover in a greenhouse experiment, we showed that upper fruits tended to be more contaminated by chlordecone than lower fruits (Clostre et al. 2014b). This resulted in good agreement with decreasing contamination levels by PCB and DDT in fruits along with an increasing distance from the roots (Whitfield Åslund et al. 2007, 2010). Globally, decreasing contaminations from the bottom to the top of the shoot at the plant scale were observed in studies on hydrophobic pollutant uptake by crops (Mattina et al. 2000, Namiki et al. 2013).

Such trends are consistent with a translocation of chlordecone through the xylem and transpiration stream. During translocation to the plant shoots along the transpiration stream, chlordecone of the xylem transpiration flow tends to adsorb onto the hydrophobic components of plant tissues and cell walls (Briggs et al. 1983, Fismes et al. 2002). For lipophilic compounds such as chlordecone, the distance of translocation will thus be limited (Barak et al. 1983, Briggs et al. 1983).

10.3.3 Uptake and Translocation Potentials Depend on the Crop

Among root vegetables, crops varied in their uptake capacity: dasheen had transfer ratios (vegetable dry weight basis) up to four or ten times higher than yam and sweet potato depending on the type of soil (Clostre et al. 2015b). The nature of the underground storage organ (corm or tuber) and periderm thickness can play a role in the differences observed (Clostre et al. 2015b, Létondor et al. 2015).

Generally speaking for hydrophobic pollutants, fruits have low risk of contamination via soil to plant route as plants poorly translocate them to shoots (Wild et al. 2005, Trapp 2007). Consistently, for chlordecone, no contamination (or low detection percentage) was observed in fruits, except in cucurbits, in studies of food and crop contamination by chlordecone (Dubuisson et al. 2007, Clostre et al. 2015a).

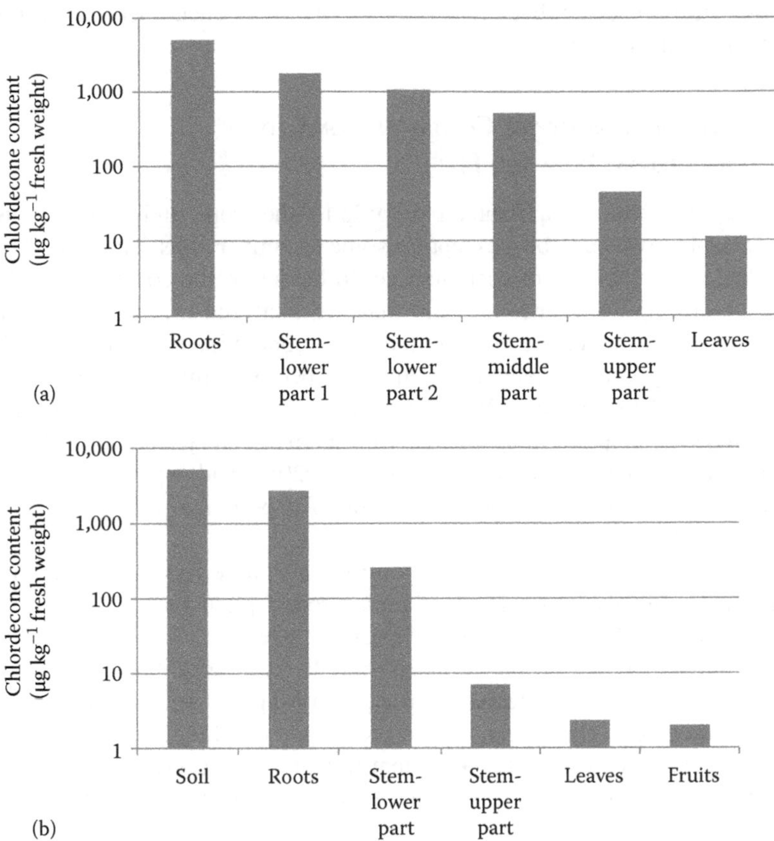

FIGURE 10.3 Contamination of (a) sugarcane (nitisol, n = 12) and (b) tomato (nitisol, n = 3) plant organs (median, logarithmic scale). For tomato, leaves and fruits were sampled in the upper part of plant.

Cucurbits (*Cucurbitaceae* family) are well known to accumulate and translocate hydrophobic pollutants, such as chlordecone, in greater quantities than other plant species (Mattina et al. 2000, Donnarumma et al. 2009). They have also been studied for their potential for soil phytoremediation of weathered DDT, DDE, DDD, and PCBs (Lunney et al. 2004, White et al. 2006a,b). Root exudates can contribute to the unique ability of cucurbits to take up pollutants from soil and translocate them to shoots and fruits (Campanella and Paul 2000).

Among cucurbits, in field conditions, zucchini and pumpkin (*Cucurbita* species) translocated more chlordecone to fruits than cucumber (*Cucumis* species), while christophine (*Sechium* species) can be classified as a non-uptaker crop (Clostre et al. 2014b). In our greenhouse experiment, contamination was significantly higher in zucchini than in cucumber: transfers (plant dry weight basis) to roots were threefold to eightfold higher depending on the type of soil; transfers to stem were 30- to 12-fold higher; transfers to leaves were not significantly different (Clostre et al. 2014b).

Likewise, zucchini accumulated tenfold more DDE than cucumber in a hydroponic experiment (Gent et al. 2007).

10.3.4 Distribution of the Contamination within an Organ: Pulp and Peel

Peel was always more contaminated than pulp for the crops studied (three root and tuber vegetables, two cucurbits) except for some cucumber plots. The lack of significant difference in the case of cucumber could be due to the contribution of seeds rather than of pulp itself since the pulp that was analyzed included seeds. Peel of underground organs was three to forty times more contaminated than pulp, with yam exhibiting the highest differences. In pumpkin, peel was threefold more contaminated than pulp (Clostre et al. 2014a). The higher chlordecone contents observed in peel compared to pulp are in agreement with an exploratory study on chlordecone contamination (Cabidoche and Lesueur-Jannoyer 2012) and experiments on uptake and distribution in plant tissues of other hydrophobic pollutants (Mattina et al. 2000, Fismes et al. 2002).

The differences in pulp and peel contamination can be linked to contamination mechanisms (adsorption to surface of underground plant parts), dilution during growth, and peel and pulp composition (lipids and fibers).

As the edible part is less contaminated than the peel, usual home preparation of contaminated fruits and vegetables (thorough washing and peeling) helps decrease consumer exposure to chlordecone. Food processing (washing, peeling, cooking) is known to generally reduce pesticide content in fruits and vegetables (Kaushik et al. 2009). In our case, cooking did not reduce the chlordecone concentration in food (Clostre et al. 2014a).

10.3.5 First Recommendations

- Avoid root vegetable cropping systems on polluted fields and promote non-uptaker crops in these conditions.
- Apply the general rule of hygiene when preparing food: thorough washing of all fruits and vegetables, and peeling of roots, tubers, and cucurbits.

10.4 CONCLUSION

Chlordecone's chemical properties explain its relatively limited uptake by plants. Contamination in plants remains generally far lower than in soils: Uptake ratios were always below 1 in the order of a few tenths or hundredths, except for zucchini roots and stem under greenhouse conditions (Clostre et al. 2014b, 2015b). The lack of chlordecone accumulation in plant organs suggests that phytoremediation would be poorly efficient, but in our experiment we did not explore the whole plant diversity.

This work was undertaken in a public health framework with the overall aim of allowing farmers to manage the risk of exceeding the regulatory threshold for chlordecone in crop products, and hence to reduce consumer exposure. Our field experiments were conducted using aged soils contaminated by chlordecone; local crop varieties and crops were cultivated under growing conditions similar to those of commercial production. Thus, our results can be used to classify the studied crops into categories according to their chlordecone uptake ability (e.g., non- (or low)-uptakers, mid-uptakers, and high-uptakers). These categories are the cornerstone to build a management tool for farmers, linking regulatory threshold (maximum residue limit) to transfers.

ACKNOWLEDGMENT

Funding was provided by the French Chlordecone National Plan and Regional Health Agency of Martinique through the "JAFA" Family Garden Health Program.

REFERENCES

Alexander, M. 2000. Aging, bioavailability, and overestimation of risk from environmental pollutants. *Environmental Science & Technology* 34(20):4259–4265. doi: 10.1021/es001069+.

Barak, E., B. Jacoby, and A. Dinoor. 1983. Adsorption of systemic pesticides on ground stems and in the apoplastic pathway of stems, as related to lignification and lipophilicity of the pesticides. *Pesticide Biochemistry and Physiology* 20(2):194–202. doi: 10.1016/0048-3575(83)90024-x.

Briggs, G. G., R. H. Bromilow, and A. A. Evans. 1982. Relationships between lipophilicity and root uptake and translocation of non-ionised chemicals by barley. *Pesticide Science* 13(5):495–504. doi: 10.1002/ps.2780130506.

Briggs, G. G., R. H. Bromilow, A. A. Evans, and M. Williams. 1983. Relationships between lipophilicity and the distribution of non-ionised chemicals in barley shoots following uptake by the roots. *Pesticide Science* 14(5):492–500. doi: 10.1002/ps.2780140506.

Brudenell, A. J. P., D. A. Baker, and B. T. Grayson. 1995. Phloem mobility of xenobiotics: Tabular review of physicochemical properties governing the output of the Kleier model. *Plant Growth Regulation* 16(3):215–231. doi: 10.1007/bf00024777.

Cabidoche, Y. M., R. Achard, P. Cattan, C. Clermont-Dauphin, F. Massat, and J. Sansoulet. 2009. Long-term pollution by chlordecone of tropical volcanic soils in the French West Indies: A simple leaching model accounts for current residue. *Environmental Pollution* 157(5):1697–1705. doi: 10.1016/j.envpol.2008.12.015.

Cabidoche, Y. M. and M. Lesueur-Jannoyer. 2012. Contamination of harvested organs in root crops grown on chlordecone-polluted soils. *Pedosphere* 22(4):562–571. doi: 10.1016/s1002-0160(12)60041-1.

Calvelo Pereira, R., M. Camps-Arbestain, B. Rodriguez Garrido, F. Macias, and C. Monterroso. 2006. Behaviour of α-, β-, γ-, and δ-hexachlorocyclohexane in the soil–plant system of a contaminated site. *Environmental Pollution* 144(1):210–217.

Campanella, B. and R. Paul. 2000. Presence, in the rhizosphere and leaf extracts of zucchini (*Cucurbita pepo* L.) and melon (*Cucumis melo* L.), of molecules capable of increasing the apparent aqueous solubility of hydrophobic pollutants. *International Journal of Phytoremediation* 2(2):145–158.

Clostre, F., P. Cattan, J.-M. Gaude, C. Carles, P. Letourmy, and M. Lesueur-Jannoyer. 2015a. Comparative fate of an organochlorine, chlordecone, and a related compound, chlordecone-5b-hydro, in soils and plants. *Science of the Total Environment* 532:292–300. doi: 10.1016/j.scitotenv.2015.06.026.

Clostre, F., P. Letourmy, and M. Lesueur-Jannoyer. 2015b. Organochlorine (chlordecone) uptake by root vegetables. *Chemosphere* 118:96–102. doi: 10.1016/j.chemosphere.2014.06.076.

Clostre, F., P. Letourmy, L. Thuriès, and M. Lesueur-Jannoyer. 2014a. Effect of home food processing on chlordecone (organochlorine) content in vegetables. *Science of the Total Environment* 490:1044–1050. doi: 10.1016/j.scitotenv.2014.05.082.

Clostre, F., P. Letourmy, B. Turpin, C. Carles, and M. Lesueur-Jannoyer. 2014b. Soil type and growing conditions influence uptake and translocation of organochlorine (chlordecone) by *Cucurbitaceae* species. *Water, Air, & Soil Pollution* 225(10):1–11. doi: 10.1007/s11270-014-2153-0.

Collins, C., M. Fryer, and A. Grosso. 2006. Plant uptake of non-ionic organic chemicals. *Environmental Science & Technology* 40(1):45–52. doi: 10.1021/es0508166.

Dawson, G. W., W. C. Weimer, and S. J. Shupe. 1979. Kepone: A case study of a persistent material. *Water American Institute of Chemical Engineers Symposium Series* 75(190):366–374.

Dettenmaier, E. M., W. J. Doucette, and B. Bugbee. 2009. Chemical hydrophobicity and uptake by plant roots. *Environmental Science & Technology* 43(2):324–329. doi: 10.1021/es801751x.

Dolfing, J., I. Novak, A. Archelas, and H. Macarie. 2012. Gibbs free energy of formation of chlordecone and potential degradation products: Implications for remediation strategies and environmental fate. *Environmental Science & Technology* 46(15):8131–8139. doi: 10.1021/es301165p.

Donnarumma, L., V. Pompi, A. Faraci, and E. Conte. 2009. Dieldrin uptake by vegetable crops grown in contaminated soils. *Journal of Environmental Science and Health, Part B* 44(5):449–454. doi: 10.1080/03601230902935113.

Dubuisson, C., F. Héraud, J.-C. Leblanc, S. Gallotti, C. Flamand, A. Blateau, P. Quenel, and J.-L. Volatier. 2007. Impact of subsistence production on the management options to reduce the food exposure of the Martinican population to chlordecone. *Regulatory Toxicology and Pharmacology* 49(1):5–16. doi: 10.1016/j.yrtph.2007.04.008.

Ficko, S., A. Rutter, and B. Zeeb. 2011. Effect of pumpkin root exudates on ex situ polychlorinated biphenyl (PCB) phytoextraction by pumpkin and weed species. *Environmental Science and Pollution Research* 18(9):1536–1543. doi: 10.1007/s11356-011-0510-8.

Fismes, J., C. Perrin-Ganier, P. Empereur-Bissonnet, and J. L. Morel. 2002. Soil-to-root transfer and translocation of polycyclic aromatic hydrocarbons by vegetables grown on industrial contaminated soils. *Journal of Environmental Quality* 31(5):1649–1656.

Fujisawa, T., K. Ichise, M. Fukushima, T. Katagi, and Y. Takimoto. 2002. Mathematical model of the uptake of non-ionized pesticides by edible root of root crops. *Journal of Pesticide Science* 27(3):242–248.

Gent, M. P. N., J. C. White, Z. D. Parrish, M. Isleyen, B. D. Eitzer, and M. I. Mattina. 2007. Uptake and translocation of p,p′-dichlorodiphenyldichloroethylene supplied in hydroponics solution to *Cucurbita*. *Environmental Toxicology and Chemistry* 26(12):2467–2475. doi: 10.1897/06-257.1.

Juraske, R., C. S. Mosquera Vivas, A. E. Velásquez, G. G. Santos, M. B. Berdugo Moreno, J. D. Gomez, C. R. Binder, S. Hellweg, and J. A. Guerrero Dallos. 2011. Pesticide uptake in potatoes: Model and field experiments. *Environmental Science & Technology* 45(2):651–657. doi: 10.1021/es102907v.

Kaushik, G., S. Satya, and S. N. Naik. 2009. Food processing a tool to pesticide residue dissipation—A review. *Food Research International* 42(1):26–40. doi: 10.1016/j.foodres.2008.09.009.

Kelsey, J. W., A. Colino, M. Koberle, and J. C. White. 2006. Growth conditions impact 2,2-bis(p-chlorophenyl)-1,1-dichloroethylene (p,p′-DDE) accumulation by *Cucurbita pepo. International Journal of Phytoremediation* 8(3):261–271. doi: 10.1080/15226510600846830.

Kenaga, E. E. 1980. Predicted bioconcentration factors and soil sorption coefficients of pesticides and other chemicals. *Ecotoxicology and Environmental Safety* 4(1):26–38.

Khan, S., L. Aijun, S. Zhang, Q. Hu, and Y.-G. Zhu. 2008. Accumulation of polycyclic aromatic hydrocarbons and heavy metals in lettuce grown in the soils contaminated with long-term wastewater irrigation. *Journal of Hazardous Materials* 152(2):506–515. doi: 10.1016/j.jhazmat.2007.07.014.

Kumar, M. and L. Philip. 2006. Adsorption and desorption characteristics of hydrophobic pesticide endosulfan in four Indian soils. *Chemosphere* 62(7):1064–1077. doi: 10.1016/j.chemosphere.2005.05.009.

Létondor, C., S. Pascal-Lorber, and F. Laurent. 2015. Uptake and distribution of chlordecone in radish: Different contamination routes in edible roots. *Chemosphere* 118:20–28. doi: 10.1016/j.chemosphere.2014.03.102.

Li, H., G. Sheng, C. T. Chiou, and O. Xu. 2005. Relation of organic contaminant equilibrium sorption and kinetic uptake in plants. *Environmental Science & Technology* 39(13):4864–4870.

Lunney, A. I., B. A. Zeeb, and K. J. Reimer. 2004. Uptake of weathered DDT in vascular plants: Potential for phytoremediation. *Environmental Science & Technology* 38(22):6147–6154. doi: 10.1021/es030705b.

Mattina, M. J., W. Iannucci-Berger, and L. Dykas. 2000. Chlordane uptake and its translocation in food crops. *Journal of Agricultural and Food Chemistry* 48(5):1909–1915. doi: 10.1021/jf990566a.

Multigner, L., P. Kadhel, F. Rouget, P. Blanchet, and S. Cordier. 2016. Chlordecone exposure and adverse effects in French West Indies populations. *Environmental Science and Pollution Research* 23:3–8. doi: 10.1007/s11356-015-4621-5.

Murano, H., T. Otani, T. Makino, N. Seike, and M. Sakai. 2009. Effects of the application of carbonaceous adsorbents on pumpkin (*Cucurbita maxima*) uptake of heptachlor epoxide in soil. *Soil Science & Plant Nutrition* 55(2):325–332.

Namiki, S., T. Otani, and N. Seike. 2013. Fate and plant uptake of persistent organic pollutants in soil. *Soil Science and Plant Nutrition* 59(4):669–679. doi: 10.1080/00380768.2013.813833.

Paraíba, L. C. and K. Kataguiri. 2008. Model approach for estimating potato pesticide bioconcentration factor. *Chemosphere* 73(8):1247–1252. doi: 10.1016/j.chemosphere.2008.07.026.

Poulsen, M. E. and J. H. Andersen. 2003. Results from the monitoring of pesticide residues in fruit and vegetables on the Danish market, 2000–01. *Food Additives & Contaminants* 20(8):742–757. doi: 10.1080/0265203031000152433.

Topp, E., I. Scheunert, A. Attar, and F. Korte. 1986. Factors affecting the uptake of 14C-labeled organic chemicals by plants from soil. *Ecotoxicology and Environmental Safety* 11(2):219–228.

Trapp, S. 2002. Dynamic root uptake model for neutral lipophilic organics. *Environmental Toxicology and Chemistry* 21(1):203–206.

Trapp, S. 2007. Fruit Tree model for uptake of organic compounds from soil and air. *SAR and QSAR in Environmental Research* 18(3–4):367–387.

US Environmental Protection Agency. 2012. Estimation Programs Interface Suite™ for Microsoft® Windows, v 4.11. United States Environmental Protection Agency, Washington, DC.

Wang, X., J. C. White, M. P. N. Gent, W. Iannucci-Berger, B. D. Eitzer, and M. I. Mattina. 2004. Phytoextraction of weathered p,p′-DDE by Zucchini (*Cucurbita pepo*) and Cucumber (*Cucumis sativus*) under different cultivation conditions. *International Journal of Phytoremediation* 6(4):363–385. doi: 10.1080/16226510490888910.

White, J. C., Z. D. Parrish, M. P. N. Gent, W. Iannucci-Berger, B. D. Eitzer, M. Isleyen, and M. I. Mattina. 2006a. Soil amendments, plant age, and intercropping impact p,p′-DDE bioavailability to *Cucurbita pepo. Journal of Environmental Quality* 35(4):992–1000.

White, J. C., Z. D. Parrish, M. Isleyen, M. P. N. Gent, W. Iannucci-Berger, B. D. Eitzer, J. W. Kelsey, and M. I. Mattina. 2006b. Influence of citric acid amendments on the availability of weathered PCBs to plant and earthworm species. *International Journal of Phytoremediation* 8(1):63–79. doi: 10.1080/15226510500507102.

Whitfield Åslund, M. L., A. I. Lunney, A. Rutter, and B. A. Zeeb. 2010. Effects of amendments on the uptake and distribution of DDT in *Cucurbita pepo* ssp. pepo plants. *Environmental Pollution* 158(2):508–513. doi: 10.1016/j.envpol.2009.08.030.

Whitfield Åslund, M. L., B. A. Zeeb, A. Rutter, and K. J. Reimer. 2007. In situ phytoextraction of polychlorinated biphenyl—(PCB)contaminated soil. *Science of the Total Environment* 374(1):1–12. doi: 10.1016/j.scitotenv.2006.11.052.

Wild, E., J. Dent, G. O. Thomas, and K. C. Jones. 2005. Direct observation of organic contaminant uptake, storage, and metabolism within plant roots. *Environmental Science & Technology* 39(10):3695–3702. doi: 10.1021/es048136a.

Woignier, T., F. Clostre, H. Macarie, and M. Jannoyer. 2012. Chlordecone retention in the fractal structure of volcanic clay. *Journal of Hazardous Materials* 241–242:224–230. doi: 10.1016/j.jhazmat.2012.09.034.

Zohair, A., A.-B. Salim, A. A. Soyibo, and A. J. Beck. 2006. Residues of polycyclic aromatic hydrocarbons (PAHs), polychlorinated biphenyls (PCBs) and organochlorine pesticides in organically-farmed vegetables. *Chemosphere* 63(4):541–553. doi: 10.1016/j.chemosphere.2005.09.012.

11 Transfer of Chlordecone from the Environment to Animal-Derived Products

Stefan Jurjanz, Catherine Jondreville, Agnès Fournier, Sylvain Lerch, Guido Rychen, and Cyril Feidt

CONTENTS

11.1 INTRODUCTION

As a consequence of the former use of chlordecone (CLD) in the French West Indies, this pesticide is still present in soils, with subsequent contamination of water, crops, animals (Dubuisson et al. 2007; Coat et al. 2011), and human impregnation through food (Guldner et al. 2010). Indeed, about one-fourth of the total agricultural acreage of these two French overseas departments (Guadeloupe, Martinique) are moderately to heavily polluted. As a consequence, CLD has been detected in about one-third of carcasses of bovines collected in the frame of the national survey plans carried out since 2008, and the concentrations exceed the maximum residue limit (MRL) of 100 μg kg^{-1} fat (regulation 839/2008/EC) in 6%–9% of them. Another point to give attention in the context of the French West Indies is the elevated proportion of self-consumption. Indeed, in the French West Indies homegrown foodstuffs are commonly consumed and may lead to non-negligible exposure to CLD for self-consumers (Dubuisson et al. 2007). While many studies have been conducted on the specific subject of vegetable contamination (Woignier et al. 2014; Clostre et al. 2014a), little information is currently available on CLD levels in products of animals reared in backyard systems by small holders.

As a result of the persistence of CLD in the environment, the contamination of animal products may persist for several hundred years (Dubuisson et al. 2007; Cabidoche et al. 2009) if practical solutions are not developed and implemented by farmers. Development of such solutions should rely on the knowledge of the ways animals are exposed to CLD during rearing and of the modalities of transfer of ingested CLD to animal-derived products. This chapter then takes stock of knowledge on the routes of farm animals' exposure, on the availability of CLD in the contaminated matrices they ingest, and on the transfer of this pesticide at the animal scale through ADME (absorption, distribution, metabolism, and excretion) processes. At the end, a regulatory frame set up to protect human health is presented, as well as its implications for implementing rearing practices leading to compliant animal-derived products.

In order to maintain the social and economic value of livestock production, there is a real need for gathering existing knowledge in order to promote rearing practices so that safe animal products are obtained even in polluted areas in both commercial and backyard contexts.

11.2 EXPOSURE OF LIVESTOCK

11.2.1 Potential Sources of Animal Exposure

Food-producing animals are exposed to CLD, mainly by oral intake of contaminated matrices such as water, feed and soil, whereas dermal contact and inhalation are negligible pathways of exposure.

Different water sources are used for the *watering* of animals. Potable water is not expected to be a source of exposure of animals since the regulation sets a maximum CLD concentration of 0.1 μg L^{-1} (Instruction 2010). The situation may be different when local watercourses, ponds, or collected rainwater is used for watering. Local sources are generally contaminated at levels below 0.4 μg L^{-1} (Godard 2007) so that their contribution to animals' exposure remains mild. Some rare sources were reported to regularly reach concentrations of 30 μg L^{-1} (ARS Guadeloupe 2011, personal information) but they are not used for watering the animals. Nevertheless, surface water may contain quite high proportions of potentially contaminated soil, as suspended matter in waterers or as sediments in troughs. Providing such sources of water to animals may considerably increase the contribution of watering to animals' exposure.

The exposure of animals to CLD *via* the intake of *plants* is more heterogeneous. As mentioned in Chapter 10 (i.e., from soil to plants), the contamination of plants grown on contaminated soils decreases from their bottom to their top (Clostre et al. 2014b). Therefore, aerial organs such as grass and similar roughage generally display very low concentrations of CLD (<1 μg kg^{-1} fresh matter [FM], i.e., <0.2 μg kg^{-1} dry matter [DM]) when grown on moderately contaminated soil (i.e., <0.1 mg kg^{-1} dry soil). Exceptionally, CLD concentrations of approximately 5 μg kg^{-1} FM (i.e., 1 μg kg^{-1} DM) were measured in grass grown on very heavily contaminated andosol (>4 mg kg^{-1} soil) (CIRAD, personal information). Therefore, the contribution of such roughage to the exposure of ruminants to CLD is low, as illustrated in Table 11.1. Here again, particulates of contaminated soil on plants may considerably increase the contribution of roughage to animal exposure.

In contrast, cucurbits and root vegetables may be contaminated to a much higher extent than roughage. Indeed, Bordet et al. (2007) reported concentrations exceeding 50 μg kg^{-1} in sweet potatoes. Clostre et al. (2014a) confirmed high concentrations of CLD not only in dasheens (>200 μg kg^{-1} FM) but also in yam (up to 130 μg kg^{-1} FM) and in sweet potatoes (up to 100 μg kg^{-1} FM) grown on a soil contaminated at a level exceeding 1 mg kg^{-1} dry soil. Therefore, the plant species and the organs fed to animals may widely impact their exposure level. Moreover, peel of fruits and vegetables is more contaminated than pulp (Clostre et al. 2014a). These plant products are likely to be a source of exposure for monogastric animals such as swine and poultry that are fed with starch-containing roots and kitchen waste rather than roughage.

Finally, animals may ingest *soil* when directly exploring outside, especially when rearing conditions are degraded (see Section 11.2.2). The concentration in certain soils may be very high (>1 mg kg^{-1}). Therefore, soil intake is considered as the main exposure pathway of animals to CLD as illustrated in Table 11.1. Thus, the main way to reduce exposure of animals to CLD consists in limiting as much as possible soil intake by outside reared animals.

TABLE 11.1
Estimates of Animals' Exposure to CLD through Water, Feed, or Soil

		Daily Exposure to CLD (μg day^{-1})			
		Via Water[a] (0.4 μg CLD L^{-1})[b]	Via Grass (0.2 μg CLD kg^{-1} DM)	Via Soil (0.1 mg CLD kg^{-1} dry soil)	
Animal Specie	**DM Intake (kg day^{-1})**			**1% of DM Intake**	**10% of DM Intake**
Beef cattle	7	11	1.4	7	70
Small ruminant	2	3.2	0.4	2	20
Growing swine	2.5	4	—[c]	2.5	25
Hens	0.1	0.16	—[c]	0.1	1

[a] Based on 4 L water per kg DM ingested (Anses 2011).
[b] Observatoire des Residus de Pesticides, n.d.
[c] Such animals would normally ingest very small amounts of grass.

Thus, soil is clearly the major source of farm animals' exposure to CLD, while water and grass, if not soiled by soil particles, are minor sources. Therefore, evaluating soil intake by animals in different rearing conditions is a prerequisite for a sound assessment of the risk of animal-derived product contamination.

11.2.2 Evaluation of Soil Intake

Soil intake can be estimated by different methods, mainly based on natural markers as acid insoluble ash (AIA) or titanium. The marker must be much more concentrated in the considered soil than in other ingested matrices, including plants and feed, for a relevant use of the Equation 11.1 proposed by Beyer et al. (1994):

$$\text{Soil intake (\% of ingested DM)} = \frac{[\text{M}]\text{feed} - [\text{M}]\text{scat} + \text{dig DM} \times [\text{M}]\text{scat}}{\text{dig DM} \times [\text{M}]\text{scat} - [\text{M}]\text{soil} + [\text{M}]\text{feed}} \quad (11.1)$$

where

[M] feed, [M] scat, and [M] soil are the concentration of the indigestible marker in feed, feces, and soil [expressed as g per kg dry matter], respectively

dig DM is the digestibility of DM of the ration including feed and grass [expressed as kg per kg]

The principle is the analysis of the concentration of the chosen marker [M] in the ingested matrices feed and soil as well as in the feces and related to them by the excreted (i.e., not digested) DM. The chemical procedure to analyze the marker AIA (e.g., van Keulen and Young 1977) is robust and not expensive. When soil does not contain enough AIA, thus when the ratio of the concentration of AIA in soil to its concentration in feed plus grass is below 30, other markers should be selected. Thus, with some Caribbean soils rich in organic matter and poor AIA concentration in plants, it is preferable to use Ti, which is analyzed by the more expensive method of inductively coupled plasma mass spectrometry. The preciseness of the whole method depends on the property of sampled feeds. Especially, plants can be contaminated by soil particles that would lead to an overestimation of the marker background and therefore an underestimation of ingested soil. Thus, special attention has to be paid to thoroughly rinse in order to be cleared of any trace of soil. Another sensitive point is the preciseness of the digestibility determination of ingested feed, especially when it is composed by several ingredients. This would concern especially monogastric animals reared outside as they generally ingest industrial feedstuffs and herbage, which would widely differ in digestibility.

There are no data in the literature on soil intake in humid tropical conditions but available data carried out in temperate conditions let us suppose that soil intake may reach 5% in cattle (Fries et al. 1982; Jurjanz et al. 2012). In poultry, basal soil ingestion in good rearing conditions is around 1%–2% of DM ingested; however, in case of degraded rearing conditions, it may reach up to 30% of DM ingested (Waegeneers et al. 2009; Jondreville et al. 2010). In free-range swine, a level of 10% was reported (Jurjanz and Roinsard 2014) and more can be expected (Rivera Ferre et al. 2001).

Nevertheless, because of the specificities of tropical conditions (soil types, climatic effects on plant growth and splash-effects, botanical composition of explored areas, grazing systems), thorough investigations are required before estimating soil intake by farm animals and, in turn, their exposure level to CLD.

11.2.3 Rearing Practices and Soil Intake

There is generally no nutritional interest to favor soil intake of free-range animals. Nevertheless, knowledge of CLD concentration in soil of the explored areas is necessary to correctly evaluate the exposure. All rearing practices should aim at holding off the animals from soil and soiled matters. Therefore, eventually distributed feed, including mineral licks, should be supplied in racks or troughs and not disposed on bare soil. A second point is the use of tuber and roots in feeding practices that would need to ensure the absence of CLD in the cultivated soil. Especially in extensive systems, a nutritionally balanced diet would reduce the motivation of animals to explore the soil as it has been shown in poultry (Jondreville et al. 2010; Almeida et al. 2012).

Watering should also be carried out in proper bowls or troughs. The use of surface water requires measurements of CLD concentrations in water and sediments. As no regulation sets up a maximal concentration for animal watering, the reference value of maximal 1.5 $\mu g\ L^{-1}$ for human consumption can be used as a threshold when surface water is used for food-producing animals.

11.2.4 Relative Bioavailability of CLD Present in Soils and Consequences on Animal-Derived Product Contamination

11.2.4.1 Relative Bioavailability

The three main soil types on which CLD was applied are volcanic andosol, ferralsol, and nitisol (Cabidoche et al. 2009). These soils, rich in organic carbon, display a high affinity for this hydrophobic compound. However, this affinity varies greatly between soils, due to the different properties of clays (Cabidoche et al. 2009; Cabidoche and Lesueur-Jannoyer 2012). Thus, CLD is strongly retained in andosol, with an availability to plants by 18 times lower than in nitisol (Cabidoche and Lesueur-Jannoyer 2012), and an estimated time for decontamination by leaching of five to seven centuries, compared to 6–10 decades in nitisol (Cabidoche et al. 2009). Given these differences, it may be hypothesized that the proportion of soil-bound CLD extracted from soil during the digestive processes in animals may be dependent of the type of contaminated soil they ingest. Andosol may display a stronger capacity of CLD retention in the digestive tract than nitisol. Subsequently, the availability of soil-bound CLD to animals and, in turn, the level of contamination of animal-derived products due to soil ingestion would be higher in nitisol than in andosol. Beside, animals likely to ingest soil during rearing belong to different categories, including birds (e.g., chickens), monogastric mammals (e.g., swine), and ruminants (e.g., goats), the regimen as well as the digestive tract of which displays huge differences such as length, retention time, and enzymatic equipment.

The potential of soil to release the pollutants it contains in the digestive tract is usually assessed by means of relative bioavailability (RBA) studies, in which the response of the animal to the ingestion of graded levels of pollutants through soil is compared to the response obtained with the same amount of pollutants ingested through a reference matrix, often spiked oil (Budinsky et al. 2008). Such studies were conducted in laying hens (Jondreville et al. 2013) and in piglets (Bouveret et al. 2013) given CLD-contaminated andosol and nitisol, and in lambs (Jurjanz et al. 2014) given CLD-contaminated andosol. The response criteria were egg and abdominal fat in laying hens, subcutaneaous fat and liver in piglets, and serum and kidney fat in lambs. In any animal species, the estimates of RBA for each soil (slope ratios) could be differentiated from 1 (Table 11.2). Thus, neither andosol nor nitisol significantly reduced the bioavailability of CLD compared to oil. In addition, at equal level of CLD ingested, andosol and nitisol elicited similar responses of CLD in tissues, demonstrating that andosol does not exhibit a stronger capacity of CLD retention in the digestive tract of these animals than nitisol.

11.2.4.2 Impact of Soil Ingestion on Animal-Derived Product Contamination

The previous results clearly show that, for the three types of farm animals, soil does not modulate CLD availability. Therefore, the risk of animal-derived product contamination through soil ingestion only depends on the amount of soil ingested and on its concentration in CLD. Knowing the accumulation ratio of ingested CLD in different animal tissues or products, one may estimate the expected concentration of CLD in edible tissues at a steady state according to the level of exposure of the animals through polluted soil ingestion. These concentrations may be compared with the regulatory maximum residue limits (MRL, Regulation 396/2005/EC). Such a calculation was performed by Jondreville et al. (2014b) for the production of eggs by laying hens (Table 11.3). It was based on the classification established for advising the population regarding the consumption of vegetables produced in familiar gardens (Lesueur-Jannoyer et al. 2012), according to which soils are considered as mildly and moderately contaminated when their CLD concentration ranges between 0.1 and 0.5 mg kg^{-1} DM and between 0.5 and 1 mg kg^{-1} DM, respectively, whereas they are considered as heavily contaminated when their CLD concentration exceeds 1 mg kg^{-1} DM. As soil ingestion by laying hens may reach up to 30 g DM daily (see also Section 11.2.2), the consumption of home-produced eggs is not recommended, even in mildly contaminated areas.

11.3 CHLORDECONE BEHAVIOR IN THE ANIMAL ORGANISM: FROM ITS ABSORPTION TO ITS EXCRETION

Experiments in dairy cows, in lambs, in birds (laying hens, Japanese quail, Muscovy ducks) and in monogastric mammals (swine) demonstrated that the chronic ingestion of CLD-contaminated matrices (feed or soil) for several days or weeks always leads to the contamination of animals and of their derived products (Naber and Ware 1965; Smith and Arant 1967; McFarland and Lacy 1969; Bouveret et al. 2013; Jondreville et al. 2013, 2014a,b; Jurjanz et al. 2014). In animals exposed at environmental CLD levels, concentrations increase proportionally to the ingested quantities (Smith and Arant 1967; Bouveret et al. 2013; Jondreville et al. 2013; Jurjanz et al. 2014).

TABLE 11.2
Parameters of the Linear Response of CLD Concentration in Tissues (μg kg^{-1} FM) to the Amount of Ingested CLD (μg kg^{-1} BW day^{-1}) Originating from Andosol, from Nitisol or from Contaminated Oil[a]

	Species	Hen		Piglet		Lamb	
	Target Tissue	Egg Yolk		Liver		Kidney Fat	
		Parameter	*P*-Value[b]	Parameter	*P*-Value[c]	Parameter	*P*-Value[c]
Intercept			*NS*		*NS*		*NS*
Ingested CLD							
	Andosol	4.64	*<0.001*	5.17	*<0.001*	15.9	*<0.001*
	Nitisol	4.59	*<0.001*	4.55	*<0.001*		
	Oil	4.78	*<0.001*	4.70	*<0.001*	14.6	*<0.001*
Ingested CLD × matrix[c]		*NS*		*NS*		*NS*	
rsd[d]		6.3		5.37		10.7	
R^2		0.79		0.81		0.85	
RBA[e]	Andosol	0.97 (0.73–1.20)		1.10 (0.81–1.39)		1.09 (0.95–1.23)	
	Nitisol	0.96 (0.76–1.16)		0.97 (0.71–1.23)			

Source: Adapted from Jondreville, C. et al., *Environ. Sci. Pollut. Res.*, 20, 292, 2013; Bouveret, C. et al., *J. Agric. Food Chem.*, 61, 9269, 2013; Jurjanz, S. et al., *Environ. Geochem. Health*, 36, 911, 2014.

[a] The equation is: chlordecone concentration in tissue = a ingested chlordecone from andosol + b ingested chlordecone from nitisol + c ingested chlordecone from oil, where a, b, and c are the estimates of the parameters attributed to the amount of CLD ingested from andosol, from nitisol, and from oil, respectively.

[b] NS, not significant ($P > 0.1$).

[c] If *P*-value > 0.05, the slopes do not significantly differ and RBA does not differ from 1.

[d] rsd, residual standard deviation.

[e] RBA, relative bioavailability of chlordecone present in soil, calculated as the ratio of the slope of the response fitted with soil to the slope of the response fitted with oil; 95% confidence limits calculated as RBA ± 2 standard error are enclosed in parentheses.

TABLE 11.3
Amount of Ingested Soil That Should Not Be Exceeded for the Production of Eggs Compliant with the MRL[a] according to the Level of Soil Contamination[b]

Soil CLD content (mg kg^{-1} DM)	0.1	0.5	1
Ingested soil not to be exceeded (g DM day^{-1})	22	4.4	2.2

Note: DM, dry matter.

[a] MRL, maximum residue limit set at 20 μg kg^{-1} FM in eggs (Regulation n° 396/2005/EC).

[b] Calculated from the steady state accumulation ratio of ingested CLD as presented by Jondreville et al. (2014b).

11.3.1 Absorption

CLD is known to be absorbed by more than 90% of animal species like rodents, Japanese quails, pigs, or goat (Eroschenko and Hackmann 1981, Guzelian 1982; Soine et al. 1983; Belfiore et al. 2007; Mahieu et al. 2014). After CLD administration, less than 48 h is sufficient to achieve a full distribution of the compound in the organism (Mutter 1980 in: Guzelian 1982).

11.3.2 Distribution

Compared to other lipophilic organochlorines, CLD displays an unusual distribution within animal body. Irrespective of the animal species, it accumulates mainly in liver, whereas lipophilic organic pollutants such as mirex or PCBs usually accumulate mainly in fat-rich tissues (especially adipose tissue) (Kavlock et al. 1980; Belfiore et al. 2007). For example, in laying hens orally exposed to CLD over 42 days, CLD concentrations in serum, adipose tissue, muscle, and egg represent 13%, 20%, 8%, and 28%, respectively, of the concentration observed in liver (fresh matter basis) (Jondreville et al., 2014b) (Table 11.4).

According to the literature on swine and rodents, the presence of a ketone group in the CLD molecule explains its preferential distribution in liver, for at least two hypothetical mechanisms, which probably act concomitantly. Firstly, the polarity of the CLD molecule, conferred by its ketone group, presumably leads to a higher affinity for polar lipids (i.e., mainly phospholipids and free cholesterol) than for neutral ones (i.e., mainly triglycerides). CLD may then share common distribution and transport pathways with polar lipids, especially cholesterol (Soine et al. 1984b), and therefore be mostly distributed in polar lipid–rich tissues, liver being one of the richest. This assumption is corroborated by the studies of Skalsky et al. (1979) and of Soine et al. (1984a), which showed that in human and pig blood, CLD is mainly associated with high-density lipoproteins (HDL) rather than with low-density lipoproteins. Indeed, HDLs are involved in the cholesterol transport from peripherical cells to the liver.

TABLE 11.4
Tissue Levels of CLD Depending on Exposure Conditions Revealed in Literature

Animals	Exposure Conditions Length in Days	Mean Tissue Levels (µg CLD kg^{-1} FM) at the End of the Exposure		Half-Life Mean (sd) in Days	Reference
Laying hens (*Gallus domesticus*)	*Exposure*: 42 day	Liver	1640	5 (0.38)	Jondreville et al. (2014b)
22 weeks old	*Dose*: 39 µg CLD kg^{-1} BW and day	Egg	460	5.5 (0.29)	
	Depuration: 35 day	Abdominal fat	331	5.3 (0.37)	
		Serum	213	5.1 (0.66)	
		Pectoral muscle	119		
		Leg muscle	127		
Growing Muscovy ducks (*Carinata moschata*)	Field experiment on contaminated soil (182 day)	Liver	1215	17	Jondreville et al. (2014a)
	Depuration: 63 day	Egg	774	19	
		Abdominal fat	212	22	
		Leg muscle[a]	122	21	
Laying hens (*Gallus domesticus*)	Spiked oil or native contaminated soils (andosol or nitisol)	Egg	35	—	Jondreville et al. (2013)
41 weeks old	*Exposure*: 23 day	Abdominal fat	90		
	Dose: 7.4 µg CLD kg^{-1} BW and day				
Growing piglets (*Sus scrofa domesticus*)	Spiked oil or native contaminated soil (andosol or nitisol)	Liver	35	—	Bouveret et al. (2013)
5 weeks old	*Exposure*: 14 day	Subcutaneous fat	70		
	Dose: 7.3 µg CLD kg^{-1} BW and day				
Growing lambs (*Ovies aries*)	Spiked oil or native contaminated soil (andosol)	Perirenal fat	30/60/80	—	Jurjanz et al. (2014)
8 weeks old	*Exposure*: 15 day	Serum	1.5/3/4		
	Dose: 2/4/6 µg CLD kg^{-1} BW and day				

(*Continued*)

TABLE 11.4 (*Continued*)
Tissue Levels of CLD Depending on Exposure Conditions Revealed in Literature

Animals	Exposure Conditions Length in Days	Mean Tissue Levels (µg CLD kg^{-1} FM) at the End of the Exposure		Half-Life Mean (sd) in Days	Reference
Dry creole dairy goats (*Capra hircus*) 2–7 years old	Unique intraveinous or oral dose: *Dose*: 1 mg CLD kg^{-1} BW *Depuration*: 200 day	—		18 (ranking from 11 to 36)	Mahieu et al. (2014)
Creole beef cows (*Bos taurus*)	Unknown field experiment *Depuration*: 168 day	—		43	Mahieu et al. (2014)
Dairy cows (*Bos taurus*)	Feed concentrate *Exposure*: 60 day *Dose*: 16 or 90 mg CLD/animal and day	Milk	100/350	20[b]	Smith and Arant (1967)
Growing male goats (*Capra hircus*) 20 weeks old	*Exposure*: 21day *Dose*: 48 µg CLD kg^{-1} BW and day *Depuration*: 21 day	Liver	4199	15 ≪ 20	Lastel et al. (2016)
		Perirenal fat	522		
		Serum	302		
		Muscle diaphragm	597		
Dry Ewes (*Ovis aries*)	*Exposure*: 35 day *Dose*: 29 µg CLD kg^{-1} BW and day *Depuration*: 21 day	Subcutaneous pericaudal fat	493	—	Lerch et al. (2016)
		Serum	297		

[a] Without skin
[b] Recalculated.

But HDLs can also induce CLD transfer from the peripheral tissues to the liver. Moreover, CLD binding to specific liver proteins, also showing cholesterol binding properties, was demonstrated in swine (Soine et al. 1984b). A second explanation related to liver sequestration of CLD was more recently developed by Belfiore et al. (2007). Based on a pharmacokinetic modeling approach, these authors proposed that the ketone group with alpha-carbon halogen (chlorine) substitutions confers to the CLD molecule a relatively high affinity for sulfhydryl groups and, therefore, a strong reversible binding to glutathione and glutathione transferases, which are abundant in liver (Belfiore et al. 2007).

11.3.3 Metabolism

Regarding CLD metabolism in organisms, data are available for human (Fariss et al. 1980), rodents (Houston et al. 1981), and swine (Soine et al. 1983), and pathways for CLD metabolism have been suggested (Fariss et al. 1980; Houston et al. 1981). Firstly, CLD can rapidly be hydrated to chlordecone hydrate before being metabolized by an aldo-ketoreductase to chlordecone alcohol (chlordecone-OH). These two forms (chlordecone hydrate and chlordecone-OH) can be transformed in glucuronide conjugates. Biliary chlordecone-OH is predominantly present, as well as its glucuronide conjugate, in humans (Fariss et al. 1980) and in swine (Soine et al. 1983) and the transformation in chlordecone-OH seems to be species-dependent since the aldo-ketoreductase does not have a similar activity in different species. The chlordecone conjugate was found in bile and plasma.

11.3.4 Excretion

Milk fat is a route of CLD excretion in dairy cows (Smith and Arant 1967) and in lactating rats (Kavlock et al., 1980), as well as egg yolk in laying hens (Jondreville et al. 2014b). Nonetheless, following a comparable level of oral exposure to CLD (i.e., approximately 500 μg kg^{-1} of DM intake) the rate of excretion was higher through hen egg than through cow milk, as the accumulation ratio (i.e., the CLD concentration in milk or egg on fat basis divided by its concentration in feed on DM basis) was 10 for egg (Jondreville et al. 2014b), and 2.8 for milk (Kavlock et al. 1980).

To our knowledge, no study has assessed the excretion routes of CLD in non-lactating or non-laying farm animals. In rats, feces represented the main route of excretion for CLD or its metabolites (assessed after [^{14}C] CLD administration), urine representing a negligible excretion way (Boylan et al. 1978; Richter et al. 1979). Studies conducted in the late 1970s on rats and also in one human equipped with a T tube implanted in the common bile duct and therefore allowing diversion of bile flow provided an opportunity to study the question of how CLD, and its metabolite chlordecone alcohol, are excreted from the organism to the gastrointestinal tract. It was shown that when bile was diverted from the duodenum, CLD fecal excretion remained unaffected in rats and increased by 6- to 10-fold in human (Boylan et al. 1979). This demonstrates that CLD excretion into the digestive lumen is, at least in part, dependent on a nonbiliary route of excretion, as proposed for many lipophilic pollutants (Rozman 1986). Conversely, when bile excretion into the duodenum was diverted, chlordecone alcohol excretion in feces was abolished, suggesting an exclusive biliary route of excretion in this latter case (Boylan et al. 1979). Attempts to increase fecal excretion, and therefore accelerate depuration from CLD, were also tested by dietary supplementation with cholestyramine or liquid paraffin. Cholestyramine, a non-absorbable anion exchange resin, interrupts the enterohepatic circulation of bile salts, and is therefore known as a hypocholesterolemic drug. Its administration at a rate of approximately 4% of the diet of chlordecone-contaminated rats led to a twofold increase in CLD fecal excretion and a 1.5- to twofold decrease in CLD concentrations in tissues (Boylan et al. 1978, 1979). Moreover in human, cholestyramine supplementation (16–24 g day^{-1}) led to a 6- to 10-fold increase in CLD fecal excretion, and after 5 months of treatment shortened the CLD half-life in blood from 165 to 80 days and in fat from 125 to 64 days (Cohn et al. 1978; Boylan et al. 1979). Conversely, dietary liquid paraffin supplementation (8% of the diet) over 24 days in rats did not increase the concentration of radioactivity in feces after administration of [^{14}C] CLD; neither did it decrease the half-life of CLD elimination from the body (Richter et al. 1979). The higher effectiveness of cholestyramine compared to liquid paraffin supplementation could be explained, at least in part, by the amphiphilic properties of cholestyramine, showing then a higher affinity for polar lipophilic pollutants (i.e., CLD), than did neutral non-absorbable lipids (i.e., liquid paraffin).

11.3.5 Half-Lives

Finally, the variations of the kinetics of CLD depuration can be due to the different processes of distribution, metabolism, and excretion, implicating different tissues of the organism as well as the excretion pathways in the products egg or milk. Table 11.4 indicates the available data in different reared species. Depending on the species, the mean half-life ranges between 5 and 20 days for controlled experiments and between 17 and 43 for field experiments. The half-lives are relatively short, surely due to an efficient metabolism and can be shortened in case of an excretion through fat products like milk or eggs.

However, no data are currently available on CLD metabolism and CLD excretion in ruminants, especially when lactating. Therefore, further research is necessary to understand CLD metabolism in such animals in order to predict concentrations in meat of the different types of ruminants reared in the French West Indies.

11.4 REGULATORY FRAME, FOOD SAFETY CONCERNS, AND PERSPECTIVES

The regulatory frame regarding CLD content in animal-derived products has been built in two main steps.

In 2003, the French Agency AFSSA (currently Anses) set up two toxicity reference values (TRV), one for chronic exposure and one for acute exposure. Based on these two TRVs and a specific study on food consumption in the French West Indies (ESCAL), AFSSA (2005) led an assessment of the risk due to oral exposure to CLD. It was concluded that two different levels in food would be protective for the general population. The chronic exposure level of 50 µg kg^{-1} FM concerned only eight food items, including poultry meat (few samples demonstrated high levels of CLD in previous control and survey plans). For all other food items, a limit of 200 μg kg^{-1} FM was set up in order to protect against acute risk that would not correspond in the case of chronic exposure.

Since 2008, a regulatory position has replaced the previous sanitary assessment. Indeed, from this date, CLD is considered as a forbidden pesticide rather than as an environmental pollutant. Therefore, regulatory values, referred to as MRLs, are derived from the European regulations n° 853/2004/EC and 396/2005/EC. This change leads to new maximum values in food items and especially in animal-derived products. Particularly, maximum concentrations in all products originating from terrestrial animals are reduced down to 100 μg kg^{-1} FM, except in poultry products (meat and eggs) for which the MRL is 200 μg kg^{-1} FM.

The national survey plans carried out since 2008 in the slaughterhouses in the French West Indies revealed unexpected contamination of animal products. CLD residues were detected in about one-third of bovine carcasses originating from CLD-contaminated areas, and 6%–9% at concentrations exceeding the MRL. These surveys focused on perirenal fat analysis (easy to sample without degrading the carcass value). The carcasses were destroyed if the value in the perirenal fat was greater than 100. But as concentrations in liver were generally much higher than in the previous tissue (Table 11.4), this foodstuff was destroyed if the value in perirenal fat was above 20.

These surveys confirmed the exposure of the populations in Martinique and in Guadeloupe to CLD through consumption of animal-derived products. However, backyard animal productions (swine, small ruminants, and poultry) were not taken into account in the survey plans, although these food-producing animals may be heavily contaminated and in turn may represent an important source of CLD exposure for local consumers. Indeed, due to specific pedoclimatic conditions and social context, self-production habits, including rearing animals in familiar backyards, are still common. A project is ongoing, led by the French institutions InVS and ANSES (KANNARI), in order to better characterize the contamination levels of animal products from backyard production.

This state will evolve due to the application of the EC regulation n°839/2008, which sets up a maximum of 100 μg kg^{-1} fat rather than FM. This change would probably significantly increase the number of animal products not suitable for human consumption.

Although the European regulations protect consumers by avoiding commercialization of contaminated products through the application of MRLs in animal products, there is a real need for innovative actions to prevent animal contamination. Risk management and risk perception are strongly linked to regulatory frame and not only to “real” contamination levels.

11.5 CONCLUSIONS

The reduction of the exposure of food-producing animals to CLD can be mainly carried out by reducing as much as possible direct and indirect soil intake. The importance of this aspect increases with the concentration of CLD in the soil of the site explored by the animals. The very high bioavailability of CLD and its absorption rate as well as its rapid distribution, especially in fatty tissues, results in a rapid increase in the concentrations of CLD in tissues of exposed animals, especially in liver. CLD is readily excreted in milk and in egg. Excretion in nonlactating or non-laying animals is scarcely studied but feces seem also to be a notable output pathway. The depuration delay in contaminated animals depends on the reached concentrations in the considered tissues, excretion activity (laying or milk yield), and likely on fat metabolism.

Further work should precisely suggest the evaluation models for soil intake in these tropical systems carried out by small holders. Refined depuration models for animal-derived products from such rearing systems would allow actualizing the regulatory framework in order to improve the protection of consumers in chronic exposure situations. Finally, the toxicity of chlordecone metabolites in animal-derived products for the consumer should be elucidated.

REFERENCES

Almeida GF, Hinrichsen LK, Horsted K, Thamsborg SM, Hermanson JE. 2012. Feed intake and activity level of two broiler genotypes foraging different types of vegetation during finishing period. *Poultry Science* 91: 2105–2113.

Afssa 2005. Première évaluation de l'exposition alimentaire de la population martiniquaise au Chlordécone. Report Afssa, ed. Bialec, Nancy, France, 40pp.

Anses 2011. Etat des lieux des pratiques et des recommandations relatives à la qualité sanitaire de l'eau d'abreuvement des animaux d'élevage. Report ANSES (saisine 2008-SA-0162), ed. Bialec, Nancy, France, April 2011, 126pp.

Belfiore CJ, Yang RSH, Chubb LS, Lohitnavy M, Lohitnavy OS, Andersen ME. 2007. Hepatic sequestration of chlordecone and hexafluoroacetone evaluated by pharmacokinetic modeling. *Toxicology* 234: 59–72.

Beyer NW, Connor EE, Gerould S. 1994. Estimates of soil ingestion by wildlife. *Journal of Wildlife Management* 58: 375–382.

Bordet F, Thieffinne A, Mallet J, Heraud F, Blateau A, Inthavong D. 2007. In-house validation for analytical methods and quality control for risk evaluation of chlordecone in food. *International Journal of Environmental Analytical Chemistry* 87: 985–998.

Bouveret C, Rychen G, Lerch S, Jondreville C, Feidt C. 2013. Relative bioavailability of tropical volcanic soil-bound chlordecone in piglets. *Journal of Agricultural and Food Chemistry* 61: 9269–9274.

Boylan JJ, Cohn WJ, Egle JL, Blanke RV, Guzelian RS. 1979. Excretion of chlordecone by the gastro-intestinal tract—Evidence for a non-biliary mechanism. *Clinical Pharmacology and Therapeutics* 25: 579–585.

Boylan JJ, Egle JL, Guzelian PS. 1978. Cholestyramine: Use as a new therapeutic approach for chlordecone (Kepone) poisoning. *Science* 199: 893–895.

Budinsky RA, Rowlands JC, Casteel S, Fent G, Cushing CA, Newsted J, Giesy JP, Ruby MV, Aylward LL. 2008. A pilot study of oral bioavailability of dioxins and furans from contaminated soils: Impact of differential hepatic enzyme activity and species differences. *Chemosphere* 70: 1774–1786.

Cabidoche YM, Achard R, Cattan P, Clermont-Dauphin C, Massat F, Sansoulet J. 2009. Long-term pollution by chlordecone of tropical volcanic soils in the French West Indies: A simple leaching model accounts for current residue. *Environmental Pollution* 157: 1697–1705.

Cabidoche YM., Lesueur-Jannoyer M. 2012. Contamination of harvested organs in root crops grown on chlordecone-polluted soils. *Pedosphere* 22: 562–571.

Clostre F, Letourmy P, Thuriès L, and Lesueur-Jannoyer M. 2014a. Effect of home food processing on chlordecone (organochlorine) content in vegetables. *Science of the Total Environment* 490: 1044–1050.

Clostre F, Letourmy P, Turpin B, Carles C, Lesueur-Jannoyer M. 2014b. Soil type and growing conditions influence uptake and translocation of organochlorine (chlordecone) by *Cucurbitaceae* species. *Water, Air, & Soil Pollution* 225: 1–11.

Coat S, Monti D, Legendre P, Bouchon C, Massat F, Lepoint G. 2011. Organochlorine pollution in tropical rivers (Guadeloupe): Role of ecological factors in food web bioaccumulation. *Environmental Pollution* 159: 1692–1701.

Cohn WJ, Boylan JJ, Blanke RV, Fariss MW, Howell JR, Guzelian PS. 1978. Treatment of chlordecone (Kepone) toxicity with cholestyramine—Results of a controlled clinical-trial. *The New England Journal of Medicine* 298: 243–248.

Dubuisson C, Héraud F, Leblanc JC, Gallotti S, Flamand C, Blateau A, Quenel P, Volatier JL. 2007. Impact of subsistence production on the management options to reduce the food exposure of the Martinican population to chlordecone. *Regulatory Toxicology and Pharmacology* 49: 5–16.

Eroschenko VP, Hackmann NL. 1981. Continuous ingestion of different chlordecone (kepone) concentrations and changes in quail reproduction. *Journal of Toxicology and Environmental Health* 8: 659–665.

Fariss MW, Blanke RV, Saady JJ. 1980. Demonstration of major metabolic pathways for chlordecone (Kepone) in Humans. *Drug Metabolism and Disposition* 8: 434–438.

Fries GF, Marrow GS, Snow PA. 1982. Soil ingestion by dairy cattle. *Journal of Dairy Science* 65: 611–618.

Godard E. 2007. Chlodécone 1999–2007—Découverte et gestion d'une pollution agricole sans précèdent. Groupe Régional des phytosanitaires (GREPHY), Fort-de-France, Martinique (France) Séance plénière du 22 octobre. 32p.

Guldner L, Multigner L, Héraud F, Monfort C, Thomé J-P, Giusti A, Kadhel P, Cordier S. 2010. Pesticide exposure of pregnant women in Guadeloupe: Ability of a food frequency questionnaire to estimate blood concentration of chlordecone. *Environmental Research* 110: 146–151.

Guzelian PS. 1982. Comparative toxicology of chlordecone (kepone) in Humans and experimental animals. *Annual Review of Pharmacology and Toxicology* 22: 89–113.

Houston TE, Mutter LC, Blanke RV, Guzellian PS. 1981. Chlordecone alcohol formation in the Mongolian Gerbil (*Meriones unguiculatus*): A model for human metabolism of chlordecone (Kepone). *Fundamental and Applied Toxicology* 1: 293–298.

Instruction DGS/EA4/2010/424 du 9 décembre 2010 relative à « la gestion des risques sanitaires en cas de dépassement des limites de qualité des eaux destinées à la consommation ». Direction Générale de la Santé, Ministère du Travail, de l'Emploi et de la Santé, Paris, France, 4pp.

Jondreville C, Bouveret C, Lesueur-Jannoyer M, Rychen G, Feidt C. 2013. Relative bioavailability of tropical volcanic soil-bound chlordecone in laying hens. *Environmental Science and Pollution Research* 20: 292–299.

Jondreville C, Fournier F, Mahieu M, Feidt C, Archimède H, Rychen G. 2014b. Kinctic study of chlordecone orally given to laying hens (*Gallus domesticus*). *Chemosphere* 114: 275–281.

Jondreville C, Lavigne A., Jurjanz S, Dalibard C, Liabeuf JM, Clostre F, Lesueur-Jannoyer M. 2014a. Contamination of free-range ducks by chlordecone in Martinique (French West Indies): A field study. *Science of the Total Environment* 493: 336–341.

Jondreville C, Travel A, Besnard J, Dziurla MA, Feidt C. 2010. Intake of herbage and soil by free-range laying hens offered a complete diet compared to a whole-wheat diet. In *XIIIth European Poultry Conference*, Duclos M, Nys Y. (eds.). Tours, France, August 23–27, 2010; *World's Poultry Science Journal* 66(Suppl.): 4.

Jurjanz S, Feidt C, Pérez-Prieto LA, Ribeiro Filho HMN, Rychen G, Delagarde R. 2012. Soil intake of lactating dairy cows in intensive strip-grazing systems. *Animal* 6: 1350–1359.

Jurjanz S, Jondreville C, Mahieu M, Fournier A, Archimède H, Rychen G, Feidt C. 2014. Relative bioavailability of soil-bound chlordecone in growing lambs. *Environmental Geochemistry and Health* 36: 911–917.

Jurjanz S, Roinsard A. 2014. Valorisation de l'herbe par des truies en plein air. *AlterAgri* 108: 25–26.

Kavlock RJ, Chernoff N, Rogers E, Whitehouse D. 1980. Comparative tissue distribution of mirex and chlordecone in fetal and neonatal rats. *Pesticide Biochemistry and Physiology* 14: 227–235.

Lastel ML, Lerch S, Fournier A, Jurjanz S, Mahieu M, Archimède H, Feidt C, Rychen G. 2016. Clordecone disappearance in tissues of growing goats after a one month decontamination period—effect of body fatness on chlordecone retention. *Environmental Science and Pollution Research* 23: 3176–3183.

Lerch S, Guidou C, Thomé JP, Jurjanz S. 2016. Non-dioxin-like Polychlorinated Biphenyls (PCBs) and Chlordecone release from adipose tissue to blood in response to body fat mobilization in ewe (Ovis aries). *Journal of Agricultural and Food Chemistry* 64: 1212–1220.

Lesueur-Jannoyer M. Cattan P, Monti D, Saison C, Voltz M, Woignier T, Cabidoche YM. 2012. Chlordécone Aux Antilles: Évolution Des Systèmes de Culture et Leur Incidence Sur La Dispersion de La Pollution. *Agronomie, Environnement et Société* 1(2): 45–58.

Mahieu M, Fournier A, Lastel ML, Feidt C, Rychen G, Archimède H. 2014. Chlordécone et élevage, variabilité individuelle des capacités d'excrétion des ruminants et conséquences sur leur contamination. *44ème congrès du Groupement Français des Pesticides*, Schoelcher, Martinique, France, May 26–29, 2014.

McFarland LZ, Lacy PB. 1969. Physiologic and endocrinologic effects of the insecticide Kepone in the Japanese quail. *Toxicology and Applied Pharmacology* 15: 441–450.

Naber EC, Ware GW. 1965. Effect of Kepone and mirex on reproductive performance in the laying hen. *Poultry Science* 44: 875–880.

Observatoire des Résidus de Pesticides n.d. http://www.observatoire-pesticides.gouv.fr, Accessed June 2016.

Richter EJ, Lay P, Klein W, Korte F. 1979. Enhanced elimination of Kepone-c-14 in rats fed liquid paraffin. *Journal of Agricultural and Food Chemistry* 27: 187–189.

Rivera Ferre MG, Edwards SA, Mayes RW, Riddoch I, Hovell FD. 2001. The effect of season and level of concentrate on the voluntary intake and digestibility of herbage by outdoor sows. *Animal Science* 72: 501–510.

Rozman K. 1986. Fecal excretion of toxic substances. In *Gastrointestinal Toxicology*, Rozman K Hänninen O. (eds.). Elsevier: Amsterdam, the Netherlands, pp. 119–145.

Skalsky HL, Fariss MW, Blanke RV, Guzelian PS. 1979. The role of plasma proteins in the transport and distribution of chlordecone (Kepone) and other polyhalogenated hydrocarbons. *Annals of the New York Academy of Sciences* 320: 231–237.

Smith JC, Arant FS. 1967. Residues of kepone in milk from cows receiving treated feed. *Journal of Economic Entomology* 60: 925–927.

Soine PJ, Blanke RV, Chinchilli VM, Schwartz CC. 1984a. High-density lipoproteins decrease the biliary concentration of chlordecone in isolated perfused pig liver. *Journal of Toxicology and Environmental Health* 14: 319–335.

Soine PJ, Blanke RV, Schwartz CC. 1983. Chlordecone metabolism in swine. *Toxicology Letter* 17: 35–41.

Soine PJ, Blanke RV, Schwartz CC. 1984b. Isolation of chlordecone binding proteins from pig liver cytosol. *Journal of Toxicology and Environmental Health* 14: 305–317.

Van Keulen J, Young BA. 1977. Evaluation of acid-insoluble ash as a natural marker in ruminant digestion studies. *Journal of Animal Science* 44: 282–287.

Waegeneers N, De Steur H, De Temmerman L, Van Steenwinkel S, Gellynk X, Viaene J. 2009. Transfer of soil contaminants to home-produced eggs and preventive measures to reduce contamination. *Science of the Total Environment* 407: 4438–4446.

Woignier T, Clostre F, Cattan P. 2014. Diagnosis and management of field pollution in the case of an organochlorine pesticide, the chlordecone. In *Environmental Risk Assessment of Soil Contamination*. Hernandez Soriano MC, (ed.). In Tech: Rijeka, Croatia, pp. 614–636.

Section V

The Public Health Issue: Exposure and Health Impacts

12 West Indian Population's Food Exposure to Chlordecone and Dietary Risk Assessment

Marie Fröchen, Carine Dubuisson, Fanny Héraud, Mathilde Merlo, and Jean-Luc Volatier

CONTENTS

12.1 INTRODUCTION

This chapter describes development of knowledge concerning the West Indian population's food exposure to chlordecone and the associated risk assessment since the early 2000s. It is divided into four parts, corresponding to the successive studies carried out by the French Food Safety Agency (AFSSA) until 2010 and by the French Agency for Food, Environmental, Occupational Health and Safety (ANSES) since then. The studies include the general population (adults, toddlers, and children) in two French overseas departments: Martinique and Guadeloupe. It also briefly describes the management measures set up by the French public administration to take into account the results and recommendations of these successive studies.

12.2 ASSESSMENT OF THE MARTINICAN POPULATION'S FOOD EXPOSURE TO CHLORDECONE, RISK ASSESSMENT, AND DIETARY RECOMMENDATIONS

12.2.1 Context

Previously used in the banana plantations of the French Antilles, chlordecone was banned in France in 1993. In 2001, several surveys revealed its widespread presence in soils and rivers (DIREN 2001, DSDS 2001) and showed possible contamination of drinking water and fishes, raising questions on food safety for the population. In 2003, the former French Food Safety Agency (AFSSA) proposed two toxicological reference values for chlordecone: a provisional tolerable daily intake (PTDI) of 0.5 mg kg^{-1} BW day^{-1} (BW for body weight) and an acute exposure limit of 10 mg kg^{-1} BW day^{-1} (AFSSA 2003). In 2005, the first food consumption data in Martinique were made available, allowing AFSSA to assess food exposure to chlordecone and to establish maximum limits for chlordecone in foods (AFSSA 2005, Dubuisson et al. 2007).

12.2.2 Data

To assess the Martinican population's dietary intake of chlordecone, a characterization of food patterns and food contamination of chlordecone was performed.

Consumption data were taken from the ESCAL ("Enquête sur la santé et les comportements alimentaires en Martinique") survey carried out from December 2003 to May 2004 in a representative sample of the Martinican population aged 3 years and over (N = 1814) (Merle et al. 2008). This survey provided information on dietary intake (Food Frequency Questionnaire [FFQ] and 24 h recalls), food supply habits (supermarkets, groceries, local markets, or subsistence production) and anthropometric characteristics of the population. Daily individual food consumption was estimated from the FFQ weighted by a median daily portion size derived from the 24-hour recalls. All data were weighted in order to provide an exposure estimate for the general population. The most eaten foods were tropical roots and tubers (sweet potato, yam, taro); bananas (fruit, plantain, or green), citrus fruits, and melon for vegetable-based foods; and chicken, creel, and net fishes for animal-based foods.

Food occurrence data were taken from specific monitoring programs implemented by the local French administration during the 2002–2004 period and from a survey on seafood conducted by the French Research Institute for Exploitation of the Sea in 2002 (Bocquené et al. 2002). These data mainly focused on the products that were thought to be more highly contaminated. The more frequently contaminated foods were seafood, milk, meat, and roots and tubers, with the highest levels observed in roots and tubers. Two levels of contamination were estimated:

1. An average contamination level (aQ), depending on the number of quantified results according to the international guidelines (European Union GEMS/Food Euro 1995): using the lower*-bound and upper†-bound hypothesis if more than 60% of results were not quantified, and using the middle‡-bound hypothesis if more than 40% of the results were quantified. For a food item, when more than 10 results were available and all of them were censored, the food item was considered to be uncontaminated.
2. A high contamination level (hQ) was calculated as the average contamination level of quantified results only.

The food consumption and food contamination data were combined taking into account the origin of food (supply habit) and the contamination status of the consumer living area, as described in Table 12.1.

12.2.3 Different Management Options Studied

The four scenarios detailed in Figure 12.1 were studied in order to assess the impact of different management options and maximum limit values.

The first theoretical exposure scenario (scenario 1) calculated a theoretical maximum daily intake (TMDI). According to this maximalist method, a maximum limit (ML) of 25 μg kg^{-1} fresh weight would have covered the risk of exceeding the provisional tolerable daily intake (PTDI) for all age groups. However, it overestimated

* All nondetectable results are set to 0, all nonquantifiable ones set to the limit of detection (LOD).

† All nondetectable results are set to the LOD, all nonquantifiable ones set to the limit of quantification (LOQ).

‡ All nondetectable results are set to half the LOD, all nonquantifiable ones set at the mid value between the LOD and the LOQ.

TABLE 12.1
Use of the Average and High (aQ and hQ) Residue Levels

	Foodstuffs from Circuit Court[a]	Foodstuffs from Mass Markets
People living in contaminated areas	hQ residue level applied	aQ residue level applied
People living in non-contaminated areas	aQ residue level applied	aQ residue level applied

[a] Small groceries, market, and self-production.

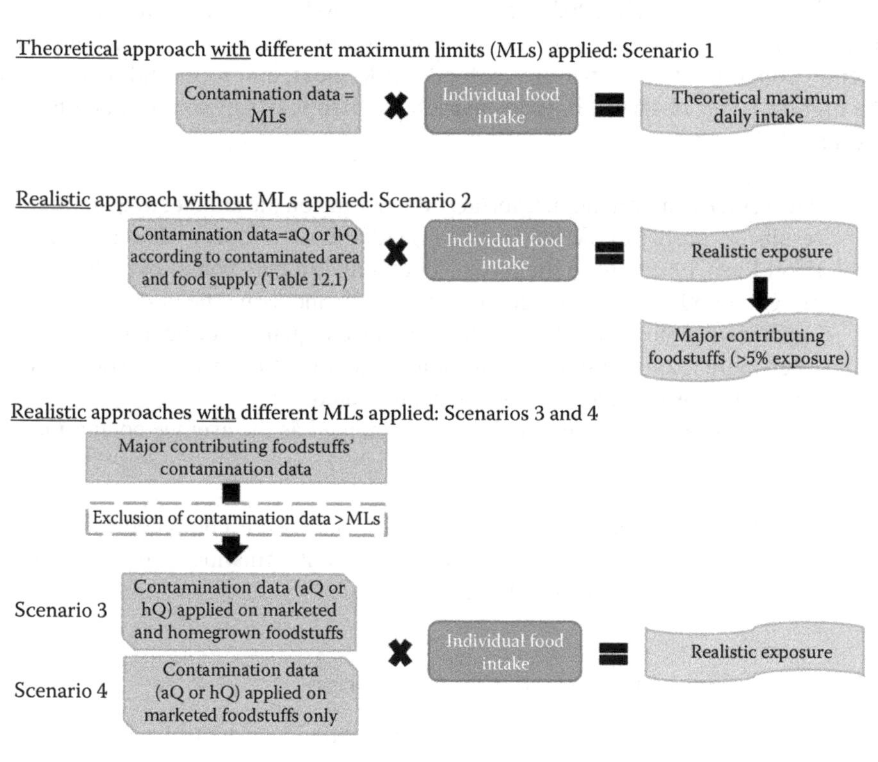

FIGURE 12.1 Description of the scenarios tested.

exposure and did not allow identification of the major foods contributing to chlordecone intake. Therefore, a more realistic exposure scenario (scenario 2) was used taking into account the contamination assumption from Table 12.1 and made it possible to identify the major food contributors. Scenario 2 assessed the probability of exceeding the PTDI of 0.5 μg kg^{-1} BW day^{-1} at, respectively, 20.9% (confidence interval CI_{95th} [6.2; 34.4]) and 15.6% (confidence interval CI_{95th} [9.6; 20.8]) for, respectively, children and adults living in a soil-contaminated area and null for the remaining population. According to this scenario, eight foods contributed to at least 5% of chronic exposure to chlordecone: dasheen (madeira), sweet potato, yam, cucumber, carrot, tomato, melon, and chicken meat.

Two scenarios were then used to test the impact of different MLs: in scenario 3 MLs were applied on all foods (marketed and homegrown), whereas in scenario 4 MLs were applied only on marketed food products since food regulations and food controls are not relevant to homegrown foods. These scenarios showed that supply habits might have significant impacts on food exposure to chlordecone.

Results correspond to the scenario using the upper-bound residue level as for the average residue level (aQ). P95 corresponds to the 95th percentile level of exposure in μg kg^{-1} BW day^{-1}. % > PTDI [95% CI] corresponds to the rate of people whose dietary intake is higher than the PTDI with the 95% confidence interval. In bold, the % > PTDI is statistically different from zero.

Scenario 3 demonstrated that a provisional maximum limit of 50–300 ppb (μg kg^{-1} fresh weight) for the eight major contributing foods was effective to protect the high-level consumers (Table 12.2). For the occasional food contributors, a limit based on the acute exposure of 200 μg kg^{-1} fresh weight would have prevented accidentally exceeding the acute toxicological reference value for adults and children and young children. However, the MLs remained of no use to reduce the probability of exceeding the PTDI when only marketed products were taken into account (scenario 4, Table 12.3). Setting maximum limits, which can be controlled only on marketed products, was therefore not sufficient to protect the population eating root vegetables grown in home gardens or directly sold by farmers (circuit court).

12.2.4 Summary

Taking into account these results, other management options were considered necessary to protect consumers and thus AFSSA recommended, as a temporary measure, to limit the consumption of homegrown sweet potato and dasheen (madeira) to approximately twice a week for families with home gardens on contaminated soil. Moreover, the French Ministries of Health, Agriculture, the Economy and Overseas Departments set two maximum limits:

1. 50 μg kg^{-1} FW for seven vegetables (dasheen, sweet potato, yam, cucumber, carrot, tomato, melon) and poultry meat;
2. 200 μg kg^{-1} FW for other vegetables and other products of animal origin.

12.3 EXTENSION OF FOOD EXPOSURE TO THE WHOLE WEST INDIAN POPULATION AND RISK ASSESSMENT

12.3.1 Material and Methods

In 2007, AFSSA updated the results of the first study taking into account the new available data (AFSSA 2007). These new data concerned the food behaviors of the population in Guadeloupe, on the one hand, and data on contamination of the representative foodstuffs of marketed products, on the other hand. Firstly,

TABLE 12.2
Impact of ML Setting on the Major Contributor on the Chlordecone Dietary Exposure (Scenario 3)

		Children (3–15 Years)			Adults (≥16 Years)		
		All	Non-Contaminated Areas	Contaminated Areas	All	Non-Contaminated Areas	Contaminated Areas
Without ML	p 95	0.416	0.189	0.956	0.391	0.102	0.76
	% > PTDI [95% CI]	**3.9 [1.7; 6.1]**	0	**20.9 [10.1; 31.7]**	**3.2 [2.0; 4.4]**	0	**15.6 [10.0; 21.2]**
ML = LOQ	p 95	0.116	0.103	0.145	0.067	0.057	0.12
	% > PTDI [95% CI]	0	0	0	0	0	0
ML = 50 ppb	p 95	0.164	0.115	0.304	0.145	0.061	0.258
	% > PTDI [95% CI]	0.1 [0; 0.5]	0	0.6 [0; 1.8]	0	0	0
ML = 100 ppb	p 95	0.215	0.134	0.37	0.182	0.073	0.32
	% > PTDI [95% CI]	0.6 [0; 1.8]	0	3.1 [0; 7.7]	0	0	0
ML = 200 ppb	p 95	0.237	0.159	0.444	0.214	0.085	0.373
	% > PTDI [95% CI]	0.8 [0; 1.8]	0	4.3 [0; 9.7]	0.2 [0; 0.5]	0	1.1 [0; 2.7]
ML = 300 ppb	p 95	0.254	0.167	0.468	0.227	0.09	0.393
	% > PTDI [95% CI]	0.8 [0; 1.8]	0	4.3 [0; 9.7]	0.4 [0; 0.8]	0	2.1 [0; 4.3]

TABLE 12.3
Impact of Subsistence Production on the Efficacy of ML Setting (Scenario 4)

		Children (3–15 Years)			Adults (≥16 Years)		
		All	Non-Contaminated Areas	Contaminated Areas	All	Non-Contaminated Areas	Contaminated Areas
Without ML	p 95	0.416	0.189	0.956	0.391	0.102	0.76
	% > PTDI [95% CI]	**3.9 [1.7; 6.1]**	0	**20.9 [10.1; 31.7]**	**3.2 [2.0; 4.4]**	0	**15.6 [10.0; 21.2]**
ML = LOQ	p 95	0.299	0.102	0.767	0.299	0.102	0.767
	% > PTDI [95% CI]	**2.5 [0.7; 3.7]**	0	**13.5 [4.4; 22.6]**	**2.5 [0.7; 3.7]**	0	**13.5 [4.4; 22.6]**
ML = 50 ppb	p 95	0.318	0.115	0.769	0.318	0.115	0.769
	% > PTDI [95% CI]	**2.8 [0.9; 4.7]**	0	**14.7 [5.3; 24.1]**	**2.8 [0.9; 4.7]**	0	**14.7 [5.3; 24.1]**
ML = 100 ppb	p 95	0.341	0.134	0.778	0.341	0.134	0.778
	% > PTDI [95% CI]	**2.9 [1.0; 4.8]**	0	**15.3 [5.8; 24.8]**	**2.9 [1.0; 4.8]**	0	**15.3 [5.8; 24.8]**
ML = 200 ppb	p 95	0.361	0.159	0.854	0.361	0.159	0.854
	% > PTDI [95% CI]	**3.2 [1.2; 5.2]**	0	**17.2 [7.2; 27.2]**	**3.2 [1.2; 5.2]**	0	**17.2 [7.2; 27.2]**
ML = 300 ppb	p 95	0.361	0.167	0.854	0.361	0.167	0.854
	% > PTDI [95% CI]	**3.2 [1.2; 5.2]**	0	**17.2 [7.2; 27.2]**	**3.2 [1.2; 5.2]**	0	**17.2 [7.2; 27.2]**

the Calbas survey (survey on Consumption and Food Supply in the Basse-Terre Region—"Consommation et Approvisionnement de la Région Basse Terre") concerns food consumption and supplies and included 683 people living in Basse-Terre, Guadeloupe. This study was carried out in children from 3 to 18 years of age as well as adults. Secondly, AFSSA also used the data on occurrence in foodstuffs from the RESO survey (Organochlorine residues in food—RÉSidus Organochlorés). This survey was conducted in Martinique and Guadeloupe (2005–2007). It provided information on the levels of contamination of food available in the food distribution systems in the West Indian population. This study had a sampling design that took into account the spending patterns and supply of the population in Martinique and Guadeloupe. It enabled the collection of 894 and 744 samples respectively.

The same evaluation methods of exposure and risks were implemented with these new data. The use of these new data made it possible to evaluate the relevance of the preliminary recommendations made by AFSSA (2005) and to adapt them.

12.3.2 Results and Discussion

The evaluation of acute exposure showed that four types of food products could still be contaminated at a level likely to represent an acute risk for the West Indian population. The food products were the malanga (*Xanthosoma* sp.), the dasheen (*Colocasia esculenta*), the sweet potato, and freshwater fish and shellfish. To protect the West Indian population from the risks related to acute exposure, AFSSA proposed a threshold of maximum contamination of food of 50 $\mu g\ kg^{-1}$ fresh weight for the dasheen, sweet potato and the malanga, and of 200 $\mu g\ kg^{-1}$ fresh weight for freshwater fish and shellfish.

The levels estimated for chronic exposure of the West Indian population were overall lower than those previously evaluated by AFSSA (2005). These results were explained by improved representativeness of the contamination data used. The RESO survey was carried out based on random sampling of food products available in the distribution systems. In contrast, the plans for monitoring and control, initially used in 2005, were more particularly targeted to the food products and areas likely to represent a risk. Chronic exposure was higher for children than adults. The maximum probability of exceeding the tolerable limit of chronic exposure was 18.5% (CI [6.3; 38.1]) in children from 3 to 5 years of age living in contaminated areas, and 0.2% (CI [0.2; 0.3]) in adults living in contaminated areas. However, the tolerable limit was not exceeded for the individuals, children, or adults, living in non-contaminated areas. Seafood and vegetable roots were the main food products contributing to total chlordecone exposure, because of their level of contamination. On the basis of these results, AFSSA recommended widening of application of the maximum threshold of 50 $\mu g\ kg^{-1}$ FW to the Caribbean cabbage (malanga) and seafood.

With regard to homegrown products, AFSSA maintained the provisional recommendation for families operating a garden on contaminated ground, to limit their consumption of dasheens (madeira), sweet potatoes, and yams coming from the garden, to twice per week. In addition, AFSSA recommended limiting the consumption of fishery products to every second day for the populations practicing leisure or subsistence fishing.

Taking into account the new European regulation on methods of establishing the MRLs of pesticides in food,* two new MRLs were defined in June 2008:

1. 100 µg kg^{-1} FW for products of terrestrial animal origin
2. 20 µg kg^{-1} FW for all other products, including vegetables, fruits, and freshwater and seafood products

This new regulation is protective compared to the previous recommendations of AFSSA.

12.4 DIETARY EXPOSURE OF 18-MONTH-OLD GUADELOUPEAN TODDLERS TO CHLORDECONE ("TIMOUN" COHORT STUDY)

12.4.1 Context

In previous studies (Merle et al. 2008, AFSSA 2003), chlordecone exposure and risk were assessed for the population aged over 3. However, since the most sensitive groups may be unborn children, infants, and toddlers, the French National Institute of Health and Medical Research (INSERM) conducted the "Hibiscus" study (2003–2004), an assessment of breast milk contamination and impregnation determinants of pregnant women. In 2008, the agency used the "Hibiscus" data to assess newborn exposure through breast milk, and the agency concluded that levels of chlordecone found in human milk were associated with a very low probability of exceeding the chronic health-based guidance value (CHGV) and the acute health-based guidance value (AHGV) for exclusively breastfed infants between 0 and 6 months. However, children's diet changes during food diversification, and they also consume more foods and water per unit of body weight than adults. The "Timoun" cohort study was conducted by the Inserm in Guadeloupe in 2006–2008 in order to characterize the impact of prenatal and postnatal exposure to chlordecone on the children's development. ANSES used the "Timoun" consumption survey data to assess dietary exposure to chlordecone, and to compare it to the other age groups (Seurin et al. 2012).

12.4.2 Data

12.4.2.1 Consumption and Contamination Data

The consumption survey was based on a questionnaire divided into three parts:

1. An FFQ, providing information for each food item: the age of its introduction in the diet, its frequency of consumption over 18 months, and its supply origin.
2. A questionnaire on milk consumption habits.
3. Two 24 h recalls, one on a weekday, and the second on a weekend. The 24 h recalls were used to quantify portion sizes.

* Regulation 396/2005 of the European Parliament and of the Council of February 23, 2005, on maximum residue levels of pesticides in or on food and feed of plant or animal origin.

Food contamination data were estimated based on the "RESO" survey, human milk contamination data came from the "Hibiscus" survey, and water contamination data came from the 2005 water monitoring program of the Departmental Public Health Services of Guadeloupe.

12.4.2.2 Estimation of Food Contamination Levels

Food contamination levels were assessed according to the hypothesis previously defined (aQ and hQ, average and high contamination levels). However, in this study, the aQ was used in the chronic exposure assessment and the hQ was used in the acute exposure assessment.

12.4.2.3 Exposure Assessment

Chronic and acute exposure was assessed for two subgroups of population, depending on whether they lived in a contaminated area or not. The exposure of 18-month-old toddlers was compared with other age groups: 3–5 years, 6–10 years, 11–15 years, and over 16 years.

12.4.2.4 Results

Consumption habits were assessed based on 240 questionnaires, 33.8% representing toddlers living in the contaminated area. The survey revealed that 73.7% of the toddlers consumed at least one type of homegrown food, mainly fruits and vegetables.

For most of the foodstuffs, contamination levels were higher in food when collected in contaminated areas. The highest *detection rate* (percentage of analysis exceeding the LOD) in contaminated areas was observed for sea products, and for root vegetables in noncontaminated areas. The highest *contamination levels* were observed in roots and tubers in both areas. The main contributors are described in Table 12.4, and Table 12.5 presents chronic exposure estimates. The CHGV was never exceeded.

Compared to the other age groups, the 18-month-old toddlers were less exposed than groups over 3 years, mainly because their diet is mostly based on milk and fruits, which are not highly contaminated by chlordecone.

The AHGV can be exceeded when consuming highly contaminated dasheen, but the level is not exceeded when considering taro contaminated at the maximum regulatory limit (20 μg kg^{-1} FW).

TABLE 12.4
Main Contributors to Total Chlordecone Intake

	Contaminated Areas		Noncontaminated Areas	
	Lower-Bound Estimate	**Upper-Bound Estimate**	**Lower-Bound Estimate**	**Upper-Bound Estimate**
Main contributors to total daily chlordecone intake	Fish (15%) Dasheen (15%) Malanga (11%) Carrot (18%)	Milk (30%) Dasheen (10%)	Shellfish (26%) Sweet potato (18%)	Milk (36%) Shellfish (11%)

TABLE 12.5
Chronic Exposure Estimates

		Mean ± SD (µg kg^{-1} BW day^{-1})	95th (µg kg^{-1} BW day^{-1})
Lower-bound estimate	Contaminated areas	0.045 ± 0.046	0.110
	Non-contaminated areas	0.018 ± 0.014	0.044
Upper-bound estimate	Contaminated areas	0.078 ± 0.051	0.144
	Non-contaminated areas	0.051 ± 0.024	0.096

12.4.3 Summary

Dietary exposure of 18-month-old Guadeloupean toddlers was lower than the other age groups, but since a developmental effect of chlordecone was observed in animals, this subgroup could have specific sensitivity. The results also confirmed that the regulatory maximum limit prevented AHGV levels from being exceeded.

12.5 KANNARI: A STUDY ON HEALTH, NUTRITION, AND CHLORDECONE EXPOSURE IN THE FRENCH WEST INDIES (2011–2016)

12.5.1 Context

12.5.1.1 Organization and Partners

The Kannari study, "Health, nutrition and chlordecone exposure in the French West Indies" (2011–2016), involves four components: the "nutrition" and "impregnation" parts, conducted by the National Institute for Health Surveillance (INVS), the "health" part conducted by the Regional Health Observatories of Martinique and Guadeloupe (ORS), and the "exposure" part conducted by the French Agency for Food, Environmental and Occupational Health and Safety (ANSES).

12.5.1.2 Objectives of the "Exposure" Part

As mentionned in Sections 12.2 and 12.3, the first exposure assessments were carried out in 2005 and 2007 based on contamination data from annual regulatory controls* and from the RESO survey. It was thus necessary to update the data and improve the knowledge based on the conclusions and recommendations of these previous studies. The study had three main objectives:

1. To update the information on *general population* eating habits and the exposure and risk assessment

* Annual controls organized by the Directorate General for Food and the Directorate General for Competition Policy, Consumer Affairs and Fraud Control.

2. To characterize the exposure and risk for different *high-risk population groups*: children aged 3–6 years, high-level consumers of local fishing products, and consumers of homegrown fruits and vegetables in contaminated areas
3. To identify the main exposure *contributors*

12.5.2 Materials and Methods

The field phase involved face-to-face meetings, 24-hour recalls, and biological sample collection carried out from September 2013 to June 2014.

12.5.2.1 Study Population

The population sampling design was determined in collaboration with the National Institute of Statistics and Economic Studies (INSEE). In order to fulfill the objectives, the sampling design was chosen so that it represented both the general population and the different high-risk population groups. Four strata were established: households composed of fishermen (strata 1), households in a defined coastal area (strata 2), households in individual dwellings in contaminated areas (strata 3), and all the other households (strata 4). Strata 1 and 2 represented high-level consumers of fishing products, strata 3 the homegrown fruit and vegetable consumers, and strata 4 represented the general population. Children were divided into three age groups (3–6, 7–10 and 11–15); individuals aged 16 and over were considered adults. In each household, one adult and one child from each age group were drawn by lot, so that one to three children were included.

12.5.2.2 Consumption and Contamination Data

Consumption data were collected during face-to-face meetings, with an FFQ and a supply habits questionnaire. The FFQ provided the annual consumption frequencies for the total diet. The supply habits questionnaire described the supply channels for locally produced food and therefore potential contributors. Additional consumption data and portion size data were collected during two 24-hour recalls made by dieticians on the phone.

Several sources of contamination data will be combined. The contamination levels of drinking water will be established from the SISE-Eaux database, administered by the Regional Health Authorities. Food contamination data will be established from three types of sources:

1. Regulatory controls organized annually by the Competition, Consumer Affairs and Prevention of Fraud authority (DGCCRF).
2. Scientific studies regarding environmental contamination conducted by the University of the French West Indies and the French Research Institute for Exploitation of the Sea (IFREMER).
3. An additional sampling campaign launched by ANSES in 2015 in order to supplement existing data. Since contamination levels of food sold on the roadside are not precisely characterized and could significantly increase exposure, this campaign focuses on this particular short supply channel.

12.5.2.3 Statistical Analysis

The sampling weights were calculated by the INSEE, taking into account non-response, out-of-scope units, and sample adjustments. All statistical analyses will be performed using Stata 12 and SAS 9.3 software, and will be carried out separately for Guadeloupe and Martinique, and for adults and children. The sampling design and weights will systematically be taken into account.

Food consumption habits will be described using mean daily consumption levels. Each food item will have a different contamination level depending on the supply channel and the area of production or marketing (contaminated or noncontaminated). The exposure will then be calculated individually, taking into account supply habits and daily consumption levels.

12.5.2.4 Summary

In addition to exposure and risk assessment, the report will establish new consumption recommendations, especially for self-grown food consumption.

The longer-term prospects are to provide guidelines for future health programs, based on the description of eating and supply habits.

12.6 DISCUSSION AND CONCLUSIONS

Chlordecone environmental contamination in the French West Indies is an exceptional environmental crisis, with potential consequences in different fields of public health: occupational and environmental health and food safety. Different agencies and research groups in agronomy and epidemiology joined their skills to launch and optimally use large surveys like dietary surveys (Escal, Calbas, Kannari), food and crop contamination surveillance surveys (Reso), and epidemiological surveys (Timoun, Kannari). These surveys have been used at the same time for research, surveillance, and risk assessment. They produced crucial information needed for risk management like the priority setting of foods contributing the most to the total exposure (vegetable roots, fish). The most exposed populations have also been identified by those studies like families eating locally grown vegetable cassavas in contaminated areas. For the management of the chlordecone crisis, dietary surveys have been decided and designed not only for nutritional purposes but also for chemical risk analysis. Usually, dietary surveys are designed by nutritionists and epidemiologists in the field of nutrition and not by chemical risk assessors and lack some information needed for dietary exposure assessment. In the case of chlordecone in the French West Indies, it has been the other way around. For instance, very useful information was recorded about the local origin of the food eaten in the dietary survey included in the Kannari project, so more precise exposure assessment and risk management is possible. These dietary surveys have also been used by the National Institute of Public Health for nutritional risk assessment, the French West Indies being concerned by a high prevalence of obesity.

Thanks to these studies, uncertainties in exposure assessment have been progressively reduced. This experience shows that such a public health crisis can be optimally managed only after years of surveillance. Risk management options need to be first very conservative when uncertainties are high. They can be adapted and detailed progressively when uncertainties decline because of the results of research and surveillance studies. It is one of the main lessons learned from the chlordecone risk assessment and management in the French West Indies in the last 15 years.

REFERENCES

AFSSA (2003) Avis de l'AFSSA relatif à l'évaluation des risques liés à la consommation de denrées alimentaires contaminées par la chlordécone en Martinique et en Guadeloupe. Saisines n°2003-SA-0330, 2003-SA-0132, 2003-SA-0091. AFSSA, Maisons-Alfort, France, 8pp.

AFSSA (2005) Première évaluation de l'exposition alimentaire de la population martiniquaise au chlordécone. Propositions de limites maximales provisoires de contamination dans les principaux aliments vecteurs.

AFSSA (2007) Actualisation de l'exposition alimentaire au chlordécone de la population antillaise, évaluation de l'impact de mesures de maîtrises des risques. Document technique AQR/FH/2007-219.

Bocquené G, Akcha F, Franco A, Grosjean P, Coat S, Godard E (2002) Bilan ponctuel de la présence et des effets des pesticides en milieu littoral martiniquais en 2002. IFREMER, Le Robert, Martinique, France, 42pp.

DIREN (2001) Le suivi de la contamination des rivières de la Martinique par les produits phytosanitaires. Bilan à l'issue des trois premières campagnes de mesure. DIREN, Fort-de-France, Martinique Island, France, 12pp.

DSDS (2001) Pesticides et alimentation en eau potable en Martinique. Etat des lieux et position sanitaire. Bilan actualisé en octobre 2001. Direction de la Santé et du Développement Social de la Martinique, Fort-de-France, Martinique Island, France, 11pp.

Dubuisson C, Héraud F, Leblanc J-C, Gallotti S, Flamand C, Blateau A, Quenel P, Volatier J-L (2007) Impact of subsistence production on the management options to reduce the food exposure of the Martinican population to Chlordecone. *Regulatory Toxicology and Pharmacology* 49, 5–16.

European Union. 1995. *GEMS/Food-EURO Second Workshop on Reliable Evaluation of Low-Level Contamination of Food. Report on a Workshop in the Frame of GEMS/Food-EURO.* Kulmbach, Federal Republic of Germany, 26-27. EUR/ICP/EHAZ.94.12/WS04 FSR/KULREP95. 8p ftp://ftp.ksph.kz/Chemistry_Food%20Safety/TotalDietStudies/Reliable.pdf. Accessed June 2016.

Merle B, Deschamps V, Merle S, Malon A, Blateau A, Pierre-Louis K et al. (2008) Enquête sur la santé et les comportements alimentaires en Martinique (ESCAL 2003–2004). Résultats du volet "consommations alimentaires et apports nutritionnels". Institut de veille sanitaire, Université Paris 13, Conservatoire national des arts et métiers, Observatoire de la santé de Martinique. Saint-Maurice, France, décembre 2008, 34pp.

Regulation (EC) NO 396/2005 of the European Parliament and of the Council of 23 February 2005 on maximum residue levels of pesticides in or on food and feed of plant and animal origin and amending Council Directive 91/414/EEC. *Official Journal of the European Union.* 16pp. http://eur-lex.europa.eu/LexUriServ/LexUriServ.do?uri=OJ:L:2005:070:0001:0016:en:pdf. Accessed June 2016.

Seurin S, Rouget F, Reninger J-C, Gillot N, Loynet C, Cordier S, Multigner L, Leblanc J-C, Volatier J-L, Héraud F (2012) Dietary exposure of 18-month-old Guadeloupian toddlers to chlordecone. *Regulatory Toxicology and Pharmacology* 63, 471–479.

13 Chlordecone Impact on Pregnancy and Child Development in the French West Indies

Sylvaine Cordier, Gina Muckle, Philippe Kadhel, Florence Rouget, Nathalie Costet, Renée Dallaire, Olivier Boucher, and Luc Multigner

CONTENTS

13.1 INTRODUCTION

Chlordecone is an organochlorine insecticide that was used intensively in the French West Indies, Guadeloupe and Martinique, from 1973 to 1993 to control the banana weevil *Cosmopolites sordidus*. It was only in 1999 that health and food authorities and regulatory agencies revealed that it was extensively distributed in soils, rivers, spring, and ground waters, aquatic biota, and crops. Given a large proportion of tap water, local bottled spring water, and local foodstuffs, animals, and vegetables were polluted and these products have been consumed, it was feared that human beings would also be contaminated (Dubuisson et al. 2007). The toxicity and harmful effects of exposure of humans to chlordecone was revealed in 1975 following a poisoning episode involving chlordecone plant workers in the industrial city of Hopewell, VA, USA (Cannon et al. 1978). The exposed workers showed evidence of toxicity involving the central nervous system and testis (Cannon et al. 1978, Cohn et al. 1978, Taylor 1982). This event involved only male workers, so no data or studies could be done for women, pregnant women, or their offspring. Meantime, experimental studies with rodents have shown that gestational and perinatal chlordecone exposure is detrimental to normal fetal development and it impairs neurobehavior during pre-weaning and post-weaning development (Faroon et al. 1995, Mactutus et al. 1982, 1984, Mactutus and Tilson 1984, 1985). Chlordecone crosses the placental barrier in pregnant rodents and is transferred to the newborn through maternal breastfeeding, thus exposing the developing organism during the earliest stages of development (Kavlock et al. 1980). In addition, it has been shown that chlordecone is an endocrine-disrupting chemical with well-established estrogenic and progestogenic characteristics (Hammond et al. 1979, Eroschenko 1981).

The first data of human contamination by chlordecone in the French West Indies among pregnant women and their newborns were obtained in a cross-sectional sample of 112 pregnant women in Guadeloupe in 2004. Chlordecone was detected in blood samples at delivery from 87% of pregnant women, in 61% of cord blood samples, and in 39% of colostrum samples of mothers who intended to breastfeed (Kadhel 2008). This raised many questions about the potential health implications of chlordecone exposure. All these observations prompted us to conduct an epidemiological study (Cordier et al. 2015) in order to evaluate the possible health consequences arising out of exposure of the general Guadeloupean population to chlordecone during pregnancy and during childhood.

13.2 POPULATION DESCRIPTION

This mother–child cohort was implemented in Guadeloupe (French West Indies), an archipelago situated in the Caribbean Sea, with a population of 450,000 inhabitants mostly of African descent. From December 2004 to December 2007, 1068 pregnant women were enrolled during third-trimester checkup visits at public health centers (University Hospital of Pointe-à-Pitre, General Hospital of Basse-Terre, and antenatal care units). To be eligible, participants had to have resided in Guadeloupe for more than 3 years. The proportion of refusal was about 7%. The study was approved by the Guadeloupean Ethics Committee for studies involving human subjects and detailed informed consent was obtained from each woman.

At enrolment, the participants answered a standardized questionnaire during a face-to-face interview with trained midwives. The questionnaire covered sociodemographic characteristics, medical and obstetrical history, and lifestyle factors. After delivery, the medical history of the pregnancy and delivery information were collected from midwives, pediatricians, and hospital medical records.

A semiquantitative food frequency questionnaire collected usual dietary intake during pregnancy. Maternal exposure in this cohort was attributed to the intake of contaminated food and water, especially seafood, root vegetables, and cucurbitaceous vegetables (Guldner et al. 2010). Maternal and cord blood samples were obtained at delivery.

A subsample of children, excluding cases of multiple birth, preterm birth, intrauterine growth restriction, neonatal serious disease or congenital malformation, and serious maternal illness before or during pregnancy, was selected for neurodevelopmental follow-up (n = 589). In total, 287 mothers could not be contacted or refused to participate. The final subsample available for follow-up consisted of 302 children. Two hospital visits were organized for the children, at the ages of 3 and 7 months. A third visit took place at the child's home, at the age of 18 months. All 302 families were invited to participate at each visit.

At each stage of the follow-up, trained research personnel completed questionnaires with the mothers about child's health, lifestyle, and dietary habits, and performed anthropometric measurements of the children. Neurobehavioral development was assessed at 7 and 18 months.

13.3 EXPOSURE ASSESSMENT

Exposure of the mother during pregnancy and prenatal exposure of the child were derived from chlordecone concentrations in blood samples collected at the end of pregnancy. Exposure during childhood (postnatal exposure) was assessed longitudinally at different ages during childhood from estimates of chlordecone intake from breast milk and from solid food. The maternal and cord blood samples were collected into tubes containing ethylenediaminetetraacetic acid as anticoagulant. After blood centrifugation, plasma samples were transferred to polypropylene Nunc® tubes, stored at –20°C and transferred in dry ice to the Center for Analytical Research and Technology (Liège, Belgium) for determination of chlordecone concentrations in plasma by gas chromatography-electron capture detection (Debier et al. 2003, Multigner et al. 2010). The limit of detection (LOD) of chlordecone was 0.06 μg L^{-1}. In the Timoun study, 88% of maternal blood samples (median: 0.39 μg L^{-1}; range: [<LOD; 19.3] μg L^{-1}) and 56% of the cord blood samples (median: 0.25 μg L^{-1}; range: [<LOD; 22.9] μg L^{-1}) had detectable levels of chlordecone. Total cholesterol and triglyceride concentrations in plasma were determined by standard enzymatic procedures (DiaSys Diagnostic Systems GmbH, Holzheim, Germany) and total lipid concentrations were calculated as described elsewhere (Bernert et al. 2007). At 3 months, a breast milk sample was collected from nursing mothers and the chlordecone concentration was determined using the same analytical method. The LOD in milk was 0.34 μg L^{-1} and 77% of the breast milk samples had detectable levels of chlordecone (median: 0.10 μg L^{-1}; range: [<LOD; 0.34] μg L^{-1}).

At the ages of 7 and 18 months, the daily dietary intake of chlordecone (μg kg^{-1} body weight day^{-1}) was estimated by combining the quantity of each food item ingested (g day^{-1})—measured by a food frequency questionnaire to the mother and its level of contamination with chlordecone (μg kg^{-1}) (Seurin et al. 2012).

13.4 OTHER CHARACTERISTICS AND ENVIRONMENTAL EXPOSURES OF MOTHERS AND CHILDREN

We collected information on maternal and child characteristics that may interfere in the relation between chlordecone exposure and health outcomes. Maternal characteristics considered were country of birth, enrolment site, age, years of education, marital status, pre-pregnancy working status, parity, prior preterm birth or miscarriage, pre-pregnancy body mass index (BMI), timing of first ultrasound, smoking or alcohol consumption during pregnancy, vitamin supplementation, weight gain during pregnancy, presence of maternal disease during pregnancy, and type of delivery and birth complications. Variables characterizing the child and its environment included sex of the newborn, number of adults living with the infant, domestic violence, maternal depression, HOME score (Home Observation for Measurement of the Environment), and environmental tobacco smoke. Potential prenatal co-exposure to other pollutants or nutrients was assessed by the determination of mercury (Hg), polychlorinated biphenyl congener 153 (PCB153), dichlorodiphenyl dichloroethene (DDE), the main metabolite of dichlorodiphenyl trichloroethane (DDT), lead (Pb), docosahexaenoic acid (DHA), and selenium (Se) concentrations in cord blood.

13.5 STATISTICAL ANALYSIS

For statistical analysis, maternal blood chlordecone concentrations were grouped in four classes to express exposure levels. Child prenatal exposure was defined from cord blood concentrations grouped in three classes. Both maternal and cord blood concentrations were used in some models as such after log transformation. A variable (four categories) combining breastfeeding status at 3 months and breast milk contamination was used to assess exposure through breastfeeding. Postnatal intake via the ingestion of contaminated food at 7 and 18 months (µg kg^{-1} body weight day^{-1}) was categorized into quartiles. All analyses of the associations between chlordecone exposure and health outcomes were conducted using multiple regression models (linear, logistic), non-linear growth models, or hazard models including, in addition to exposure and health measures, all covariates likely to modify these associations.

13.6 STUDY OF THE ASSOCIATIONS BETWEEN CHLORDECONE EXPOSURE AND PREGNANCY COMPLICATIONS AND OUTCOMES

13.6.1 Pregnancy Complications

Hypertensive disorders and diabetes mellitus are common problems during pregnancy, and are associated with significant short-term and long-term adverse health outcomes for both mothers and their children. They are particularly widespread in the French West Indies and in the Timoun cohort (Rouget et al. 2013).

In the Timoun cohort study, diagnoses of gestational hypertension, preeclampsia, and gestational diabetes mellitus were established from the medical history of pregnancy. Chlordecone exposure was found to be associated with a significantly lower risk of gestational hypertension (Saunders et al. 2014). Conversely, no statistically significant association was found between chlordecone exposure and the risk of preeclampsia and gestational diabetes mellitus.

A similar association has been reported between exposure to DDT and other organochlorine compounds, and a lower risk of gestational hypertension in a U.S. cohort study of 1933 women pregnant in the years 1959–1965 (Savitz et al. 2014). The negative association observed between exposure to chlordecone and risk of gestational hypertension could be due to a hypotensive effect mediated by the sympathomimetic and/or progestin properties of the molecule (Hammond et al. 1979, Tilson et al. 1987). However, the possibility of unrecognized bias or reverse causation (i.e., gestational hypertension would affect the chlordecone concentration in blood) cannot be excluded.

13.6.2 Preterm Birth

A number of studies have reported associations between maternal or cord blood concentrations of persistent organochlorine pollutants (especially DDE, the main metabolite of DDT), and decreased duration of gestation and/or increased risk of preterm birth. The potential effects of maternal chlordecone exposure on length of gestation and preterm birth (birth before 37 completed weeks of gestation) have also been examined in the Timoun cohort study (Kadhel et al. 2014). Gestational age (in weeks) was estimated by the obstetricians in charge of follow-up. It was based on the first day of the last menstrual period and was confirmed or corrected by ultrasound, available for 97% of the pregnancies. Preterm births were classified as spontaneous or medically induced. Higher maternal exposure to chlordecone was associated with shorter gestation, at least 3 days shorter for the 40% of women with blood chlordecone levels of more than 0.52 $\mu g\ L^{-1}$. As a consequence, higher levels of exposure were associated with an increased risk of preterm birth (condition with clinical significance and consequences on child health) in a dose–response-type association for all preterm births, and whatever the mode of onset of labor.

Although toxicological studies in rats and mice do not suggest a negative impact of chlordecone exposure during gestation on fetal growth (Chernoff and Rogers 1976), the association observed with length of gestation in the Timoun study has some biological plausibility. Parturition is triggered by the shortening and dilatation of the cervix associated with uterine contractions. Progesterone plays a key role in maintaining pregnancy, and treatment of pregnant women with progesterone-receptor antagonists induces labor at any stage of pregnancy (Chwalisz and Garfield 1994). Chlordecone binds estrogen receptors (ER) and stimulates the synthesis *in vivo* of the progesterone receptor in rat uterine tissues (Hammond et al. 1979), and this process is mediated by ERs. These observations suggest that the observed associations between exposure to chlordecone and decreased gestational length and increased risk of preterm birth, if real, may be due to estrogenic and/or progestin activities of chlordecone.

13.7 STUDY OF THE ASSOCIATIONS BETWEEN CHLORDECONE EXPOSURE AND CHILD DEVELOPMENT

Chlordecone crosses the placental barrier in pregnant rodents and transfers to the newborn through maternal breastfeeding, thus exposing the developing organism during the earliest stages of development (Kavlock et al. 1980).

13.7.1 GROWTH

Among the 222 babies with cord plasma chlordecone measurements available, 182 (82%) attended the 3-month visit, 165 (74%) the 7-month visit, and 167 (75%) the 18-month visit. All measurements (height [cm] and weight [g]) performed during the visits, and those recorded in the child's health records up to 18 months, were collected. Body mass index (BMI) was calculated as weight divided by height squared (kg m^{-2}). The total number of measurements recorded per infant varied from 2 to 11. Structured Jenss-Bayley growth models were fitted to individual height and weight growth trajectories. The impact of exposure on growth curve parameters was estimated directly with adjusted mixed nonlinear models. Weight, height, BMI, and instantaneous height and weight growth velocities at specific ages were also analyzed related to exposure.

Higher prenatal exposure was associated with a higher BMI at 3 months in boys and with higher BMI, and lower height and height growth velocity at 8 and 18 months in girls (Costet et al. 2015). Higher postnatal exposure was associated with lower weight and height at 3 months in boys, and with lower weight, height, and BMI at 18 months, in boys and girls. Height growth velocity was more particularly affected in girls. The underlying mechanism requires further investigation. Bone growth is a long process that starts in the intrauterine life and continues into early adulthood. Among other factors, estrogens play an important role in skeletal growth and maturation (Patandin et al. 1998). It remains unknown whether chlordecone intervenes in these processes, through its interactions with ERs or the activation of alternative estrogen signaling pathways (Lee and Witchel 1997, Jackson et al. 2010).

13.7.2 DEVELOPMENT

Follow-ups of the Timoun cohort at 7 and 18 months of age were conducted to document whether prenatal and postnatal exposures to chlordecone were related to neurobehavioral development in a cohort where exposure was not associated with any clinical sign of intoxication.

The infants were assessed at 7 months with the *Fagan Test of Infant Intelligence* (FTII), the *Teller Visual Acuity Card Test* (TAC; Teller et al. 1986), and two subscales from the *Revised Brunet-Lezine Scale of Psychomotor Development of Early Childhood* (BLS-R). The FTII (Fagan and Singer 1983) was administered to assess visual recognition memory. The infant is shown two identical photos depicting a person's face for a fixed period and is then shown the familiar target paired with a novel one; there are 10 problem sets. At this age, this test assesses "pre-explicit" recognition memory, which involves the construction of a memory trace of the visual stimulus, recall of information about the stimulus, and comparison of the memory trace with the visual input (Jacobson et al. 2008). Two measures are computed: (1) novelty preference, defined as proportion of looking time devoted to the novel stimulus; and (2) fixation duration, defined as the average duration of the infant's visual fixations to the stimuli. Shorter visual fixations indicate more rapid encoding of the memory trace (Colombo et al. 1991); novelty preference reflects the strength of the dishabituation to the familiar stimulus (Bakker et al. 2003). Both FTII measures have been shown to be moderately predictive of childhood IQ (McCall and Carriger 1993, McGrath et al. 2004). The TAC was used to assess binocular visual acuity in infants. It consists of a series of rectangular cards on which a patch of black and white vertical stripes with different spatial frequency is imposed to the left or right of a central 4 mm peephole. Acuity is defined as the narrowest grid for which the child shows a visual preference. Two subscales from the *BLS-R* were used to assess fine and gross motor development. In order to reduce the duration of testing, only 3 out of 11 fine motor items and 6 out of 13 gross motor items were assessed through maternal report.

At 18 months of age, infant development was assessed using an adapted version of the parent-completed *Ages and Stages Questionnaire* (ASQ), a test aimed at identifying children at risk for developmental delay (Bricker and Squires 1999). This 30-item questionnaire covers five developmental areas: personal-social, communication, problem-solving, fine motor, and gross motor. Each item corresponds to a developmental behavior or skill, which the parent must answer "yes" (10 points) if the child performs the given behavior, "sometimes" (5 points) if the behavior is performed occasionally or is emerging, and "not yet" (0 point) if the child does not yet perform the behavior. For each area, cutoff values, set at 2 standard deviations below the mean using reference norms, are proposed for identifying children at risk for developmental delay. Administration procedures were adapted for the purpose of this study since a pilot phase of our study indicated that parents tended to respond positively to each item: Each item from the communication, problem-solving, and fine motor and gross motor areas was administered directly to the child by a trained research personnel, and their administration was standardized according to the Bayley Scales of Infant Development-II (BSID-II) procedures (Bayley 1993). When a child failed an item, the examiner asked the mother whether her child "can," "sometimes can," or "cannot yet" perform the behavior or the skill. Thereafter, 10 points were allocated following a success to the administered item or to an item answered "can" by the mother, 5 points when the mother responded "sometimes can," and 0 point for a "cannot yet" answer. Points were then summed to yield a total score for the communication, problem-solving, and gross motor and fine motor areas. For the personal-social area, the items from the 18-month and the 20-month questionnaires were administered according to the ASQ's original procedure.

Increasing cord chlordecone concentration was found to be associated with poorer novelty preference at 7 months. Furthermore, higher postnatal intake of contaminated food also tended to be associated with poorer novelty preference and slower processing speed of visual information. By contrast, visual acuity during infancy was unrelated to chlordecone exposure. Finally, infants with a detectable level of chlordecone in cord were at greater risk of lower score of fine motor development (Dallaire et al. 2012). At 18 months of age, prenatal exposure to chlordecone was associated with poorer fine motor function, and stratified analyses by sex revealed that this effect was exclusively found in boys. Chlordecone exposure was not associated with the other developmental domains (personal-social, communication, problem-solving) assessed in the ASQ (Boucher et al. 2013).

The preference for novelty score from the FTII assesses short-term visual memory during infancy, and the fixation duration score reflects the time required to process the visual information into memory. Results from the Timoun prospective cohort study are consistent with reports of poorer short-term memory among Hopewell workers (Cannon et al. 1978) and results from rodent studies where pre-weaning rats exposed to chlordecone during the neonatal period exhibited memory impairment (Mactutus et al. 1982, Mactutus and Tilson 1984). Visual impairment was the third most reported symptom in the neurological syndrome associated with chlordecone intoxication in adult workers (Cannon et al. 1978). This deficit was characterized by incapacities to fixate and focus and the presence of a horizontal or multidirectional nystagmus. Visual acuity was unaffected, although blurred vision and visual hallucinations have been reported (Taylor 1982). In the Timoun cohort where pregnant mothers were exposed to lower doses of chlordecone than the Hopewell workers, and where exposure occurs during the prenatal period and first months of life, visual acuity during infancy was not related to prenatal or postnatal exposure to chlordecone.

Results from the 18-month follow-up corroborated the finding of impaired fine motor skills as a function of prenatal chlordecone exposure observed when infants were tested at 7 months of age (Dallaire et al. 2012). The results from the 18-month assessment are important in that they reveal that the previously observed fine motor impairments are persistent during infancy, and they identify boys as more especially vulnerable to the neurotoxic effects of chlordecone. The observed effects of prenatal chlordecone exposure on fine motor function at both 7 and 18 months of age are in accordance with the observation of severe and consistent tremors among intoxicated Hopewell employees: Presence of tremor in the upper limbs was the main complaint, with functional impact on daily activities (Cannon et al. 1978). Tremors are also a key characteristic of chlordecone neurotoxicity in experimental studies with rodents exposed either during the neonatal period or in adulthood (Dietz and McMillan 1979, Fujimori et al. 1982, 1986, Gerhart et al. 1982, 1985, Mactutus et al. 1984, Benet et al. 1985). That the adverse effects on fine motor function at 18 months were only observed among boys is compatible with data from experimental studies suggesting sex-dependent neurotoxic effects following developmental chlordecone exposure (Squibb and Tilson 1982, Mactutus and Tilson 1985).

13.8 THYROID HORMONES

No previous data, human or animal, are available on the action of chlordecone on the thyroid hormone system. In the Timoun cohort, thyroid stimulating hormone (TSH), free tri-iodothyronine (FT3), free thyroxine (FT4) were determined in child blood at 3 months, and these levels were related to prenatal and postnatal exposure to chlordecone. Sex-related associations were observed: Cord chlordecone was associated with an increase in TSH in boys, whereas postnatal exposure was associated with a decrease in FT3 overall, and in FT4 among girls (Cordier et al. 2015).

According to results of animal studies (Desaiah 1982), exposure to chlordecone may affect the hypothalamo-pituitary-thyroid axis at central and peripheral level in a sex-specific manner: TSH production at the central level among boys and FT3-FT4 at the peripheral level among girls, although we cannot exclude the possibility that the increase in TSH secretion observed in boys may mask an eventual decreased FT3-FT4 production.

Another observation in the Timoun cohort was that higher TSH level at 3 months was positively associated with the ASQ score of fine motor development at 18 months among boys, but TSH did not modify the association between prenatal chlordecone exposure and poorer ASQ fine motor score. This suggests that chlordecone developmental neurotoxicity may be mediated by the estrogen signaling pathway rather than by the thyroid endocrine system or by an interaction between these two hormonal systems (Duarte-Guterman et al. 2014).

13.9 STRENGTHS AND LIMITATIONS OF THIS STUDY

The strengths of the Timoun study include its prospective design, the evaluation of prenatal and postnatal exposure to chlordecone from several sources of exposure (transplacental transfer, consumption of contaminated water and food, and breast milk), and the consideration of a number of co-exposures to other toxicants (Hg, Pb, DDE, and PCB 153) and beneficial nutrients (DHA, Se). Single determinations of maternal and cord plasma chlordecone concentration provide an accurate reflection of the load of this compound in the body (Cohn et al. 1978, Guzelian 1982). Its half-life in blood is around 6 months, so a single measure at the end of pregnancy can be considered to be reasonably representative of prenatal exposure. The estimation of lactational exposure was based on measures of breast milk contamination and postnatal food chlordecone intake estimates benefited from recent survey (French Agency for Food Safety) on contamination level of foodstuffs in Guadeloupe.

This study also presents some limitations. Recruitment was mainly in the University Hospital of Guadeloupe, thus it may have favored the inclusion of pregnant women with higher morbidity and thus at risk of preterm birth. The prevalence of preterm birth in our study population was 14.1%, consistent with the high rate of preterm births estimated in overseas French territories (including French West Indies) and in populations of African descent (Rouget et al. 2013). However, this rate may be overestimated, but any such oversampling should not impair internal comparisons. Additionally, several participants were excluded from the follow-up during childhood because of missing information on cord chlordecone concentration, failure to contact, or for medical purposes (only children born at term, of normal weight, and without major birth defects were selected). They were therefore free from conditions previously related to chlordecone exposure (gestational hypertension, preterm birth) that may have influenced neurodevelopment or growth. There were no statistical difference between excluded and included participants with regard to maternal and newborns characteristics. With regard to the assessment of neurobehavioral development, we were not able to perform the whole BLS-R because of time constraints and there was no standardized test administered directly to 18-month-old children without relying on parental report. Finally, residual confounding cannot be completely excluded and nondifferential exposure misclassification might have underestimated the strength of the association with childhood health outcomes.

13.10 CONCLUSIONS

Findings from the Timoun prospective cohort study support the hypothesis that *in utero* and postnatal exposure to environmental levels of chlordecone is associated with less optimal pregnancy outcomes, growth, cognitive, and fine motor development during infancy. These adverse effects are observed at blood levels a thousand times lower than those measured among the intoxicated Hopewell workers. This confirms that the prenatal period is a period of increased sensitivity to the endocrine effects of this pesticide. Increased sensitivity during the perinatal period has also been documented for other neurotoxic chemicals such as Pb, PCBs, and Hg, and probably reflects the enhanced vulnerability of the developing brain to neurotoxicants due to interference with developmental processes, leading to permanent alterations in central nervous system structure and function. Follow-up assessments later during development are required to investigate long-term developmental deficits, and evaluate if these early subtle impairments are predictive of poorer neurobehavioral development at school age, and alterations of growth curves and obesity. The follow-up of children from the Timoun cohort at 7 years of age ended in 2015, and results from this follow-up will be available in the ensuing years.

REFERENCES

Bakker EC, Ghys AJA, Kester ADM et al. 2003. Long-chain polyunsaturated fatty acids at birth and cognitive function at 7y of age. *Eur J Clin Nutr* 57: 89–95.

Bayley N. 1993. *Bayley Scales of Infant Development*, 2nd edn. San Antonio, TX: The Psychological Corporation.

Benet H, Fujimori K, Ho IK. 1985. The basal ganglia in chlordecone-induced neurotoxicity in the mouse. *Neurotoxicology* 6: 151–158.

Bernert JT, Turner WE, Patterson DG, Jr. et al. 2007. Calculation of serum "total lipid" concentrations for the adjustment of persistent organohalogen toxicant measurements in human samples. *Chemosphere* 68: 824–831.

Boucher O, Simard M-N, Muckle G et al. 2013. Exposure to an organochlorine pesticide (chlordecone) and development of 18-month-old infants. *Neurotoxicology* 35: 162–168.

Bricker D, Squires J. 1999. *Ages and Stages Questionnaire (ASQ): A Parent-Completed, Child Monitoring System*, 2nd edn. Baltimore, MD: Brookes Publishing.

Cannon SB, Veazey JM, Jr., Jackson RS et al. 1978. Epidemic kepone poisoning in chemical workers. *Am J Epidemiol* 107: 529–537.

Chernoff N, Rogers EH. 1976. Fetal toxicity of kepone in rats and mice. *Toxicol Appl Pharmacol* 38: 189–194.

Chwalisz K, Garfield RE. 1994. Antiprogestins in the induction of labor. *Ann N Y Acad Sci* 734: 387–413.

Cohn WJ, Boylan JJ, Blanke RV et al. 1978. Treatment of chlordecone (Kepone) toxicity with cholestyramine. Results of a controlled clinical trial. *N Engl J Med* 298: 243–248.

Colombo J, Mitchell DW, Coldren JT, Freeseman LJ. 1991. Individual differences in infant visual attention: Are short lookers faster processors or feature processors. *Child Dev* 62: 1247–1257.

Cordier S, Bouquet E, Warembourg C et al. 2015. Perinatal exposure to chlordecone, thyroid hormone status and neurodevelopment in infants: The Timoun cohort study in Guadeloupe (French West Indies). *Environ Res* 138: 271–278.

Costet N, Pelé F, Comets E et al. 2015. Perinatal exposure to chlordecone and infant growth. *Environ Res* 142: 123–134.

Dallaire R, Muckle G, Rouget F et al. 2012. Cognitive, visual and motor development of infants exposed to chlordecone in Guadeloupe. *Environ Res* 118: 79–85.

Debier C, Pomeroy PP, Dupont C et al. 2003. Quantitative dynamics of PCB transfer from mother to pup during lactation in UK grey seals *Halichoerus grypus*. *Mar Ecol Prog Ser* 247: 237–248.

Desaiah D. 1982. Biochemical mechanisms of chlordecone neurotoxicity: A review. *Neurotoxicology* 3: 103–110.

Dietz DD, McMillan DE. 1979. Comparative effects of Mirex and Kepone on schedule-controlled behavior in the rat. I. Multiple fixed-ratio 12 fixed-interval 2-min schedule. *Neurotoxicology* 1: 369–385.

Duarte-Guterman P, Navarro-Martín L, Trudeau VL. 2014. Mechanisms of crosstalk between endocrine systems: Regulation of sex steroid hormone synthesis and action by thyroid hormones. *Gen Comp Endocrinol* 203C: 69–85.

Dubuisson C, Heraud F, Leblanc JC et al. 2007. Impact of subsistence production on the management options to reduce the food exposure of the Martinican population to chlordecone. *Regul Toxicol Pharmacol* 43: 5–16.

Eroschenko VP. 1981. Estrogenic activity of the insecticide chlordecone in the reproductive tract of birds and mammals. *J Toxicol Environ Health* 8: 731–742.

Fagan J, Singer L. 1983. Infant recognition memory as a measure of intelligence. In *Advances in Infancy Research*, Lipsitt L, Rovee-Collier C (eds.). Ablex, Norwood, NJ, pp. 31–78.

Faroon O, Kueberuwa S, Smith L, De Rosa C. 1995. ATSDR evaluation of health effects of chemicals. II. Mirex and chlordecone: Health effects, toxicokinetics, human exposure, and environmental fate. *Toxicol Ind Health* 11: 1–203.

Fujimori K, Benet H, Mehendale HM, Ho IK. 1986. In vivo and in vitro synthesis, release, and uptake of [3-H]-dopamine in mouse striatal slices after in vivo exposure to chlordecone. *J Biochem Toxicol* 1: 1–12.

Fujimori K, Nabeshima T, Ho IK, Mehendale HM. 1982. Effects of oral administration of chlordecone and mirex on brain biogenic amines in mice. *Neurotoxicology* 3: 143–148.

Gerhart JM, Hong JS, Tilson HA. 1985. Studies on the mechanism of chlordecone-induced tremor in rats. *Neurotoxicology* 6: 211–230.

Gerhart JM, Hong JS, Uphouse LL, Tilson HA. 1982. Chlordecone-induced tremor: Quantification and pharmacological analysis. *Toxicol Appl Pharmacol* 66: 234–243.

Global availability of information on agrochemicals, University of Hertfordshire. IUPAC, chlordecone (ref ENT 16391), http://sitem.herts.ac.uk/aeru/iupac/Reports/1293.htm. Accessed June 2016.

Guldner L, Multigner L, Héraud F et al. 2010. Pesticide exposure of pregnant women in Guadeloupe: Ability of a food frequency questionnaire to estimate blood concentration of chlordecone. *Environ Res* 110: 146–151.

Guzelian PS. 1982. Comparative toxicology of chlordecone (Kepone) in humans and experimental animals. *Annu Rev Pharmacol* 22: 89–113.

Hammond B, Katzzenellenbogen BS, Krauthammer N, McConnell J. 1979. Estrogenic activity of the insecticide chlordecone (Kepone) and interaction with uterine estrogen receptors. *Proc Natl Acad Sci USA* 76: 6641–6659.

Jackson LW, Lynch CD, Kostyniak PJ, McGuinness BM, Louis GMB. 2010. Prenatal and postnatal exposure to polychlorinated biphenyls and child size at 24 months of age. *Reprod Toxicol* 29: 25–31.

Jacobson JL, Jacobson SW, Muckle G, Kaplan-Estrin M, Ayotte P, Dewailly É. 2008. Beneficial effects of a polyunsaturated fatty acid on infant development: Evidence from the Inuit of Arctic Quebec. *J Pediatr* 152: 356–364.

Kadhel P. 2008. Pesticides in the Antilles, impact on the function of reproduction [in French]. PhD thesis. Université des Antilles et de la Guyane, Guadeloupe, French West Indies.

Kadhel P, Monfort C, Costet N et al. 2014. Chlordecone exposure, length of gestation, and risk of preterm birth. *Am J Epidemiol* 179: 536–544.

Kavlock RJ, Chemoff N, Rogers E, Whitehouse D. 1980. Comparative tissue distribution of mirex and chlordecone in fetal and neonatal rats. *Pestic Biochem Physiol* 14: 227–235.

Lee PA, Witchel SF. 1997. The influence of estrogen on growth. *Curr Opin Pediatr* 9: 431–436.

Mactutus CF, Tilson HA. 1984. Neonatal chlordecone exposure impairs early learning and retention of active avoidance in the rat. *Neurobehav Toxicol Teratol* 6: 75–83.

Mactutus CF, Tilson HA. 1985. Evaluation of long-term consequences in behavior and/or neural function following neonatal chlordecone exposure. *Teratology* 31: 177–186.

Mactutus CF, Unger KL, Tilson HA. 1982. Neonatal chlordecone exposure impairs early learning and memory in the rat on a multiple measure passive avoidance task. *Neurotoxicology* 3: 27–44.

Mactutus CF, Unger KL, Tilson HA. 1984. Evaluation of neonatal chlordecone neurotoxicity during early development: Initial characterization. *Neurobehav Toxicol Teratol* 6: 67–73.

McCall RB, Carriger MS. 1993. A meta-analysis of infant habituation and recognition memory as predictors of later IQ. *Child Dev* 64: 57–79.

McGrath E, Wypij D, Rappaport LA, Newburger JW, Bellinger DC. 2004. Prediction of IQ and achievement at age 8 years from neurodevelopmental status at age 1 year in children with d-transposition of the great arteries. *Pediatrics* 114: e572–e576.

Multigner L, Ndong JR, Giusti A et al. 2010. Chlordecone exposure and risk of prostate cancer. *J Clin Oncol* 28: 3457–3462.

Patandin S, Koopman-Esseboom C, De Ridder MAJ, Weisglas-Kuperus N, Sauer PJJ. 1998. Effects of environmental exposure to polychlorinated biphenyls and dioxins on birth size and growth in Dutch children. *Pediatr Res* 44: 538–545.

Rouget F, Lebreton J, Kadhel P et al. 2013. Medical and sociodemographic risk factors for preterm birth in a French Caribbean population of African descent. *Matern Child Health J* 17: 1103–1111.

Saunders L, Kadhel P, Costet N et al. 2014. Hypertensive disorders of pregnancy and gestational diabetes mellitus among French Caribbean women chronically exposed to chlordecone. *Environ Int* 68: 171–176.

Savitz DA, Klebanoff MA, Wellenius GA, Jensen ET, Longnecker MP. 2014. Persistent organochlorines and hypertensive disorders of pregnancy. *Environ Res* 132: 1–5.

Seurin S, Rouget F, Reninger J-C et al. 2012. Dietary exposure of 18-month-old Guadeloupian toddlers to chlordecone. *Regul Toxicol Pharmacol* 63: 471–479.

Squibb RE, Tilson HA. 1982. Effects of gestational and perinatal exposure to chlordecone (Kepone®) on the neurobehavioral development of Fischer-344 rats. *Neurotoxicology* 3: 17–26.

Taylor JR. 1982. Neurological manifestations in humans exposed to chlordecone and follow-up results. *Neurotoxicology* 3: 9–16.

Teller DY, McDonald M, Preston K, Sebris SL, Dobson V. 1986. Assessment of visual acuity in infants and children: The acuity card procedure. *Dev Med Child Neurol* 28: 779–789.

Tilson HA, Hong JS, Gerhart JM, Walsh TJ. 1987. Animal models in neurotoxicology: The neurobehavioral effects of chlordecone (Kepone). In *Neurobehavioral Pharmacology*, Thompson T, Dews PB, Barrett JE (eds.). Lawrence Erlbaum Association, Hillsdale, NJ, pp. 249–273.

Section VI

Remediation

14 Theoretical Approach to Chlordecone Biodegradation

Hervé Macarie, Igor Novak, Isabel Sastre-Conde, Yoan Labrousse, Alain Archelas, and Jan Dolfing

CONTENTS

14.1 INTRODUCTION

Nowadays, the sanitary, economic, and social crisis caused by chlordecone (CLD) in the French West Indies (FWI) may be considered as essentially contained thanks to the efforts undertaken by the authorities to avoid the population's dietary exposure to it (see chapters 17 and 18). A final solution to the problem would, however, involve eliminating the source of CLD responsible for the diffuse pollution of all the FWI environmental compartments and related food resources and so to destroy the stock of CLD still present in the soils. One of the cheapest and most environmental-friendly destruction methods corresponds to microbial degradation. Such a mode of destruction seems to be particularly appropriate in the case of FWI since it can often be implemented *in situ*, using techniques (e.g., watering; addition of nutrients, labile organic matter, microorganisms) that are fairly easy to incorporate into existing agricultural practices. The latter is important as the pollution is estimated to cover some 19,000 ha of arable lands (Le Déaut and Procaccia, 2009). Until now, however, there is no evidence of natural attenuation in the environments impacted by CLD and therefore of the possibility to stimulate the rate of the process. For instance, in 1989, Hugget stated

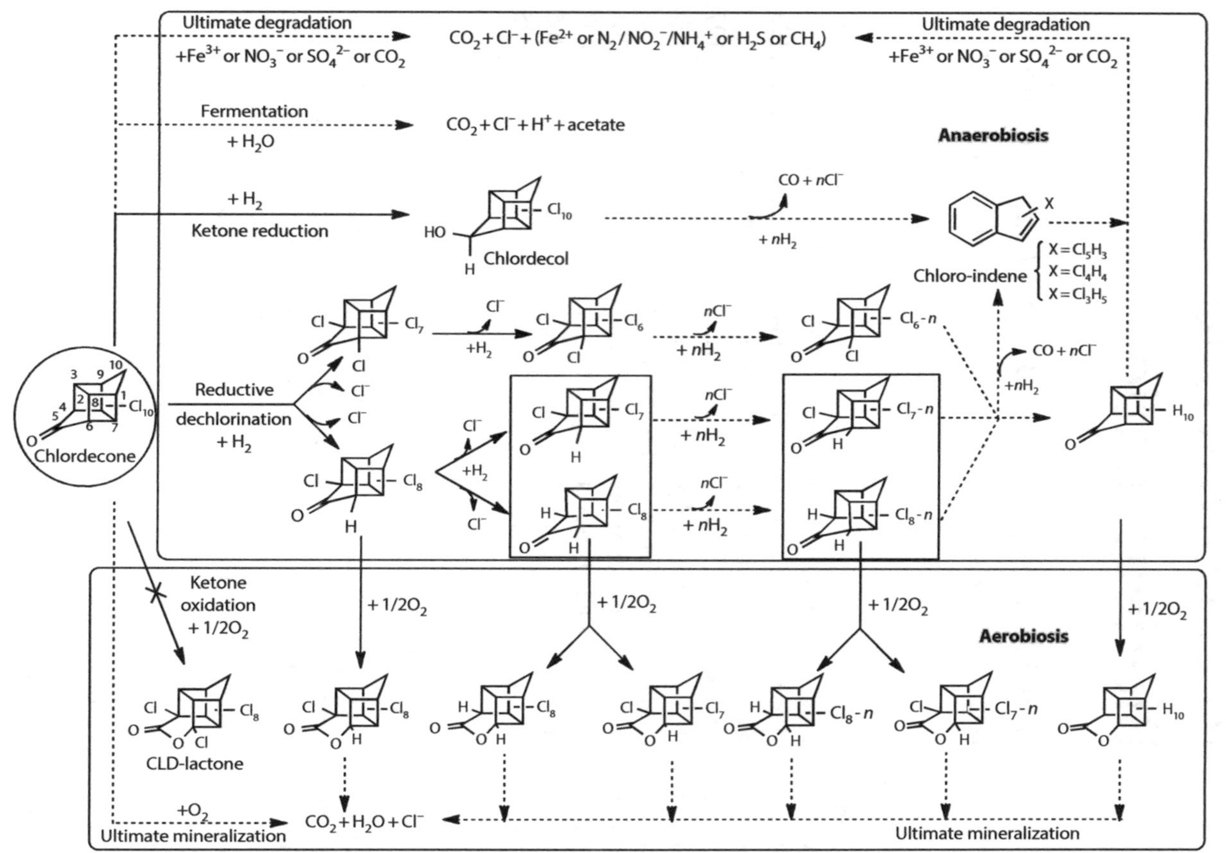

FIGURE 14.1 Different hypothetical and/or experimental pathways of CLD transformation.

with respect to the James River pollution by CLD that 13 years of observations did not demonstrate CLD degradation. In the same way, 20 years later Cabidoche et al. (2009) concluded the same about the fate of CLD in the FWI soils using a simple water-leaching model to simulate soil CLD content. Old and more recent research consisting of incubating freshwater sediments or FWI soils in the presence of freshly spiked CLD under controlled conditions in the laboratory for short (1 month) or longer term (7 months) and following CLD concentration over time and/or the production of CO_2 seem to confirm in the first analysis that the natural microbial populations present in these ecosystems are unable to, or at least have very poor capacity to, attack CLD both under aerobic than more reduced redox conditions (Skaar et al. 1981, Portier and Meyers 1982, Gambrell et al. 1984, Fernandez-Bayo et al. 2013a). Similar results were obtained with 103 strains of aerobic fungi isolated from FWI soils (Merlin et al. 2014). The reason usually advocated to explain this apparent absence of degradation in the environment is the peculiar chemical structure of CLD (bishomocubane "cage" structure with a high steric hindrance caused by the 10 chlorine atoms bound to the cage; see Figure 14.1), coupled with its low aqueous solubility (3 mg L^{-1} at 20°C) and high hydrophobicity (Log $K_{ow20°C,pH7} = 4.5$)* (IUPAC 2016) that would make it refractory to degradation. Indeed, such characteristics indicate intuitively that CLD must be poorly available to microorganisms that necessarily thrive in water and that the access of their enzymes to the CLD carbon skeleton to open the cage will not be easy. CLD is also known to be toxic to microorganisms (e.g., Orndorff and Colwell 1980b) and was even patented as an antimicrobial agent against Gram-positive bacteria and dermatophytic fungi in 1969 (US Patent 3,448,194).

In this chapter, we will demonstrate through a thermodynamic approach that there is no energetic reason why the structure of CLD should not be amenable to microbial degradation and we will propose some possible reasons for the apparent absence of CLD degradation in the FWI and what could be done to reverse the situation.

14.2 GIBBS FREE ENERGY OF POTENTIAL REACTIONS OF CLD TRANSFORMATIONS AND REDUCTION POTENTIAL OF CLD/CLD-$Cl_{10-n}H_n$ COUPLES

Through the action of microorganisms in the environment CLD could—hypothetically—be fully mineralized or just partially transformed (Figure 14.1). The ultimate degradation could proceed under different redox conditions (aerobiosis, iron reduction, denitrification, sulfate reduction, methanogenesis, etc.) depending on the main electron acceptors (O_2, Fe^{3+}, NO_3^-, SO_4^{2-}, CO_2, etc.) present in the environmental compartments (soil, surface water, groundwater, freshwater or marine sediments) impacted by CLD. CLD partial transformations that can be contemplated correspond to (1) the oxidation or reduction of its ketone group to form the corresponding lactone or alcohol, (2) its sequential dechlorination (removal of 1, 2, 3, 4, ..., 10 chlorine atoms) that can generate up to 484† different partially dechlorinated intermediate

* K_{ow}: octanol water partition coefficient.

† Including 92 mesomeric compounds and 196 pairs of enantiomers.

products until the fully dechlorinated congener (Dolfing et al. 2012), or (3) its transformation by fermentation (Figure 14.1).

Among the different thermodynamic functions, the change in Gibbs free energy or ΔG associated with a chemical reaction is the one that allows predicting the direction of that reaction. A negative value indicates that the reaction is exergonic and should occur spontaneously, while a positive value indicates that the reaction is endergonic and cannot proceed spontaneously under given conditions. It must be remembered, however, that even when a reaction is exergonic this does not necessarily imply that it will occur at an observable rate. The way to calculate the ΔG of the reactions is described in most biochemical or thermochemistry textbooks. The reader may also refer to Dolfing (2003) and Dolfing et al. (2012). The calculation of the $\Delta G^{\circ\prime}$ for all the hypothetic reactions of CLD microbial transformation requires the knowledge of the ΔG_f° (Gibbs free energy of formation) of CLD and related CLD derivatives (e.g., dechlorination products) in their aqueous state.* Since these values were not available in the literature, they have been estimated by *ab initio quantum* calculations using the G3(MP2)/B3LYP method implemented in Gaussian 03 software. The tabulated ΔG_f° have been reported previously (Dolfing et al. 2012).

The $\Delta G^{\circ\prime}$ values calculated from these ΔG_f° for the hypothetic reactions mentioned earlier are presented in Table 14.1. They show that all these reactions are extremely favorable ($\Delta G^{\circ\prime}$ very negative) and that even the removal of a single chlorine atom or the oxidation or reduction of the CLD ketone function should liberate enough energy to allow the synthesis of ATP and therefore microbial growth.

The standard conditions used for $\Delta G^{\circ\prime}$ calculations being far away from those found in the environment (reactants and products are generally not present at a concentration of 1 M or 1 atm!), it was necessary to evaluate whether the reactions would remain favorable under more realistic conditions. This was possible for some of the reactions and environmental compartments (groundwater, soil water solution) for which the requested *in situ* concentrations were available. The resulting *in situ* ΔG values showed that the reactions that could be tested (CLD aerobic mineralization and removal of one Cl) are as favorable or even more favorable than under standard conditions (Table 14.1), a situation that should be similar for the other reactions (Dolfing et al. 2012).

$\Delta G^{\circ\prime}$ calculations show also that CLD dechlorination is thermodynamically more favorable than the formation of chlordecol resulting from the reduction of the CLD ketone group. This suggests that in natural environments where the reducing equivalents (e.g., H_2) required by the two reactions are present in limited amounts, both reactions will compete for them and that the dechlorination should proceed preferentially to chlordecol formation. When the reducing equivalents are not limiting, it is not impossible, however, that some microorganisms may be able to dechlorinate CLD and simultaneously reduce its ketone group in order to maximize energy recovery. A last option that must be considered is that in some ecosystems,

* (°) superscript indicates that ΔG calculations are done under standard conditions, which means that the concentration of all aqueous reactants and products is 1 M or a partial pressure of 1 atm for gaseous compounds. (°′) superscript means that ΔG calculations are done under standard conditions except for pH equal to 7. In both cases the temperature is equal to 298.15 K.

TABLE 14.1
Gibbs Free Energy of Potential Reactions of Transformation of CLD and Some of Its Degradation Products under Standard (pH 7) and *In Situ* Conditions Compared to That of ATP Synthesis

Reaction	$\Delta G^{o\prime}$	ΔG *In Situ*
	kJ mol^{-1} CLD	
Mineralization or ultimate degradation		
Aerobic conditions		
$C_{10}Cl_{10}O + 15H_2O + 7O_2 \rightarrow 10HCO_3^- + 20H^+ + 10Cl^-$	−4443	−5344[gw]
Iron(III) reducing conditions		
$C_{10}Cl_{10}O + 29H_2O + 28Fe^{3+} \rightarrow 10HCO_3^- + 48H^+ + 10Cl^- + 28Fe^{2+}$	−4204	—
Nitrate-reducing conditions 1		
$5C_{10}Cl_{10}O + 61H_2O + 28NO_3^- \rightarrow 50HCO_3^- + 72H^+ + 50Cl^- + 14N_2$	−4146	—
Nitrate-reducing conditions 2		
$C_{10}Cl_{10}O + 15H_2O + 14NO_3^- \rightarrow 10HCO_3^- + 20H^+ + 10Cl^- + 14NO_2^-$	−3291	—
Sulfate-reducing conditions		
$2C_{10}Cl_{10}O + 7SO_4^{2-} + 30H_2O \rightarrow 20HCO_3^- + 33H^+ + 20Cl^- + 7HS^-$	−1541	—
Methanogenic conditions		
$2C_{10}Cl_{10}O + 37H_2O \rightarrow 13HCO_3^- + 33H^+ + 20Cl^- + 7CH_4$	−1483	—
Dechlorination		
CLD + $H_2 \rightarrow$ monohydrochlordecone + H^+ + Cl^-	[−160; −155]	[−160; −138][gw] [−164; −142][sw]
monohydrochlordecone + $H_2 \rightarrow$ dihydrochlordecone + H^+ + Cl^-	[−165; −142]	
CLD + $10H_2 \rightarrow$ decahydrochlordecone + $10H^+$ + $10Cl^-$	−1448 (−145)[a]	
Reduction of CLD ketone group		
CLD + $H_2 \rightarrow$ chlordecol	−70	—
Oxidation of CLD ketone group		
CLD + ½$O_2 \rightarrow$ CLD-lactone	−148	—
Fermentation		
$2C_{10}Cl_{10}O + 30H_2O \rightarrow 6HCO_3^- + 33H^+ + 20Cl^- + 7CH_3COO^-$	−1375	—
ATP synthesis from ADP and AMP	kJ/reaction	
ADP + PO_4^{3-} + $H^+ \rightarrow$ ATP + H_2O	+30.5	~ +70[b]
AMP + $2PO_4^{3-}$ + $2H^+ \rightarrow$ ATP + $2H_2O$	+61	—

"gw" and "sw" superscripts stand for "groundwater" and "soil water" respectively. Concentrations and temperature used for the groundwater *in situ* ΔG calculations correspond to average values found in Martinique aquifers (Gourcy et al. 2009, Gourcy L. 2011 and Bristeau S. 2012, BRGM, personal communications): 4.97 μg CLD L^{-1}, 0.22 μg (8-monohydrochlordecone) L^{-1}, 44.6 mg Cl^- L^{-1}, 102 mg HCO_3^- L^{-1}, 2.7 mg O_2 L^{-1}, pH 6.8, 28°C. Concentrations used for the soil water *in situ* ΔG calculation correspond to the average values measured in the leaching water of a lysimeter implemented in an andosol of Guadeloupe containing 5.4 mg CLD kg^{-1} soil dry weight and considered as representative of the soil solution: 2.2 μg CLD L^{-1}, 43 mg Cl^- L^{-1}, pH 7.2 (Cabidoche Y.-M. 2010, personal communication). The concentration of monohydrochlordecone was estimated from the average 8-monohydrochlordecone/CLD mass ratio value found in the FWI dry soils (19.7‰, Devault et al. 2016) that also have an average temperature of 25°C within the first 30 cm (Mouvet C. 2015, BRGM, personal communication). The ΔG_f° used for the calculations for all the compounds were taken from Thauer et al. (1977) and Stumm and Morgan (1996) except CLD and its derivatives that were taken from Dolfing et al. (2012). The H_2 concentration used for the calculation of the groundwater and soil *in situ* dechlorination ΔG covers the range of concentrations (0.01–70 nM equivalent to a partial pressure of 0.001–10 Pa) usually found at steady state in natural environments poised by O_2, Fe^{3+}, Mn^{4+}, NO_3^-, SO_4^{2-}, and CO_2 according to Lovley and Goodwin (1988), Conrad (1996), and Heimann et al. (2009).

[a] Value within brackets correspond to kJ mol^{-1} Cl removed.

[b] Amount of energy required *in vivo* for the synthesis of 1 mol ATP taking into account that the efficiency of energy conservation is not 100% (see Schink, 1997).

the microorganisms present may have only the metabolic capacity to reduce CLD into chlordecol.

CLD dechlorination being a series of oxidation-reduction reactions, it is interesting to know the reduction potential ($E°'$) corresponding to the redox couples formed by CLD and its dechlorinated products that can be schematized as follows: $C_{10}Cl_nH_{(10-n)}O/C_{10}Cl_{(n-1)}H_{(11-n)}O$ with n = 1–10 (e.g., $C_{10}Cl_{10}O/C_{10}Cl_9HO$; $C_{10}Cl_9HO/C_{10}Cl_8H_2O$ ··· $C_{10}ClH_9O/C_{10}H_{10}O$). The $E°'$ of the resulting couples can

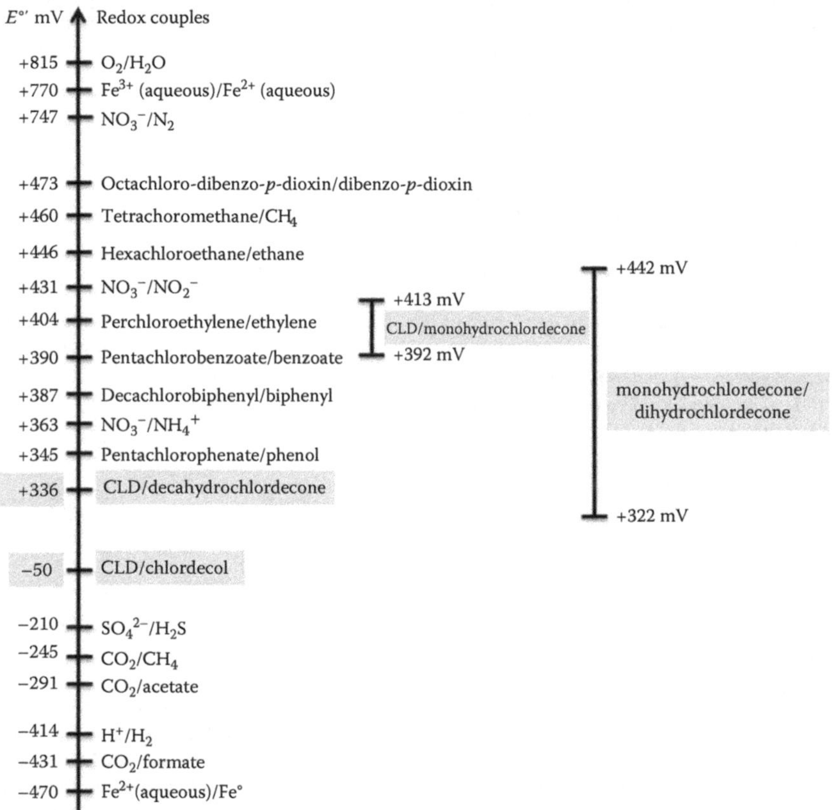

FIGURE 14.2 Comparison of the redox potential ($E°'$) of the $C_{10}Cl_nH_{(10-n)}O/C_{10}Cl_{(n-1)}H_{(11-n)}O$ couples (n = 1–10) with those of the main environmental electron acceptors and other organochlorine compounds. *Note*: The $E°'$ were calculated according to the Nernst equation (see text) using the $\Delta G_f°$ tabulated by Dolfing and Novak (2015) for pentachlorophenate, phenol, and pentachlorobenzoate; Shock (1995) for benzoate (experimental value); Holmes et al. (1993) for decachlorobiphenyl and biphenyl; Dolfing and Jansen (1994) for tetrachloromethane, hexachloroethane, tetrachloroethylene, methane, ethane, and ethylene; Huang et al. (1996) for octachloro-dibenzo-*p*-dioxin and dibenzo-*p*-dioxin; Dolfing et al. (2012) for CLD and dechlorinated compound; Thauer et al. (1977) for H_2, H^+, H_2O, Fe^{3+}, Fe^{2+}, NO_3^-, NO_2^-, NH_4^+, N_2, SO_4^{2-}, H_2S, CO_2, CH_4, acetate, and formate; and Stumm and Morgan (1996) for O_2. The following compounds were considered under their gaseous state: H_2, N_2, H_2S, CO_2, CH_4, and O_2; all others were considered under their aqueous state.

be calculated from the $\Delta G^{\circ\prime}$ of the dechlorination reaction with H^+/H_2 couple as electron donor (i.e., $C_{10}Cl_nH_{(10-n)}O + H_2 \rightarrow C_{10}Cl_{(n-1)}H_{(11-n)}O + H^+ + Cl^-$) using the Nernst Equation ($\Delta G^{\circ\prime} = -nF\Delta E^{\circ\prime}$ see Dolfing 2003).* Starting in 1984 with the isolation of *Desulfomonile tiedjei*, the first dechlorinating bacterium, a new group of microorganisms with the capacity to "respire" halogenated organic compounds, thus to use them as terminal electron acceptors through their dehalogenation coupled to energy conservation, was discovered (Mohn and Tiedje 1992). Nowadays, a large range of alkyl (perchlorethylene, tetrachloromethane, hexachloroethane, etc.) and aryl (chlorobenzoates, chlorophenols, chlorobenzenes, dioxins, PCB, etc.) halogenated compounds are known to be "respired" by a wide diversity of bacteria (e.g., Maphosa et al. 2010). As can be seen in Figure 14.2, the reduction potentials of CLD and its partially dechlorinated intermediates indicate that they correspond to electron acceptors comparable to the other organochlorine compounds and as strong as NO_3^-. Indeed, the $E^{\circ\prime}$ of the CLD/monohydrochlordecone, monohydrochlordecone/dihydrochlordecone, and CLD/decahydrochlordecone couples range from +322 to +442 mV against +363 mV for the NO_3^-/NH_4^+ and +431 mV for the NO_3^-/NO_2^- couples. The amount of energy that can be recovered per chlorine atom removed in the case of CLD ($\Delta G^{\circ\prime}_{\text{reaction}}$ = −142 to −165 kJ mol^{-1} Cl, Table 14.1) is also well within the range of energy that can be recovered for the dechlorination of the aforementioned organochlorine compounds ($\Delta G^{\circ\prime}_{\text{reaction}}$ = −130 to −180 kJ mol^{-1} Cl, Dolfing 2003). All this shows that in spite of its peculiar cage structure, CLD is thermodynamically very similar to other organochlorine compounds and should behave like them.

14.3 EXPERIMENTAL CONFIRMATION OF CLD'S SUSCEPTIBILITY TO MICROBIAL TRANSFORMATIONS

The previous considerations show that there are apparently no thermodynamic impediments to several chemical transformations of CLD that could be mediated by microorganisms. Three of the papers published after the poisoning of the workers of the Life Science CLD production site in the city of Hopewell, Virginia, USA, in 1975 and the associated pollution of the James River clearly confirmed that CLD is at least susceptible to dechlorination under the action of microorganisms or some of their coenzymes. For instance, Orndorff and Colwell (1980a) demonstrated that a pure culture of a *Pseudomonas aeruginosa* strain isolated from the water of a lagoon used for the storage of a CLD-contaminated sewage sludge at Hopewell was able to convert 30% of the initial CLD (5 mg L^{-1}) into a mixture of 8-monohydrochlordecone† and 2,8-dihydrochlordecone after 1 week of incubation

* In the Nernst equation $\Delta E^{\circ\prime} = E^{\circ\prime}$ (electron acceptor couple) − $E^{\circ\prime}$ (electron donor couple) = $E^{\circ\prime}$ $[C_{10}Cl_nH_{(10-n)}O/C_{10}Cl_{(n-1)}H_{(11-n)}O]$ − $E^{\circ\prime}$ (H^+/H_2); n = number of electrons transferred in the reaction = 2; F = Faraday constant = 0.096485 kJ/mV; $E^{\circ\prime}(H^+/H_2)$ = redox potential of H^+/H_2 couple = −414 mV (Thauer et al. 1977). After rearrangement this gives that $E^{\circ\prime}[C_{10}Cl_nH_{(10-n)}O/C_{10}Cl_{(n-1)}H_{(11-n)}O]$ in mV = $(-\Delta G^{\circ\prime}/nF) + E^{\circ\prime}(H^+/H_2) = (-\Delta G^{\circ\prime}/0.193) - 414$.

† Carbon numbering according to IUPAC nomenclature throughout this chapter—see Figure 14.1 and Dolfing et al. (2012).

in the dark under aerobic conditions at 25°C and in the presence of peptone, acetone (solvent for CLD stock solution), and yeast extract. This bacterial strain was, however, unable to use CLD as sole carbon and energy source or to modify further its carbon skeleton. In fact, its action on CLD was apparently limited to the removal of two chlorine atoms under conditions of cometabolism. Two years before the work of Orndorff and Colwell (1980a), Schrauzer and Katz (1978) had demonstrated *in vitro* that CLD was highly susceptible to dechlorination in presence of vitamin B_{12s} under reducing conditions. Vitamin B_{12s} is a cobalt transition metal coenzyme very common among prokaryotes and known since a long time to be involved in nonspecific fortuitous reductive dechlorination of chlorine atoms bound to alkyl (aliphatic compounds) or aryl (aromatic compounds) carbon atoms (Mohn and Tiedje 1992). Depending on the experimental conditions, vitamin B_{12s} revealed the capacity to remove up to four chlorine atoms from the CLD bishomocubane "cage" but more interestingly to induce the opening of the "cage" resulting in the formation of indene compounds of formula $C_9Cl_{8-n}H_n$ (with $n = 3$–5; see Figure 14.1 for structure). The last article in the aforesaid series of three papers showed *in vivo* that a culture of the methanogenic *Archaeum*, *Methanosarcina thermophila*, could convert 86% of the initial CLD into products that gave on thin layer chromatography (TLC) a profile identical to the one obtained with vitamin B_{12s} and that two transition metal complexes of this *Archaea* containing Co and Ni respectively were involved in the process since they gave on TLC the same profile of CLD transformation products as the whole cells (Jablonski et al. 1996). The previous old reports are complemented by the recent work of Belghit et al. (2015), who showed that CLD could easily lose up to five chlorine atoms when put into contact with a micrometric elemental iron powder in an aqueous phase at room temperature in the dark. Although this reaction was not biologically mediated, it confirms that CLD is not chemically inert but is susceptible to chemical transformations under mild conditions compatible with biochemical reactions. Therefore, it is not surprising that again recently Devault et al. (2016) could demonstrate indirectly that the 8-monohydrochlordecone detected in the soils of Martinique and which had been considered for a long time to be an impurity formed during the synthesis of CLD and brought into the soils as an accompanying product during the spread of the CLD pesticide commercial formulations is in fact a dechlorination product of CLD, although the biotic or abiotic nature of the dechlorination process in the soils could not be established.

The formation of chlordecol and chlordecol dechlorinated products were also recorded by Orndorff and Colwell (1980a) and Schrauzer and Katz (1978). Since the reactions leading to the formation of the CLD alcohol products described by Schrauzer and Katz (1978) were performed in the presence of methanol, it cannot be totally excluded that these products may correspond in this case to analytical artefacts formed in the injector of the gas chromatograph used for the analysis as this was observed by Soine et al. (1983). The biological formation of chlordecol from CLD has, however, been unambiguously demonstrated in the liver of humans and some other mammals (gerbil, pig, rabbit) where it is catalyzed by a specific aldo-keto reductase (Molowa et al. 1986).

14.4 LIMITS OF THE THERMODYNAMIC APPROACH

In principle, thermodynamics should also help rationalize degradation pathways. In the case of dechlorination, for instance, it has often been observed that the dechlorination products that are formed result from the reactions that liberate the most energy (e.g., Dolfing 2003). For CLD, this would imply that among the four possible monohydrochlordecone isomers, the formation of 8-monohydrochlordecone should be favored over the others and that in turn this compound should be preferentially dechlorinated into 4,8-dihydrochlordecone (Table 14.2).

While 8-monohydrochlordecone was actually found as the sole monohydrochlordecone formed by CLD photolysis (Alley et al. 1974) or aerobic microbial attack by *P. aeruginosa* (Orndorff and Colwell 1980a), it was only a minor product in the reactions performed with vitamin B_{12s} and Fe° where it was accompanied by another major monohydrochlordecone isomer, 10-monohydrochlordecone in the case of vitamin B_{12s} (Katz 1978) and 9- or 10-monohydrochlordecone (absolute identification impossible with the analytical tools used) in the case of Fe° (Belghit et al. 2015). Incongruities with the thermodynamic predictions are also evident at the level of the dihydrochlordecone isomers formed by photolysis (2,8-dihydrochlordecone; Wilson and Zehr 1979) or by vitamin B_{12s} (*cis*-8,10-dihydrochlordecone; Katz 1978, Schrauzer and Katz 1978), which were not the first to be expected to be formed from 8- or 10-monohydrochlordecone (see Table 14.2). In the cases discussed earlier, the observed inconsistencies may result from the fact that the differences between the $\Delta G^{\circ\prime}$ of the reactions of formation of the different dechlorination isomers are most often between 0.4 and 7.3 kJ reaction^{-1}, which is below the accuracy (within 4–8 kJ mol^{-1}) that can be achieved in the estimation of the ΔG_f° values (Dolfing et al. 2012), which causes considerable uncertainty in their use to discriminate among pathways. The discrepancy may also come from the fact that all calculations were done using ΔG_f° of the carbonyl forms of CLD and of its dechlorinated products while all these compounds are known to convert easily into diols in presence of water and in polar solvents such as acetone and acetonitrile (Wilson and Zehr 1979) which could be therefore their main form in some of the above mentioned experiments. The diols could be even deprotonated in some of the experiments with vitamin B_{12s} performed at pH 9.6 (Schrauzer and Katz 1978). In any case, it is important to keep in mind that a thermodynamic approach based on the energy of the initial and final products does not take into account the kinetics of the reactions and the activation energy (barrier) that has to be overcome for a reaction to proceed. For identical initial and final energetic states (i.e., same $\Delta G_{reaction}$), it is indeed the reaction with the lower activation energy that will be favored over the others. The calculation of the activation energy is theoretically possible for each isomer, but it would require knowing the mechanism involved in the reactions, which is presently not the case for CLD and its derivatives.

The limits of the thermodynamic approach also appear when considering the formation of the CLD-lactone that could be the first intermediate to be formed by direct reaction of O_2 with the keto group of CLD during a potential aerobic degradation of the insecticide and which is thermodynamically favorable according to our calculations

TABLE 14.2
$\Delta G^{\circ\prime}$ of the Different Reactions of CLD Dechlorination up to the Removal of 2 Chlorine Atoms[a]

Original Compound	Monohydrochlordecone	$\Delta G^{\circ\prime}$ kJ mol⁻¹	DihydroCLD	$\Delta G^{\circ\prime}$ kJ mol⁻¹
Chlordecone	8-monohydrochlordecone*	−159.7	**4,8-dihydroCLD***	**−161.1**
	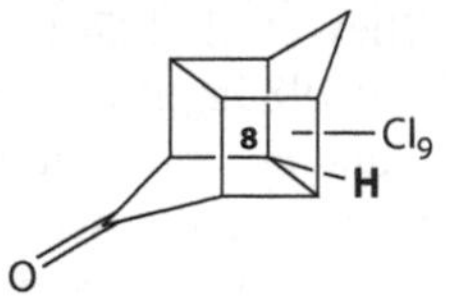		6,8-dihydroCLD*	−160.4
			2,8-dihydroCLD*	−159.5
			8,9-dihydroCLD*	−158.6
			cis-8,10-dihydroCLD*	−157.6
			trans-8,10-dihydroCLD*	−156.8
			1,8-dihydroCLD*	−155.3
			7,8-dihydroCLD	−153.7
			3,8-dihydroCLD	−142.1
	9-monohydrochlordecone	−159.3	**6,9-dihydroCLD**	**−160.2**
	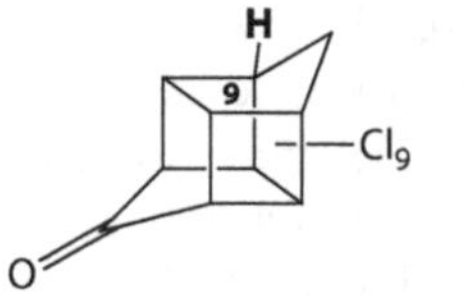		4,9-dihydroCLD	−159.3
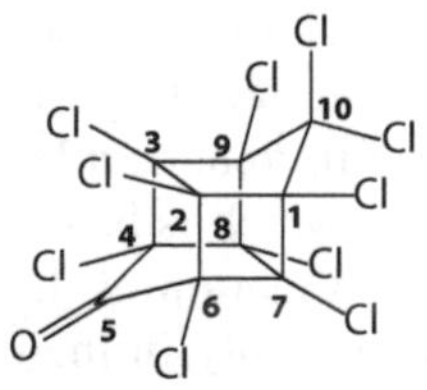			8,9-dihydroCLD*	−159.0
			9,10-dihydroCLD*	−156.3
			7,9-dihydroCLD*	−155.7
			1,9-dihydroCLD	−155.4
	10-monohydrochlordecone	−158.4	**6,10-dihydroCLD***	**−161.1**
	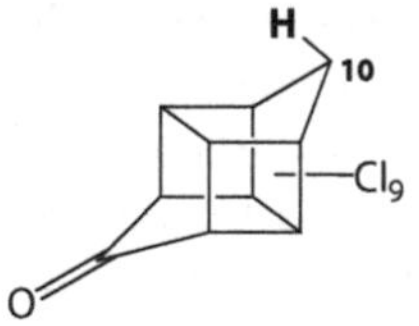		*cis*-8,10-dihydroCLD*	−158.9
			trans-8,10-dihydroCLD*	−158.1
			9,10-dihydroCLD*	−157.2
			10,10-dihydroCLD	−150.1
	6-monohydrochlordecone	−155.5	**6,7-dihydroCLD***	**−165.3**
	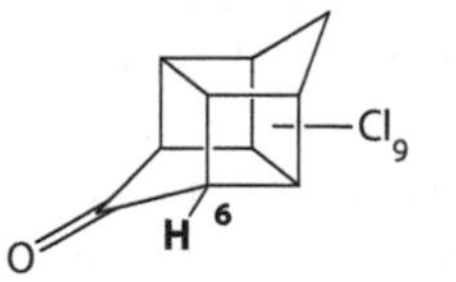		6,8-dihydroCLD*	−164.6
			6,10-dihydroCLD*	−164.0
			6,9-dihydroCLD	−164.0
			4,6-dihydroCLD	−163.9
			1,6-dihydroCLD	−163.1

Note: Compounds in bold correspond to those whose formation is thermodynamically most favorable; compounds with gray background were detected after CLD photolysis; underlined compounds were detected after reaction with B_{12s}.

[a] $\Delta G^{\circ\prime}$ values were calculated according to the reactions $C_{10}Cl_nH_{(10-n)}O + H_2 \rightarrow C_{10}Cl_{(n-1)}H_{(11-n)}O + Cl^- + H^+$. Compounds with an asterisk have an enantiomer. Carbon numbering have been selected to show direct affiliation to parent compounds and do not follow necessarily the standard IUPAC numbering rule that gives the highest number to the carbon bearing a hydrogen.

(Figure 14.1; Table 14.1). *In vivo*, the oxidation of carbonyl groups to lactones are catalyzed by flavin-dependent-Bayer-Villiger monooxygenases that are active even when the carbonyl group belongs to polycyclic structures (e.g., adamantanone) that are energetically as constrained as CLD (Selifonov 1992). Since 1976, moreover, Metha et al. have shown that a lactone could be formed from 1,4-bishomocubanone, the fully dechlorinated CLD, by reaction with the strong oxidant ceric ion. To our knowledge, such an oxidation, however, does not seem to occur when the carbons adjacent to the carbonyl group bear a chlorine atom. Indeed, the migrating ability of the alkyl group to form the lactone is decreased if an electron-withdrawing substituent such as Cl is placed to the adjacent alkyl group (Grein et al. 2006). This suggests that lactone formation from CLD will probably be possible only after the removal of at least one of the Cl atoms carried by the carbons 4 and 6 (Figure 14.1).

14.5 POSSIBLE REASONS FOR THE APPARENT ABSENCE OF CLD'S NATURAL ATTENUATION IN FWI ENVIRONMENTS

Both the thermodynamic approach used here and the old and recent experimental results discussed in the previous sections clearly show that the chemical structure of CLD should not be refractory to a microbial attack and so should not be the reason *per se* why its degradation has not been observed so far—apparently at least—in the FWI environments. Other reasons that may be advocated to explain this are the following:

- The absence of autochthonous microorganisms in FWI environments with the capacity to attack CLD, although such organisms may exist somewhere else in the world and this despite the fact that four decades have elapsed since CLD was used for the first time in the FWI, which would have allowed for a long enrichment process.
- Inadequate environmental conditions to permit the expression of the catabolic capacities of the CLD-degrading autochthonous microorganisms that could be present.
- Trapping of CLD within the clay-organic complex of FWI soils making it inaccessible to the potential degrading autochthonous microorganisms or to their enzymes, a process that could become more prominent with soil aging, making CLD less and less bioavailable with time and which is well known to reduce in soils the rate and extent of biodegradation of organic compounds even one as readily biodegradable as citrate (Chenu and Stotzky 2002).

The extremely high propensity of CLD to partition from water to organic matter and so the importance of the third point is well exemplified by the experiments of Cimetiere et al. (2014), who showed that in less than 5 min, 85% of CLD at 1 mg L^{-1} in water disappeared from solution after the addition of 0.5 g L^{-1} of inactivated sewage sludge. High experimental K_{oc}* values (580–16,522 mL g^{-1}) reported in the

* Compound soil water partition coefficient normalized to the soil organic content.

literature for CLD on clay materials (Iyengar et al. 1983) and the FWI's andosols and nitisols (Fernandez-Bayo et al. 2013b) further support such a possibility.

The recent finding by Devault et al. (2016) that 8-monohydrochlordecone was apparently naturally formed in FWI soils suggests, however, that the second point certainly warrants further scrutiny. Although thermodynamics indicate that in principle CLD should be mineralizable under a wide range of redox conditions, it is now well established that polyhalogenated compounds will be much more readily attacked anaerobically rather than aerobically, while the contrary is observed for less halogenated compounds (e.g., Vogel et al. 1987, Field et al. 1995, Holliger et al. 2003, Löffler et al. 2003). This is probably due to the oxidized nature of the chloro substituents, which hinder their electrophilic attack by oxygenases (Löffler et al. 2003). Under anaerobic conditions, the main mechanism involved in organochlorine attack corresponds to reductive dechlorination, which results in the replacement of a chlorine atom by hydrogen. As mentioned previously, in this scheme, organochlorine compounds play the role of electron acceptors instead of electron donors. This means also that their dechlorination will require the presence of electron donors, usually molecular hydrogen, generated during fermentative processes of readily biodegradable organic compounds performed by nondehalogenating microorganisms. According to the redox scale presented in Figure 14.2, O_2 ($E^{\circ\prime}$ O_2/H_2O = +815 mV) is a so strong electron acceptor that under aerobiosis, the reducing equivalents will be used so efficiently for O_2 reduction by aerobes that they will be unavailable for a reductive dechlorination process.

In FWI volcanic islands, banana crop and its associated CLD spreading was performed on three main types of volcanic soils—andosol, nitisol, and ferralsol—all known to be particularly well draining (Dorel et al. 2000, Vidal-Torrado and Cooper 2008) and therefore to be well aerated. This characteristic is appropriate for banana trees, whose roots do not develop in poorly aerated conditions (Aguilar et al. 2003) but will necessarily prevent the occurrence of CLD reductive dechlorination according to the previous comments. Such a restriction should be even stronger in the case of andosols, whose organic matter is known to be extremely poorly biodegradable (e.g., Chevallier et al. 2010) and so probably unable to play efficiently the role of electron donor required by a reductive dechlorination process. Therefore, it is not surprising that a Guadeloupean andosol incubated up to seven months under aerobic conditions showed an extremely low capacity to mineralize or even just to partially transform freshly spiked ^{14}C-CLD (Fernandez-Bayo et al. 2013a).

A similar situation is expected to occur in the ground waters of Martinique that have been reported to contain in average 2.7 ± 0.9 mg dissolved O_2 L^{-1},which is indicative of oxic conditions and an amount of organic carbon (1.4 ± 0.6 mg L^{-1}) probably too low to be an adequate source of electron donors for the removal of chloro substituents (values from Gourcy et al. 2009, Gourcy L. 2011, personal communication).

Adequate conditions for CLD reductive dechlorination may exist, however, in FWI ecosystems dominated by electron acceptors weaker than CLD like sulfates ($E^{\circ\prime}$ SO_4^{2-}/H_2S = −210 mV) or CO_2 ($E^{\circ\prime}$ CO_2/CH_4 = −245 mV; methanogenic conditions). Such ecosystems correspond to the sediments of coastal marshes, mangroves, lake, and seashore bottoms that can be the receptacle of CLD from banana plantations and where the flooding regime favors the development of anaerobic conditions as confirmed by very negative measured redox potentials (e.g., Imbert and Delbé 2006). It is well known also that anoxic microniches may form in the interior of soil aggregates, allowing the functioning of anaerobic processes in otherwise macroscopically oxic soils (e.g., Sexstone et al. 1985). N_2O emissions from andosols and nitisols under forest cover or used for the cultivation of bananas in Costa Rica and Panama, with climatic conditions similar to FWI's, and that were clearly mediated by denitrification (an anaerobic process) provide circumstantial evidence that such anoxic microniches may occur in FWI soils in absence of flooding (Veldkamp and Keller 1997, Corre et al. 2014).* Therefore, anoxic microniches could well be at the origin of the increased 8-monohydrochlordecone/CLD ratios observed in the soils from Martinique compared to the ratios found in CLD commercial formulations spread on these soils (Devault et al. 2016).

14.6 CONCLUDING REMARKS AND PERSPECTIVES

The information discussed in this chapter shows that in principle there should be no thermodynamic impediment to a wide range of biologically mediated chemical transformations of CLD and that nature has already evolved some of the essential building blocks (e.g., corrinoids, other transition metal complexes) that are necessary for microorganisms to perform at least its dechlorination. In order to achieve an extended biological transformation of CLD structure, four initial conditions, however, seem to be required:

1. Anaerobic conditions
2. Presence of an electron donor and of an additional carbon source (they can be the same)
3. Presence of microorganisms with CLD-dehalogenating activity
4. Bioavailability of CLD

Due to the aforesaid characteristics of the FWI soils impacted by CLD, direct human action will be at least necessary to fulfill the first two conditions.

In order to obtain anoxic conditions, two strategies may be followed. The first one consists of reducing the oxygen diffusion from the atmosphere to the soil. This should be achievable by soil compaction which reduces the soil porosity at the surface and therefore decreases physically the capacity of atmospheric gases to freely move into the soil structure. Oxygen transfer from the atmosphere to the soil could be further reduced by increasing the soil humidity through watering—if possible at least until field capacity—in order to fill the soil pores with water and take advantage of the low solubility of oxygen into water. The second strategy, which may be combined with

* In the work by Corre et al. (2014) denitrification was proven by monitoring of $^{15}N_2O$ emission upon addition of $^{15}NO_3^-$ or $^{15}NH_4^+$ to the soil. In the work by Veldkamp and Keller (1997), denitrification was likely because N_2O peaks occurred at WFPS (Water Filled Pore Space) ≥70%, a value widely accepted as the threshold between nitrification/denitrification as the dominant source of N_2O. *Note*: Andosol and nitisol appear as *andisol* and *inceptisol*, respectively, in the USDA soil taxonomy used by these last authors.

the first one, consists of increasing the oxygen consumption by soil microorganisms above the soil aeration capacity. This can be achieved through the addition of easily biodegradable organic matter that could be chosen among organic residues available on the islands and whose characteristics are compatible with the soil needs.* Particularly, the selected organic matter should allow maintaining an adequate equilibrium between the main nutrients C and N in order to avoid any microbial disequilibrium that would prevent a possible CLD biodegradation. The organic matter addition will also provide the electron donors required for a dechlorination process. Data from the literature and practical experience confirm that both approaches can indeed result in the formation of anoxic conditions, sometimes clearly evidenced by a substantial decrease of the soil redox potential (e.g., Dorel 1993, Stepniewski et al. 1994, de Cockborne et al. 1999). In terms of CLD bioavailability, a test of the ISCR (*in situ* chemical reduction) process performed by BRGM (French geological survey office) consisting of adding a mixture of micrometric Fe° and plant organic matter (crushed dried alfalfa) to the soil to promote dechlorination has shown at the lab scale that 90%–95% of the CLD present in historically polluted FWI ferralsols and nitisols could be removed by such a process (Dictor et al. 2011). This indicates that despite its high K_{oc} and so its tight binding to the soil matrix, most of the CLD of these two types of soils was accessible to the added chemical agents, which should also be the case for biological agents. The efficiency of the ISCR process was much less when applied to andosols (44% removal; Dictor et al. 2011) revealing a lower bioavailability of CLD in this type of soil compared to the others. The CLD removed by the ISCR process could correspond, however, to the CLD fraction that is mobile in such a soil and whose removal could be sufficient to avoid the contamination of crops and percolating waters but probably not of the livestock since the digestive tracts of birds and mammals seem to be very efficient in extracting CLD from soil particles independently of the soil type (see chapter 11).

Although anaerobic conditions are nowadays widely recognized as almost indispensable for achieving the primary attack on polychlorinated compounds, it is also recognized that their full mineralization will usually not occur under such conditions and that their degradation will result in the accumulation of less chlorinated compounds (e.g., Mohn and Tiedje 1992, Field et al. 1995) that could be more toxic than the parent compound as observed in some cases (e.g., vinyl chloride resulting from tetra- and trichloroethylene dechlorination; Rosner et al. 1997). Such a risk does not appear, however, to be significant in the case of CLD. Indeed, the only two comparative toxicity studies that exist in the literature and that were performed with the few CLD congeners existing as pure compounds have shown that CLD alcohol toxicity ≥ CLD > 8-monohydrochlordecone ⋙ 2,8-dihydrochlordecone for rat liver mitochondria (Soileau and Moreland 1983) and mysid shrimps (Carver and Griffith 1979). This suggests that the toxicity of CLD congeners for biological targets decreases with decreasing degree of chlorination. Another good news is that usually

* For example, sugarcane vinasse for a quick effect combined with biosolids from wastewater treatment plants to maintain anoxic conditions for a longer period but not sugarcane bagasse that is too bulky and would promote soil aeration or compost which is a stabilized organic matter and so necessarily a poor electron donor.

the partially dechlorinated compounds formed under anaerobic conditions have been observed to be more accessible to aerobic microorganisms than the original parent compounds and to degrade with fast kinetics aerobically (Guiot et al. 1994, Field et al. 1995). As suggested by Orndorff and Colwell (1980a), a succession of anaerobic and aerobic conditions could thus be optimal to achieve the ultimate mineralization of CLD as this has been observed for other polychlorinated pesticides such as DDT (Beunink and Rehm 1988), methoxychlor (Fogel et al. 1982), or 2,3,6-trichlorobenzoic acid (Gerritse and Gottschal 1992). The establishment of aerobic conditions after an anaerobic phase could be obtained by a tilling of the soil. This would also allow decompacting the soil when compaction was used to create anoxic conditions since this practice could otherwise negatively impact future crops due to possible root anoxia and difficulty for the roots to colonize a compacted environment (e.g., Dorel 1993). The present management of banana plantations in the FWI involves destroying the banana plants every 4–5 years by direct injection of glyphosate at the base of the pseudostem followed by a one-year fallow and subsequent plantation of pest-free banana plants produced by tissue culture (Chabrier and Quénéhervé 2003). Such a management results in a more rational harvest since all banana plants will be productive at the same time and allows a good control of nematodes and also indirectly of the black weevil populations that could otherwise cause a significant banana yield reduction. In terms of soil remediation, the enforcement of the one-year fallow period for those polluted lands that are still used for banana production is excellent since it would allow sufficient time to perform all the operations that could be necessary (e.g., soil compaction, watering, amendment with labile organic matter and source of microorganisms, re-aeration by tilling) to achieve CLD destruction without the need to stop the agronomic use of otherwise productive parcels with the associated economic loss. Hopefully, the length of the fallow would be long enough also to allow the soil to rest for sufficient time after the remediation activities in order to be ready for the next 5 years' banana crop cycle.

ACKNOWLEDGMENTS

Part of the work presented in this chapter was performed in the frame of the ABACHLOR project financed by the DEMICHLORD program implemented by INRA with funds of the CLD National Action Plan (PNAC) set up by the French Government. It was also financially supported through an IRD incentive budget and by the European Union (FEDER Martinique 2007–2013). Half of the $\Delta G_f°$ were calculated with the computing facilities of the CRCMM, "Centre Régional de Compétences en Modélisation Moléculaire" from Aix Marseille University made available by Prof. Didier Siri. HM is grateful to the editors for their invitation to contribute to the present book.

DEDICATION

This chapter is dedicated to the memory of Dr. Gerhard N. Schrauzer, who passed away in September 2014. His pioneering work on CLD degradation by vitamin B_{12s} remains an inspiration for those currently working on this subject.

REFERENCES

Alley, E. G., B. R. Layton, and J. P. Minyard. 1974. Identification of photoproducts of insecticides mirex and kepone. *Journal of Agricultural and Food Chemistry* 22:442–445.

Aguilar, E. A., D. W. Turner, D. J. Gibbs, W. Armstrong, and K. Sivasithamparam. 2003. Oxygen distribution and movement, respiration and nutrient loading in banana roots (*Musa* spp. L.) subjected to aerated and oxygen-depleted environments. *Plant and Soil* 253:91–102.

Belghit, H., C. Colas, S. Bristeau, C. Mouvet, and B. Maunit. 2015. Liquid chromatography–high-resolution mass spectrometry for identifying aqueous chlordecone hydrate dechlorinated transformation products formed by reaction with zero-valent iron. *International Journal of Environmental Analytical Chemistry* 95:93–105.

Beunink, J. and H. J. Rehm. 1988. Synchronous anaerobic and aerobic degradation of DDT by an immobilized mixed culture system. *Applied Microbiology and Biotechnology* 29:72–80.

Cabidoche, Y-.M., R. Achard, P. Cattan, C. Clermont-Dauphin, F. Massat, and L. Sansoulet. 2009. Long-term pollution by chlordecone of tropical volcanic soils in the French West Indies: A simple leaching model accounts for current residue. *Environmental Pollution* 157:1697–1705.

Carver, R. A. and F. D. Griffith. 1979. Determination of kepone dechlorination products in finfish, oysters, and crustaceans. *Journal of Agricultural and Food Chemistry* 27:1035–1037.

Chabrier, C. and P. Queneherve. 2003. Control of the burrowing nematode (*Radopholus similis* Cobb) on banana: Impact of the banana field destruction method on the efficiency of the following fallow. *Crop Protection* 22:121–127.

Chenu, C. and G. Stotzky. 2002. Interactions between microorganisms and soil particles: An overview. In *Interactions between Soil Particles and Microorganisms: Impact on the Terrestrial Ecosystem*, P. M. Huang, J.-M. Bollag, and N. Senesi (eds.). Chichester, U.K.: John Wiley & Sons, Ltd., pp. 3–40.

Chevallier, T., T. Woignier, J. Toucet, and E. Blanchart. 2010. Organic carbon stabilization in the fractal pore structure of andosols. *Geoderma* 159:182–188.

Cimetiere, N., S. Giraudet, M. Papazoglou, H. Fallou, A. Amrane, and P. Le Cloirec. 2014. Analysis of chlordecone by LC/MS–MS in surface and wastewaters. *Journal of Environmental Chemical Engineering* 2:849–856.

Conrad, R. 1996. Soil microorganisms as controllers of atmospheric trace gases (H_2, CO, CH_4, OCS, N_2O and NO). *Microbiological Reviews* 60:609–640.

Corre, M. D., J. P. Sueta, and E. Veldkamp. 2014. Nitrogen-oxide emissions from tropical forest soils exposed to elevated nitrogen input strongly interact with rainfall quantity and seasonality. *Biogeochemistry* 118:103–120.

de Cockborne, A. M., V. Vallès, L. Bruckler, G. Sevenier, B. Cabibel, P. Bertuzzi, and V. Bouisson. 1999. Environmental consequences of apple waste deposition on soil. *Journal of Environmental Quality* 28:1031–1037.

Devault D. A., C. Laplanche, H. Pascaline, S. Bristeau, C. Mouvet, and H. Macarie. 2016. Natural transformation of chlordecone into 5b-hydrochlordecone in French West Indies soils: Statistical evidence for investigating long-term persistence of organic pollutants. *Environmental Science and Pollution Research* 23:81–97.

Dictor, M. C., A. Mercier, L. Lereau, L. Amalric, S. Bristeau, and C. Mouvet. 2011. Decontamination of soils polluted by CLD. Validation of physico-chemical and biological decontamination processes, study of the degradation products and improvement of analytical sensitivity for CLD in soils. Final report BRGM/RP-59481-FR, BRGM, Orléans, France, 201pp. http://infoterre.brgm.fr/rapports/RP-59481-FR.pdf (in French). Accessed 24, May 2016.

Dolfing, J. 2003. Thermodynamic considerations for dehalogenation. In *Dehalogenation: Microbial Processes and Environmental Applications*, M. M. Häggblom and I. D. Bossert (eds.). Boston, MA: Kluwer Academic Publishers, pp. 89–114.

Dolfing, J. and D. B. Janssen. 1994. Estimates of Gibbs free energies of formation of chlorinated aliphatic compounds. *Biodegradation* 5:21–28.

Dolfing, J. and I. Novak. 2015. The Gibbs free energy of formation of halogenated benzenes, benzoates and phenols and their potential role as electron acceptors in anaerobic environments. *Biodegradation* 26:15–27.

Dolfing, J., I. Novak, A. Archelas, and H. Macarie. 2012. Gibbs free energy of formation of chlordecone and potential degradation products: Implication for remediation strategies and environmental fate. *Environmental Science and Technology* 46:8131–8139.

Dorel, M. 1993. Growing banana on an andosol in Guadeloupe: Effect of soil compaction. *Fruits* 48:83–88 (in French).

Dorel, M., J. Roger-Estrade, H. Manichon, and B. Delvaux. 2000. Porosity and soil water properties of Caribbean volcanic ash soils. *Soil Use and Management* 16:133–140.

Fernandez-Bayo, J. D., C. Saison, C. Geniez, M. Voltz, H. Vereecken, and A. E. Berns. 2013b. Sorption characteristics of chlordecone and cadusafos in tropical agricultural soils. *Current Organic Chemistry* 17:2976–2984.

Fernández-Bayo, J. D., C. Saison, M. Voltz, U. Disko, D. Hofmann, and A. E. Berns. 2013a. Chlordecone fate and mineralisation in a tropical soil (andosol) microcosm under aerobic conditions. *Science of the Total Environment* 463–464:395–403.

Field, J. A., A. J. Stams, M. Kato, and G. Schraa. 1995. Enhanced biodegradation of aromatic pollutants in cocultures of anaerobic and aerobic bacterial consortia. *Antonie van Leeuwenhoek* 67:47–77.

Fogel, S., R. L. Lancione, and A. E. Sewall. 1982. Enhanced biodegradation of methoxyclor in soil under sequential environmental conditions. *Applied and Environmental Microbiology* 44:113–120.

Gambrell, R. P., C. N. Reddy, V. Collard, G. Green, and W. H. Patrick. 1984. The recovery of DDT, kepone and permethrin added to soil and sediment suspensions incubated under controlled redox potential and pH conditions. *Journal of Water Pollution Control Federation* 56:174–182.

Gerritse, J. and J. C. Gottschal. 1992. Mineralization of the herbicide 2,3,6-trichlorobenzoic acid by a coculture of anaerobic and aerobic bacteria. *FEMS Microbiology Ecology* 101:89–98.

Gourcy, L., N. Baran, and B. Vittecoq. 2009. Improving the knowledge of pesticide transfer processes using age-dating Tools (CFC, SF_6, ^3H) in a volcanic Island (Martinique, French West Indies). *Journal of Contaminant Hydrology* 108:107–117.

Grein, F., A. C. Chen, D. Edwards, and C. M. Crudden. 2006. Theoretical and experimental studies on the Baeyer-Villiger oxidation of ketones and the effect of α-halo substituents. *The Journal of Organic Chemistry* 71:861–872.

Guiot, S. R., H. Macarie, J. C. Frigon, and M. F. Manuel. 1994. Aerobic and anaerobic synchronous treatment of PCP-contaminated wastewater. In *Proceedings 23rd Annual Technical Symposium of the Water Environment Association of Ontario*, Windsor, Ontario, Canada, April 17–19, 1994, pp. 29–38.

Heimann, A., R. Jakobsen, and C. Blodau. 2009. Energetic constraints on H_2-dependent terminal electron accepting processes in anoxic environments: A review of observations and model approaches. *Environmental Science and Technology* 44:24–33.

Holliger, C., C. Regeard, and G. Diekert. 2003. Dehalogenation by anaerobic bacteria. In *Dehalogenation: Microbial Processes and Environmental Applications*, M. M. Häggblom and I. D. Bossert (eds.). Boston, MA: Kluwer Academic Publishers, pp. 115–157.

Holmes, D. A., B. K. Harrison, and J. Dolfing. 1993. Estimation of Gibbs free energies of formation for polychlorinated biphenyls. *Environmental Science and Technology* 27:725–731.

Huang, C. L., B. K. Harrison, J. Madura, and J. Dolfing. 1996. Gibbs free energies of formation of PCDDs: Evaluation of estimation methods and application for predicting dehalogenation pathways. *Environmental Toxicology and Chemistry* 15:824–836.

Hugget, R. J. 1989. Kepone and the James River. In *Contaminated Marine Sediments: Assessments and Remediation*. Washington, DC: National Academic Press, pp. 417–424.

Imbert, D. and L. Delbé. 2006. Ecology of fire-influenced *Cladium jamaicense* marshes in Guadeloupe, Lesser Antilles. *Wetlands* 26:289–297.

IUPAC. 2016. Global availability of information on agrochemicals, chlordecone, ref ENT 16391, University of Hertfordshire. http://sitem.herts.ac.uk/aeru/iupac/Reports/1293.htm. Accessed June 2016.

Iyengar, S. S., M. D. Treblow, and J. C. Wright. 1983. Attenuation of chlorocarbon compounds by clay liner materials of a waste disposal facility. In *Hazardous and Industrial Solid Waste Testing Second Symposium*, ASTM STP 805, R. A. Conway and W. P. Gulledge (eds.). Conshohocken, PA: American Society for Testing and Materials, pp. 265–282.

Jablonski, P. E., D. J. Pheasant, and J. G. Ferry. 1996. Conversion of kepone by *Methanosarcina thermophila. FEMS Microbiology Letters* 139:169–173.

Katz, R. N. 1978. Studies in cobalamin chemistry: The mechanism of enzymatic diol dehydration and the non-enzymatic dechlorination of kepone and mirex. PhD dissertation, University of California, San Diego, CA.

Le Déaut, J.-Y. and C. Procaccia. 2009. Pesticide use in the Antilles: Current situation and perspectives for change. OPECST report n° 487 (2008–2009). French Senat, Paris, France: OPECST, 223pp. ISBN: 9782111267688. http://www.senat.fr/rap/r08-487/r08-4871.pdf (in French). Accessed 24, May 2016.

Löffler, F. E., J. R. Cole, K. M. Ritalahti, and J. M. Tiedje. 2003. Diversity of dechlorinating bacteria. In *Dehalogenation: Microbial Processes and Environmental Applications*, M. M. Häggblom and I. D. Bossert (eds.). Boston, MA: Kluwer Academic Publishers, pp. 53–87.

Lovley, D. R. and S. Goodwin. 1988. Hydrogen concentrations as an indicator of the predominant terminal electron accepting reactions in aquatic sediments. *Geochimica et Cosmochimica* 52:2993–3003.

Maphosa, F., W. M. de Vos, and H. Smidt. 2010. Exploiting the ecogenomics toolbox for environmental diagnostics of organohalide-respiring bacteria. *Trends in Biotechnology* 28:308–316.

Merlin, C., M. Devers, O. Crouzet, C. Heraud, C. Steinberg, C. Mougin, and F. Martin-Laurent. 2014. Characterization of chlordecone-tolerant fungal populations isolated from long-term polluted tropical volcanic soil in the French West Indies. *Environmental Science and Pollution Research* 21:4914–4927.

Metha, G., P. N. Pandey, and T.-L. Ho. 1976. Regiospecific Baeyer-Villiger oxidation of polycyclic ketones with ceric ion. *The Journal of Organic Chemistry* 41:953–956.

Mohn, W. W. and J. M. Tiedje. 1992. Microbial reductive dehalogenation. *Microbiology Reviews* 56:482–507.

Molowa, D. T., A. G. Shayne, and P. S. Guzelian. 1986. Purification and characterization of chlordecone reductase from human liver. *Journal of Biological Chemistry* 261:12624–12627.

Orndorff, S. A. and R. R. Colwell. 1980a. Microbial transformation of kepone. *Applied and Environmental Microbiology* 39:398–406.

Orndorff, S. A. and R. R. Colwell. 1980b. Distribution and characterization of kepone-resistant bacteria in the aquatic environment. *Applied and Environmental Microbiology* 39:611–622.

Portier, R. J. and S. P. Meyers. 1982. Monitoring biotransformation and biodegradation of xenobiotics in simulated aquatic microenvironmental systems. *Developments in Industrial Microbiology* 23:459–475.

Rosner, B. M., P. L. McCarty, and A. M. Spormann. 1997. In vitro studies on reductive vinyl chloride dehalogenation by an anaerobic mixed culture. *Applied and Environmental Microbiology* 63:4139–4144.

Schink, B. 1997. Energetics of syntrophic cooperation in methanogenic degradation. *Microbiology and Molecular Biology Reviews* 61:262–280.

Schrauzer, G. N. and R. N. Katz. 1978. Reductive dechlorination and degradation of mirex and kepone with vitamin B_{12s}. *Bioinorganic Chemistry* 9:123–142.

Selifonov, S. A. 1992. Microbial oxidation of adamantanone by *Pseudomonas putida* carrying the camphor catabolic plasmid. *Biochemical and Biophysical Research Communications* 186:1429–1436.

Sexstone, A. J., N. P. Revsbech, T. B. Parkin, and J. M. Tiedje. 1985. Direct measurement of oxygen profiles and denitrification rates in soil aggregates. *Soil Science Society of America Journal* 49:645–651.

Shock, E. L. 1995. Organic acids in hydrothermal solutions: Standard molal thermodynamic properties of carboxylic acids and estimates of dissociation constants at high temperatures and pressures. *American Journal of Science* 295:496–580.

Skaar, D. R., B. T. Johnson, J. R. Jones, and J. N. Huckins. 1981. Fate of kepone and mirex in a model aquatic environment: Sediment, fish, and diet. *Canadian Journal of Fisheries and Aquatic Sciences* 38:931–938.

Soileau, S. D. and D. E. Moreland. 1983. Effects of chlordecone and its alteration products on isolated rat liver mitochondria. *Toxicology and Applied Pharmacology* 67:89–99.

Soine, W. H., T. R. Forrest, and J. D. Smith. 1983. Thermal reduction of chlordecone in the presence of alcohol. *Journal of Chromatography A* 281:95–99.

Stepniewski, W., B. C. Ball, B. D. Soane, and C. V. Ouwerkerk. 1994. Effects of compaction on soil aeration properties. In *Soil Compaction in Crop Production*, B. D. Soane and C. van Ouwerkerk (eds.). Amsterdam, the Netherlands: Elsevier Science B.V., pp. 167–189.

Stumm, W. and J. J. Morgan. 1996. *Aquatic Chemistry*, 3rd edn. New York: John Wiley & Sons.

Thauer, R. K., K. Jungermann, and K. Decker. 1977. Energy conservation in chemotrophic anaerobic bacteria. *Bacteriology Reviews* 41:100–180.

Veldkamp, E. and M. Keller. 1997. Nitrogen oxide emissions from a banana plantation in the humid tropics. *Journal of Geophysical Research* 102:15889–15898.

Vidal-Torrado, P. and M. Cooper. 2008. Ferralsols. In *Encyclopedia of Soil Science*, W. Chesworth (ed.). Dordrecht, the Netherlands: Springer, pp. 237–240.

Vogel, T. M., C. S. Criddle, and P. L. McCarty. 1987. ES&T critical reviews: Transformations of halogenated aliphatic compounds. *Environmental Science and Technology* 21:722–736.

Wilson, N. K. and R. D. Zehr. 1979. Structures of some kepone photoproducts and related chlorinated pentacyclodecanes by carbon-13 and proton nuclear magnetic resonance. *The Journal of Organic Chemistry* 44:1278–1282.

15 Reduced Pesticide Bioavailability in Soil by Organic Amendment

Thierry Woignier, Florence Clostre, Paula Fernandes, Alain Soler, Luc Rangon, and Magalie Lesueur Jannoyer

CONTENTS

15.1 INTRODUCTION

The fate and behavior of organic pollutants in soils are determined by many different factors, including soil characteristics and chemical properties (Mottes et al. 2013). If the bioavailable fraction of pesticide is reduced, so are the risks of crop contamination and human and ecosystem exposure. A key to managing contaminated sites could thus be reducing pesticide bioavailability in the soil. Bioavailability is related to the intrinsic physical-chemical properties of pollutants. Water solubility is one key parameter along with the formation of bound residues, which is more extensive in case of hydrophobic pollutants (Semple et al. 2001). The porous structure of the matrix with which pollutants are associated may also influence the bioavailability of pollutants (Chung and Alexander 2002, Peters et al. 2007, Woignier et al. 2012). As soil organic matter plays the most important role in the sorption and bioavailability of persistent organic pollutants (Pignatello 1998, Vlčková and Hofman 2012), the use of compost may reduce the risk of contamination by reducing the bioavailable fraction through increased adsorption.

Chlordecone, a persistent organic pollutant, undergoes diffuse pollution in agricultural soils, which, in turn, become new sources of contamination for cultivated crops and ecosystems (Cabidoche et al. 2009, Woignier et al. 2012). Exposure to chlordecone through food has been linked to different effects on health (see Chapter 4 for more information), including impaired child development (Boucher et al. 2013). Chlordecone pollution affects large areas, and chlordecone-polluted sites require efficient *in situ* treatments. As no such treatment is currently available, we tested the incorporation of compost in soils as a possible way to reduce contamination of crops and water with the overall aim of preventing consumer exposure (Fernandes et al. 2010, Woignier et al. 2013). Exogenous materials (peat manure, sewage sludge, compost, etc.) have already been added to soil not only to improve soil fertility but also to increase the adsorption, retention, or degradation of organic pollutants (Fernandes et al. 2006, Hernández-Soriano et al. 2007, Cabrera et al. 2011, Marchal et al. 2013). In this chapter we have gathered data on soil to crop transfers of chlordecone for different kinds of crops (roots, cucurbit, leaves, etc.) and for two kinds of volcanic soils (andosol and nitisol) composed with two kinds of clays (allophane and halloysite respectively). This chapter has two objectives: (1) to show that compost addition significantly reduces chlordecone transfer to crops and (2) to report the results of our investigation of the role of pore changes in soil microstructure caused by the addition of compost in trapping pesticides in the soil. Indeed, the pore structure of soils has come to be recognized as a critical factor in the sorption of pesticides. Sorption delays diffusion of solutes and chemicals and has an influence on the transport of organic compounds (Pignatello 1998).

15.2 SOIL TO CROP TRANSFER

We compared soil to plant transfers of chlordecone with and without compost incorporation in two types of soil (andosol, nitisol) and for different crops. We used historically contaminated field soils to account for the aging process. The chosen compost (Vegethumus® from Phalippou-Frayssinet) was a blend of sheepfold manure, fruit pulp, and cake (olive, cacao, coffee, and sunflower), wool stuffing, and magnesium, with 47% organic content (Fernandes et al. 2010). Transfers were calculated as the ratio of chlordecone content in the plant sample (μg kg^{-1} fresh weight) to soil chlordecone content (μg kg^{-1} dry soil) and normalized to transfer in control soil.

Figure 15.1a through c shows that the incorporation of compost reduced contamination of the edible part for lettuce, radish, and cucumber in andosol and nitisol under controlled conditions (Woignier et al. 2013).

To confirm this difference in sequestration, we compared the transfer of chlordecone from the soil to radish and cucumber in field conditions in a nitisol plot (Clostre et al. 2014). For radish, we tested an additional rate of amendment: 1% w/w of compost, and we analyzed fine root, tuber, and leaves, separately.

In field conditions, organic amendment significantly reduced transfer to plant organs for both crops (Figure 15.2a, b), which validates the results of the pot experiments. Moreover, the radish experiment brings additional results regarding the effect of compost addition according to its application rate: the higher the rate of compost, the higher the reduction of transfer. A 1% amendment doesn't increase sufficiently the natural sequestration to have a significant effect, while a 5% amendment does. These crop experiments confirmed the efficiency of compost in reducing the bioavailability of chlordecone for a range of crop. This is in accordance with the results of leaching experiments (Woignier et al. 2012, Clostre et al. 2014), which showed that compost addition to soil reduces contamination in the water after percolation through chlordecone-polluted soil columns. Hence, in soil amended with

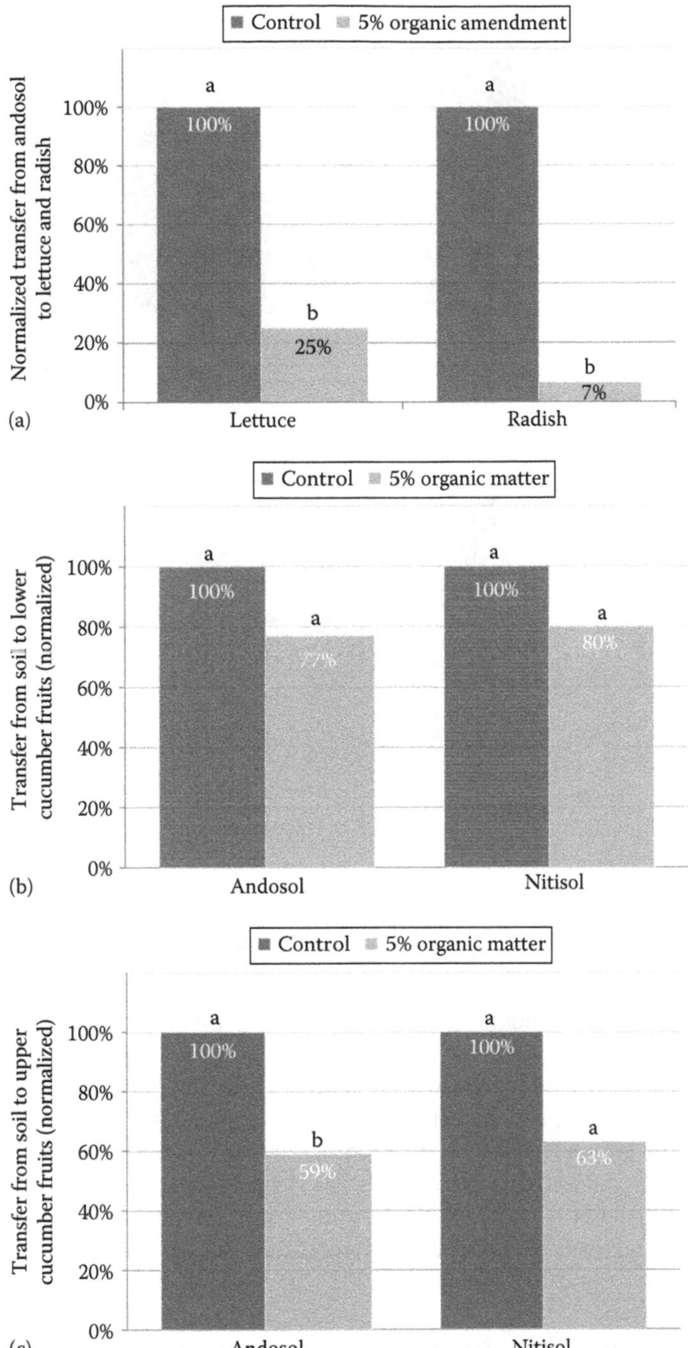

FIGURE 15.1 Effect of organic amendment on normalized chlordecone transfer from soil to lettuce leaves, radish tuber (a), and cucumber lower and upper fruits (b and c, pot experiment). Different letters for the same crop indicate significant differences ($p < 0.05$).

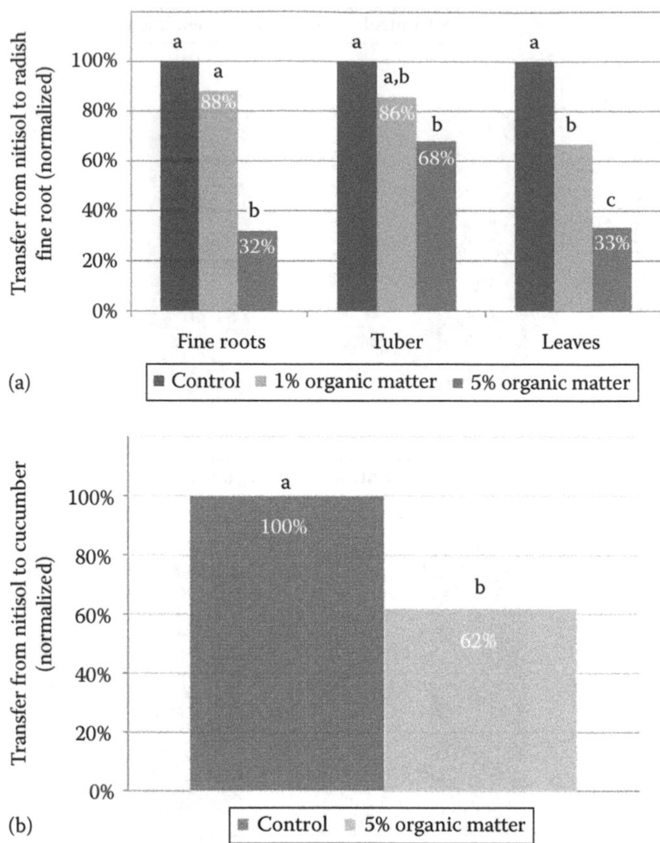

FIGURE 15.2 Effect of organic amendment on normalized chlordecone transfer from soil to (a) radish (fine roots, tubers, and leaves) and (b) cucumber fruits (field experiments). Different letters for the same crop indicate significant differences ($p < 0.05$).

compost, soil solution should be less contaminated than without, and the risk of transfer to water resources and to crops should be reduced as well.

In the literature, similar results were obtained after applying exogenous organic matter to soil. After organic matter addition, pesticide sorption to soil was increased for a range of pesticides such as propachlor, several triazine, 2,4-D, and fluometuron (Cox et al. 2000, Albarran et al. 2004, Delgado-Moreno et al. 2007, Cabrera et al. 2011, Zhang et al. 2012). Several studies showed that this reduced mobility of pesticides could lead to a reduced potential for leaching (Fernandes et al. 2006, Majumdar and Singh 2007, Zhang et al. 2012) and crop uptake (Murano et al. 2009, Saito et al. 2011).

In a larger nitisol field experiment (unpublished data), we tested the effect of organic matter addition on plant transfer for three crops—radish, cucumber, and sweet potato—to investigate the effect of compost addition over a longer time. Results showed a lesser plant uptake for all crops with organic amendment compared to control soil: Uptake was divided respectively by 4, 3, and 2 for radish, cucumber, and sweet potato, 6 months after organic matter addition.

15.3 CHANGE IN SOIL POROSITY DURING ORGANIC MATTER INCUBATION

Organic carbon content and also its composition influence soil sorption capacity (Cabrera et al. 2011, Vlčková and Hofman 2012). Reports in the literature agree on the preferential role of humin and especially on the sorption of hydrophobic contaminants (Hernández-Soriano et al. 2007, Vlčková and Hofman 2012). However, the porous structure of the matrix with which the pollutant is associated may also influence the bioavailability of pollutants (Alexander 2000, Chung and Alexander 2002, Woignier et al. 2012). Few studies have focused on the impact of compost addition on soil porosity, and precise knowledge of changes in the microstructural features resulting from the incorporation of compost is lacking. Moreover, soil microstructure is rarely taken into account in studies on organic carbon soil content and pesticide uptake by plants.

Porosity exists at all levels in the soil column: micropores between particles, mesopores within aggregates, and macropores between aggregates. Most of the soil surface area exists within the micropores and smaller mesopores. One would therefore expect the pore structure to have an influence on the mechanism of pesticide diffusion and sorption (Pignatello 1998). Accordingly, any change in the soil pore volume and especially in the microstructure and mesopore structure should affect the diffusion and the desorption process that limit pesticide bioavailability. We hypothesized that the reduction in water content observed after the addition of compost in the andosol also affected pore features, including specific surface area and the mesopore volume. The specific surface area (S) was measured by N_2 adsorption techniques (Brunauer–Emmett–Teller analysis). The pore surface distribution and mesopore volume were calculated from desorption isotherms using the Barrett–Joyner–Halenda model (Chevallier et al. 2008).

Figure 15.3 shows a clear decrease in the specific surface area and specific mesopore volume with time after compost addition. Thirty days after compost addition, most of the soil mesopore volume and specific surface area was reduced respectively by 40% and 60%, and after 90 days, the loss of the mesopore volume and specific surface area ranged between 60% and 80%.

However, changes in pore volume and specific surface area give only a macroscopic description of changes in the pores. Figure 15.4 compares the pore volume distribution of the two samples (control and amended soil) 90 days after compost addition. The pore volume distribution of the control soil ranged from 5 to 60 nm. These results indicate a hierarchical aggregation, where larger and larger clay aggregates lead to wide pore-size distribution.

These observations confirmed the hierarchical organization of the allophane aggregates (Chevallier et al. 2008). Ninety days after compost addition, mesoporosity was strongly reduced, indicating a loss of pores in the size range 10–80 nm, 40%–60% loss of mesopore surface area and volume.

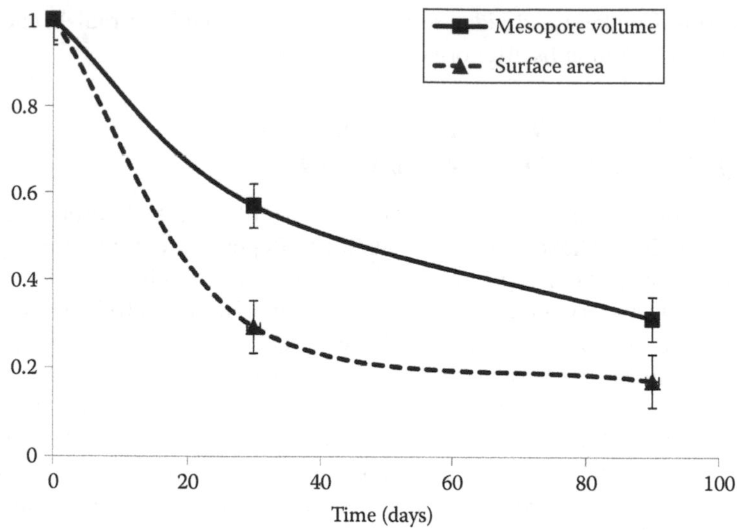

FIGURE 15.3 Changes in the specific surface area (S/S_0), with mesopore volume (V/V_0) as a function of the length of incubation. $S_0 = 61 \pm 5$ m² g⁻¹ is the specific surface area and $Vp_0 = 0.1 \pm 0.01$ cm³ g⁻¹ is the mesopore volume of the control sample.

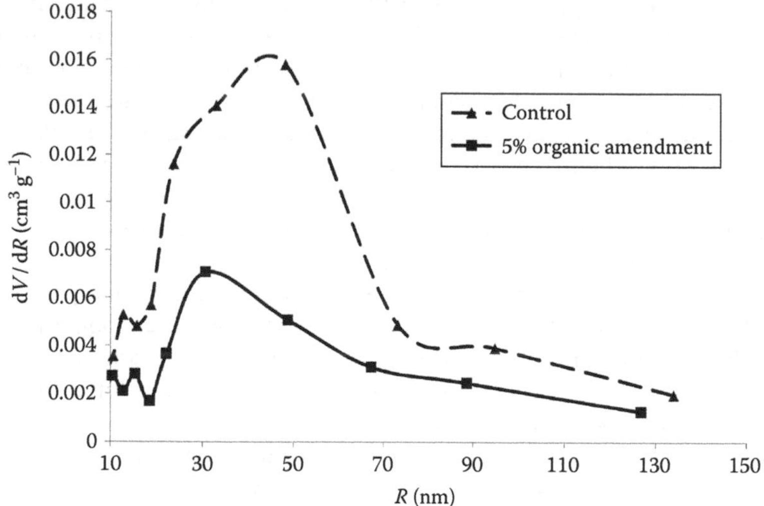

FIGURE 15.4 Changes in pore volume distribution: control (▲); compost (■).

In the literature, it has been already been reported that addition of organic amendments reduced soil porosity (Fernandes et al. 2006), mainly by a reduction in the largest pore size studied (10–100 microns). Similar results were obtained by Albarran et al. (2004). Our results showed that the compost treatment altered the features of the mesopores (10–80 nm) during time. We thus hypothesized that the

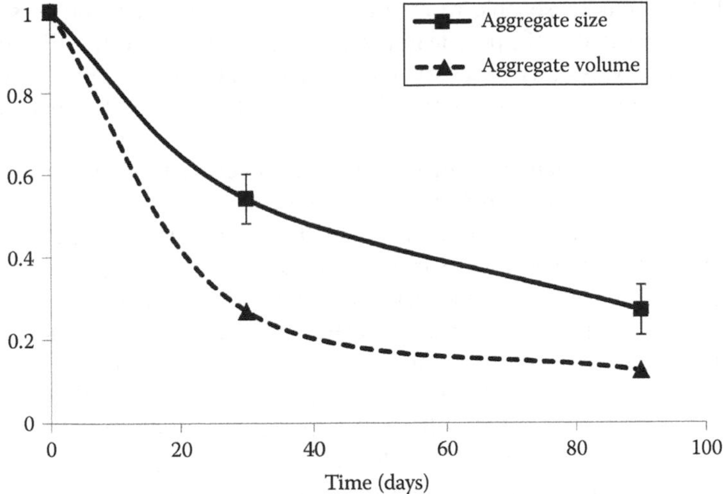

FIGURE 15.5 Changes in aggregate size L/L_0 and aggregate volume Q/Q_0 as a function of the length of incubation. $L_0 = 48 \pm 4$ nm is the mean size of the allophane aggregate of the control sample and $Q_0 = 58 \times 10^3$ nm^3 is calculated from L_0.

microstructure of clay and associated pores is fragile. The microstructure of allophane aggregates, which are present in our volcanic soil, is a very open (fractal) structure with poor mechanical properties (Chevallier et al. 2008). The nanoscale structure was studied by small-angle X-ray scattering (SAXS). SAXS experiments provided various information: the mean size of the aggregates (L) and the compactness of the aggregates, which is characterized by the fractal dimension D (Marlière et al. 2001). Figure 15.5 shows the marked decrease in aggregate size as a function of the incubation time. Experimental data showed that during incubation of the compost, the clay aggregates collapsed but the fractal dimension, $D = 2.5$–2.7 did not change (Woignier et al. 2012), meaning that the compactness of the clay aggregates was not affected. From the aggregate size decrease, we calculated the evolution of the aggregate volume $Q = 4/3\pi(L/2)^3$ (Figure 15.5). The aggregate volume Q strongly decreases with the incubation time, which shows that the allophane aggregates collapsed. This shrinkage is analogous to the result of compression stresses: shrinkage of the aggregate size and collapse of the inter-aggregate volume but preservation of the connectivity inside the clay aggregates. The main result is the progressive collapse of the clay aggregates associated with closure of the mesoporosity.

A corollary result of pore collapse is a reduction of transport properties like diffusion in the porous matrix (Archie 1942). Thus, the poor soil to water and soil to crop transfers could also be attributed to the difficulty of water and chemicals to diffuse through the clay porosity. Accessibility at the scale of the allophane is difficult due to the confined structure and small pore size of allophane clay. Regarding chlordecone, located in or near the allophane aggregates, possible exchanges or chemical reactions with water, chemical, or biological species, are likely to be hindered. It is thus reasonable to assume that the ability of a soil to sequester chlordecone is related to the pore features and to physical properties like diffusion inside the porosity. As it is not possible to measure the transport properties at the scale of allophanic aggregates, a modeling approach was used to calculate these properties (Woignier et al. 2014).

The modeling approach shows that the diffusion coefficient (*Di*,) decreased by 10 when pore size decreased from 50 to 4 nm. Thus, the clay microstructure is highly porous but has low transport properties, and fluids and chemical species migrate with difficulty inside the porosity. Some authors have postulated a protective role for amorphous minerals in soil organic matter with the formation of stable micro aggregates where soil organic matter is encapsulated in the long term (Hernández-Soriano et al. 2007). Allophanic soils also have the capacity to immobilize phosphorus, thereby decreasing its availability to plants (Calabi-Floody et al. 2012). Acid phosphatase was found to be located within micro aggregates of allophane; these results suggest that acid phosphatase is immobilized by encapsulation (Calabi-Floody et al. 2012). Thus, the high porosity of allophane may result in micropore adsorption of pesticides.

We provided evidence that the clay aggregates progressively shrink after compost addition. Given this shrinkage, we calculated the lowering of the diffusion coefficient (Woignier et al. 2014). These calculations characterized the change in the transport properties due to the collapse of porosity. The results of the calculations demonstrated a marked decrease in *Di* and hence the effect of the change in the soil microstructure on the pesticide trapping. Three months after compost addition, the calculated coefficient was the half of the initial diffusion coefficient. Thanks to the modeling approach of *Di*, we estimated that chlordecone would be confined and trapped in the porosity and that the chemical exchange reactions would be poor. This reduced accessibility clearly plays a role in the sequestration process. The chemical affinity for organic matter associated with reduced transport properties inside clay aggregates should account for the whole trapping effect of pesticides. The organic content of the compost could play a role in the microstructural change in porosity and associated diffusion processes.

Considering the peculiar clay microstructure of andosol, we hypothesize that pore collapse during incubation of the compost is the result of capillary stresses associated with the poor mechanical properties of the clay aggregates. Capillary stresses are the result of different degrees of wettability, which may be modified through adsorbed organic matter. Sorption of organic matter might thus have a significant impact on the soil–water distribution in soils. The wettability of the surface of the particles is a precondition for processes in water-filled soil pores such as flow or solute transport (Bachmann et al. 2008). The impact of heterogeneous water distribution caused by reduced wettability can be effective at different spatial scales down to that of primary particles (Puentes Jácome and Van Geel 2013). After addition of organic matter, water repellency could promote the displacement of water film that coats the mineral grains and the fluid displacements caused capillary stresses within the clay aggregates. The tensile strength of the clay aggregate is in the range 80–120 kPa (Bachmann et al. 2008). Consequently, the capillary strain will collapse the clay microstructure. Pesticides are mechanically trapped in the collapsed structure of allophane aggregates, sequestration resulted from the combination of the chemical affinity of chlordecone for organic matter, and its entrapment in the mesopores. Given this physical entrapment, which is partially irreversible, the sequestration process should be partially preserved in the long term despite mineralization of organic matter. For nitisols, the layer-like microstructure of halloysite clay is not fractal and is less sensitive to capillary forces than the spongy structure of allophane. Consequently, if chlordecone sequestration in andosol may be partly due to the physical impact of adding compost, this is not the case with nitisol.

15.4 CONCLUSIONS

Soils store and accumulate high levels of organic pollutants where they can serve as long-term sources of exposure to the surrounding environment (Marchal et al. 2013). Contaminants pose a risk only if they are available in a form that can impact human or ecosystem health. Although complete destruction of pollutants by mineralization may be desirable, in practice this destruction may not be possible. So, a risk-based approach that incorporates the bioavailability of contaminants may be more relevant than an approach based on total contaminant loading (Harmsen and Naidu 2013).

The present results show that organic amendment greatly affects the fate of pesticides: The end result is a compost-soil mixture that contains a matrix that provides less pollutant in a bioavailable form.

It should be borne in mind that while the compost traps pollutants, the long-term stability of such stabilized matrix is uncertain. Thus, it is necessary to investigate both the process involved in the retention and release of pesticides by soils and the factors influencing these processes. Trapping of pollutants in the collapsed porosity plays a role in the sequestration process and should improve its long-term stability. Addition of organic matter modifies the wetting behavior of the surface of the pores and has a significant impact on soil microstructure and diffusion properties, thus supporting pesticide stabilization in the soil.

This example shows that agro-ecological engineering offers different opportunities to control and reduce the release of pollutants into the environment. Moreover, the addition of organic matter will be less disturbing for the soil ecosystem, fertility, and biophysical properties than the incorporation of chemicals, selected bacteria, or soil digging. However, successful implementation of these risk-based approaches requires acceptance by the local community and agreement for *in situ* management of contaminants rather than the complete removal of the contaminated soil (Harmsen and Naidu 2013).

ACKNOWLEDGMENTS

Funding was provided by the French Chlordecone National Plan ("JAFA" project), the French National Research Agency ("Chlordexco" project), and the French Ministry for Overseas development (MOM).

REFERENCES

Albarran, A., R. Celis, M. C. Hermosin, A. Lopez-Pineiro, and J. Cornejo. 2004. Behaviour of simazine in soil amended with the final residue of the olive-oil extraction process. *Chemosphere* 54(6):717–724. doi: 10.1016/j.chemosphere.2003.09.004.

Alexander, M. 2000. Aging, bioavailability, and overestimation of risk from environmental pollutants. *Environmental Science & Technology* 34(20):4259–4265. doi: 10.1021/es001069+.

Archie, G. E. 1942. The electrical resistivity log as an aid in determining some reservoir characteristics. *Transactions of the American Institute of Mining and Metallurgical Engineers* 146(1):54–62. doi: 10.2118/942054-G.

Bachmann, J., G. Guggenberger, T. Baumgartl, R. H. Ellerbrock, E. Urbanek, M. O. Goebel, K. Kaiser, R. Horn, and W. R. Fischer. 2008. Physical carbon-sequestration mechanisms under special consideration of soil wettability. *Journal of Plant Nutrition and Soil Science-Zeitschrift fur Pflanzenernahrung und Bodenkunde* 171(1):14–26. doi: 10.1002/jpln.200700054.

Boucher, O., M.-N. Simard, G. Muckle, F. Rouget, P. Kadhel, H. Bataille, V. Chajès et al. 2013. Exposure to an organochlorine pesticide (chlordecone) and development of 18-month-old infants. *NeuroToxicology* 35:162–168. doi: 10.1016/j.neuro.2013.01.007.

Cabidoche, Y. M., R. Achard, P. Cattan, C. Clermont-Dauphin, F. Massat, and J. Sansoulet. 2009. Long-term pollution by chlordecone of tropical volcanic soils in the French West Indies: A simple leaching model accounts for current residue. *Environmental Pollution* 157(5):1697–1705. doi: 10.1016/j.envpol.2008.12.015.

Cabrera, A., L. Cox, K. A. Spokas, R. Celis, M. C. Hermosín, J. Cornejo, and W. C. Koskinen. 2011. Comparative sorption and leaching study of the herbicides fluometuron and 4-chloro-2-methylphenoxyacetic acid (MCPA) in a soil amended with biochars and other sorbents. *Journal of Agricultural and Food Chemistry* 59(23):12550–12560. doi: 10.1021/jf202713q.

Calabi-Floody, M., G. Velásquez, L. Gianfreda, S. Saggar, N. Bolan, C. Rumpel, and M. L. Mora. 2012. Improving bioavailability of phosphorous from cattle dung by using phosphatase immobilized on natural clay and nanoclay. *Chemosphere* 89(6):648–655. doi: 10.1016/j.chemosphere.2012.05.107.

Chevallier, T., T. Woignier, J. Toucet, E. Blanchart, and P. Dieudonné. 2008. Fractal structure in natural gels: Effect on carbon sequestration in volcanic soils. *Journal of Sol-Gel Science and Technology* 48(1):231–238. doi: 10.1007/s10971-008-1795-z.

Chung, N. and M. Alexander. 2002. Effect of soil properties on bioavailability and extractability of phenanthrene and atrazine sequestered in soil. *Chemosphere* 48(1):109–115. doi: 10.1016/s0045-6535(02)00045-0.

Clostre, F., T. Woignier, L. Rangon, P. Fernandes, A. Soler, and M. Lesueur-Jannoyer. 2014. Field validation of chlordecone soil sequestration by organic matter addition. *Journal of Soils and Sediments* 14(1):23–33. doi: 10.1007/s11368-013-0790-3.

Cox, L., R. Celis, M. C. Hermosín, J. Cornejo, A. Zsolnay, and K. Zeller. 2000. Effect of organic amendments on herbicide sorption as related to the nature of the dissolved organic matter. *Environmental Science & Technology* 34(21):4600–4605. doi: 10.1021/es0000293.

Delgado-Moreno, L., L. Sánchez-Moreno, and A. Peña. 2007. Assessment of olive cake as soil amendment for the controlled release of triazine herbicides. *Science of the Total Environment* 378(1–2):119–123. doi: 10.1016/j.scitotenv.2007.01.023.

Fernandes, M. C., L. Cox, M. C. Hermosin, and J. Cornejo. 2006. Organic amendments affecting sorption, leaching and dissipation of fungicides in soils. *Pest Management Science* 62(12):1207–1215. doi: 10.1002/ps.1303.

Fernandes, P., M. Jannoyer-Lesueur, A. Soler, R. Achard, and T. Woignier. 2010. Effects of clay microstructure and compost quality on chlordecone retention in volcanic tropical soils: Consequences on pesticide lability and plant contamination. In *19th World Congress of Soil Science, Soil Solutions for a Changing World,* Brisbanne, AU.

Harmsen, J. and R. Naidu. 2013. Bioavailability as a tool in site management. *Journal of Hazardous Materials* 261:840–846. doi: 10.1016/j.jhazmat.2012.12.044.

Hernández-Soriano, M. C., A. Peña, and M. D. Mingorance. 2007. Retention of organophosphorous insecticides on a calcareous soil modified by organic amendments and a surfactant. *Science of the Total Environment* 378(1–2):109–113. doi: 10.1016/j.scitotenv.2007.01.011.

Majumdar, K. and N. Singh. 2007. Effect of soil amendments on sorption and mobility of metribuzin in soils. *Chemosphere* 66(4):630–637. doi: 10.1016/j.chemosphere.2006.07.095.

Marchal, G., K. E. C. Smith, A. Rein, A. Winding, S. Trapp, and U. G. Karlson. 2013. Comparing the desorption and biodegradation of low concentrations of phenanthrene sorbed to activated carbon, biochar and compost. *Chemosphere* 90(6):1767–1778. doi: 10.1016/j.chemosphere.2012.07.048.

Marlière, C., T. Woignier, P. Dieudonné, J. Primera, M. Lamy, and J. Phalippou. 2001. Two fractal structures in aerogel. *Journal of Non-Crystalline Solids* 285:175–180. doi: 10.1016/S0022-3093(01)00450-1.

Mottes, C., M. Lesueur-Jannoyer, M. Bail, and E. Malézieux. 2013. Pesticide transfer models in crop and watershed systems: A review. *Agronomy for Sustainable Development* 34(1):229–250. doi: 10.1007/s13593-013-0176-3.

Murano, H., T. Otani, T. Makino, N. Seike, and M. Sakai. 2009. Effects of the application of carbonaceous adsorbents on pumpkin (*Cucurbita maxima*) uptake of heptachlor epoxide in soil. *Soil Science & Plant Nutrition* 55(2):325–332.

Peters, R., J. W. Kelsey, and J. C. White. 2007. Differences in *p,p'*-DDE bioaccumulation from compost and soil by the plants *Cucurbita pepo* and *Cucurbita maxima* and the earthworms *Eisenia fetida* and *Lumbricus terrestris*. *Environmental Pollution* 148(2):539–545. doi: 10.1016/j.envpol.2006.11.030.

Pignatello, J. J. 1998. Soil organic matter as a nanoporous sorbent of organic pollutants. *Advances in Colloid and Interface Science* 76–77:445–467. doi: 10.1016/S0001-8686(98)00055-4.

Puentes Jácome, L. and P. Van Geel. 2013. An initial study on soil wettability effects during entrapped LNAPL removal by surfactant flooding in coarse-grained sand media. *Journal of Soils and Sediments* 13(6):1001–1011. doi: 10.1007/s11368-013-0673-7.

Saito, T., T. Otani, N. Seike, H. Murano, and M. Okazaki. 2011. Suppressive effect of soil application of carbonaceous adsorbents on dieldrin uptake by cucumber fruits. *Soil Science and Plant Nutrition* 57(1):157–166. doi: 10.1080/00380768.2010.551281.

Semple, K. T., B. J. Reid, and T. R. Fermor. 2001. Impact of composting strategies on the treatment of soils contaminated with organic pollutants. *Environmental Pollution* 112(2):269–283. doi: 10.1016/s0269-7491(00)00099-3.

Vlčková, K. and J. Hofman. 2012. A comparison of POPs bioaccumulation in *Eisenia fetida* in natural and artificial soils and the effects of aging. *Environmental Pollution* 160:49–56. doi: 10.1016/j.envpol.2011.08.049.

Woignier, T., F. Clostre, H. Macarie, and M. Jannoyer. 2012. Chlordecone retention in the fractal structure of volcanic clay. *Journal of Hazardous Materials* 241–242:224–230. doi: 10.1016/j.jhazmat.2012.09.034.

Woignier, T., P. Fernandes, M. Jannoyer-Lesueur, and A. Soler. 2012. Sequestration of chlordecone in the porous structure of an andosol and effects of added organic matter: An alternative to decontamination. *European Journal of Soil Science* 63(5):717–723. doi: 10.1111/j.1365-2389.2012.01471.x.

Woignier, T., P. Fernandes, M. Lesueur-Jannoyer, and A. Soler. 2014. La séquestration des pesticides, une alternative à la dépollution. Résultats expérimentaux et modélisation. In *44e congrès du Groupe Français des Pesticides*, Schoelcher, France.

Woignier, T., P. Fernandes, A. Soler, F. Clostre, C. Carles, L. Rangon, and M. Lesueur-Jannoyer. 2013. Soil microstructure and organic matter: Keys for chlordecone sequestration. *Journal of Hazardous Materials* 262:357–364. doi: 10.1016/j.jhazmat.2013.08.070.

Zhang, J., L. Yang, L. Wei, X. Du, L. Zhou, L. Jiang, Q. Ding, and H. Yang. 2012. Environmental impact of two organic amendments on sorption and mobility of propachlor in soils. *Journal of Soils and Sediments* 12(9):1380–1388. doi: 10.1007/s11368-012-0561-6.

16 Remediation of Chlordecone-Contaminated Waters Using Activated Carbons

Sarra Gaspard, Ronald Ranguin, Ulises Jaurégui Haza, Nady Passé-Coutrin, and Axelle Durimel

CONTENTS

16.1 INTRODUCTION

Banana crop, which has been the main agricultural product in the French West Indies (Guadeloupe and Martinique), requires the intensive use of pesticides to prevent attacks by insect pests. Chlorinated pesticides, such as hexachlorocyclohexane (HCH), chlordecone (CLD), and dieldrin, were used from the 1960s until the beginning of the 1990s. CLD has a strong persistence in natural environments and high resistance to chemical reactions and microbiological degradations. This pesticide was used until 1993, resulting in a generalized diffuse contamination of the soil and water in the areas of banana production. Around 8%–9% of the cultivation areas of Guadeloupe contain CLD concentrations higher than 1 mg kg^{-1} in topsoil (Cabidoche et al. 2009). CLD is known for its endocrine-disrupting character (Tapiero and Nguyen 2002, Multigner et al. 2015), carcinogenic potential (EPA 2009, Landau-Ossondo et al. 2009), and its accumulation in the food chain (Borga et al. 2001). Banana plantations are generally localized in humid areas that are often the drinking water production zones. Drinking water production plants of the French West Indies were equipped with activated carbon (AC) filters, as ACs are commonly used to remove pesticides from contaminated water (Bembnowska et al. 2003, Faur et al. 2008, Njoku and Hameed 2011, Salman et al. 2011). Many biomass can be used as feedstock for AC preparation such as agricultural wastes, such as sugarcane bagasse (SCB), coir pith, which is a soft biomass obtained from the coconut husk during its preparation in the coconut industry, banana pith, sago waste, silk cotton hull, corn cob, maize cob, straw, rice husk, rice hulls, fruit stones nutshells, pinewood, sawdust, coconut tree sawdust, bamboo and cassava peel, and marine plant residues (Gaspard et al. 2014). ACs can be prepared by physical or chemical activation process as well as by alternative synthesis methods for energy and chemical saving such as microwave heating and hydrothermal carbonization treatment (Gaspard et al. 2014). Textural and chemical properties of the materials govern adsorptive properties of ACs. Adsorption properties depend on surface chemistry that gives the charge of the surface, its hydrophobicity, and the electronic density of the graphene layers.

There is only one work published in 2013 on CLD adsorption by ACs. Indeed, Durimel et al. (2013) prepared ACs with different textural properties and surface chemistry from sugarcane bagasse (SCB ACs) to remove CLD from contaminated water. SCB is known as a precursor for the production of ACs with interesting adsorption properties (Mall et al. 2007, Kalderis et al. 2008). It is produced by sugarcane industries that are one of the main industries in the French West Indies, such providing a sustainable cleaning process, because a high-value product is obtained from a low-cost material, bringing as well solutions to the problem of wastes and local water pollution. Data regarding the influence of parameters such as solution pH, AC surface properties and textural characteristics on the adsorption of CLD onto SCB ACs are presented. These results are correlated to molecular modeling of CLD interactions with surface functional groups of AC. Nevertheless, after water treatment CLD-contaminated ACs can be considered as a hazardous waste that need to be treated. A method for CLD desorption from AC surface and its following degradation is proposed.

16.2 ACTIVATED CARBON PREPARATION AND CHARACTERIZATION

16.2.1 Activated Carbon Preparation

ACs can be prepared by the classical experimental procedures, physical and chemical activation, of new renewable feedstock that includes wood, fruits stones or shells, and waste from agro-industries as well as by using nonconventional methods such as microwave heating and hydrothermal carbonization treatment. During physical activation, the precursor is first carbonized and then activated using an oxidizing gas. During the carbonization process, the starting material is converted to carbon. The second step—physical activation—consists of a gasification of the resulting carbon at high temperature with steam, carbon dioxide, air, or a mixture of these leading to a final AC having a well-developed porosity.

For realizing the chemical activation process, the precursor is generally impregnated with chemicals, such as KOH, NaOH, H_3PO_4, $ZnCl_2$, H_2SO_4, $(NH_4)_2SO_4$, HCl, $MgCl_2$, HNO_3, or $CaCl_2$, followed by heating under a nitrogen flow at 450°C–900°C. Hydrolysis reactions with loss of volatile matter occur during impregnation, leading to a weakening of the precursor structure, increasing of elasticity, and swelling of the particle. The amount of chemical agent incorporated in the precursor and the impregnation duration govern the porosity of the carbon. This activation method is considered as very flexible for the preparation of ACs with different pore size distributions.

Microwave heating processing can also be used for AC preparation, and presents the advantage to have lower environmental impact due to time and space saving (Appleton et al. 2005). Microwave irradiation interacts with biomass at specific temperatures, leading to a structural change within the biomass (Appleton et al. 2005). Due to microwave energy transformation into heat inside the particles by dipole rotation and ionic conduction, the high-temperature gradient from the interior of the char particle to its cool surface allows the microwave-induced reaction to proceed more quickly and effectively when compared with the conventional process (Gaspard et al. 2014).

Hydrothermal carbonization (HTC) of biomass is another low-cost method with low environmental impact for AC production. It consists of heating raw materials dispersed in an aqueous solution and autoclaving at temperatures between 150°C and 350°C for about 2–24 h at saturated pressures. During this process, dehydration and decarboxylation reactions, which are exothermic, render the process self-sufficient (Gaspard et al. 2014).

16.2.2 Activated Carbon Characterization

The surface chemistry of the ACs essentially depends on oxygen groups at their surface, for example, carboxyls, phenols, lactones, carbonyls as well as quinones. Adsorption properties depend on surface chemistry that are determined using different methods for chemical group characterization, which include thermal techniques such as thermal programmed adsorption or spectrophotometric methods such as X-ray photoelectron spectroscopy (XPS) or Fourier transform infrared (FTIR) spectroscopy.

AC surface functional groups performed by FTIR gives qualitative information of characteristic functional groups on the surface by band assignment based on published data (Gaspard et al. 2014). XPS analysis allows determination of elemental surface composition obtained over a depth of about a few nanometers. From XPS spectra, curve fitting as well as changes in the chemical bonding states and concentrations of the surface functional groups of ACs can be determined.

Another technique, temperature programmed desorption, coupled with mass spectroscopy that detects gases produced from functional groups decomposition upon temperature increase can be used for the determination of the nature of surface functional groups on ACs. Gases evolved are carbon monoxide, carbon dioxide, or water vapor that are correlated with the presence of a given functional group onto the AC surface by their specific temperature of decomposition (Brender et al. 2012, Durimel et al. 2013).

According to the classification adopted by the International Union of Pure and Applied Chemistry (IUPAC) (Rouquerol et al. 1999), porous materials contain micropores (<2 nm), mesopores (2–50 nm), and macropores (>50 nm). Textural characteristics of ACs that include surface area and pore size distribution are important data for studying the porous texture formation during AC synthesis and for determining the potential applications of the material (Figueiredo et al. 1999, Rodriguez-Reinoso, and Molina-Sabio 1998). Gas adsorption techniques allow the determination of the textural parameters, such as specific surface area and microspore and mesopore volumes. Specific surface area values of ACs are generally between 500 and 1500 $m^2 g^{-1}$ and AC samples are considered as being microporous or mesoporous when they contain either a high microporous or microporous volume, respectively.

16.3 INFLUENCE OF PHYSICOCHEMICAL PARAMETERS ON CHLORDECONE ADSORPTION BY ACTIVATED CARBONS

16.3.1 Role of Solution pH

Adsorption isotherm studies of CLD on four bagasse ACs with different textural and chemical characteristics were performed in order to understand the role of AC surface functional groups on CLD adsorption and to get information on CLD adsorption mechanism. Adsorption capacity was shown to be higher for the most acidic samples. CLD adsorption isotherm shape obtained for acidic AC samples was characteristic of strong competition from the solvent molecules for the acidic surface sites and to moderate intermolecular interaction between the CLD molecules, as it will be demonstrated later as a result of molecular modeling of the CLD–CLD and CLD–surface functional group interactions.

Moreover, a high amount of C and acidic groups, and consequently low pH_{pzc} (Figure 16.1), were shown to favor CLD adsorption, whereas high contents of oxygen, basic groups, and the presence of ether and hydroxyl groups were detrimental to CLD adsorption (Figure 16.2). It was proposed that CLD might bound the AC surface by hydrogen bounds between carboxyl groups at the AC surface and the chlorine atoms. Functionalities, such as ethers or hydroxyls, may interact with water through hydrogen bounds. Adsorption capacity of CLD reached a maximum at pH values close to the pH_{pzc} of acidic and basic AC samples. On the other hand, for acidic AC sample a second maximum was then reached around pH 9, indicating that CLD adsorption was favored when the AC surface was not charged (Figure 16.1). At a pH greater than 9, CLD formed CLD hydroxide (C_{10} Cl_{10} $(OH)_2$) (Dawson et al. 1979, Guzelian 1982) by substitution of the carbonyl function with two alcohol groups. The chlordecone hydrate formed, which is much more hydrophilic

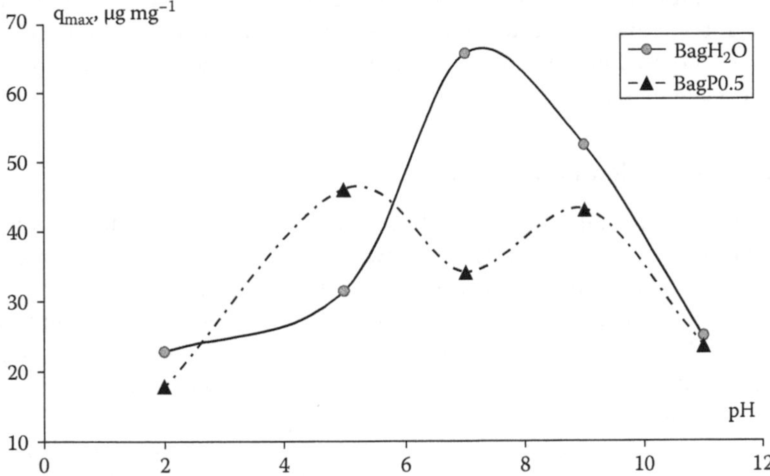

FIGURE 16.1 Influence of pH solution on total CLD adsorption capacity (q_{max}) of BagH_2O, AC samples, sample prepared by water activation and BagP0.5, sample prepared by chemical activation with phosphoric acid/bagasse ratio, Xp = 0.5. (Modified from Durimel, A. et al., *Chem. Eng. J.*, 229, 239, 2013.)

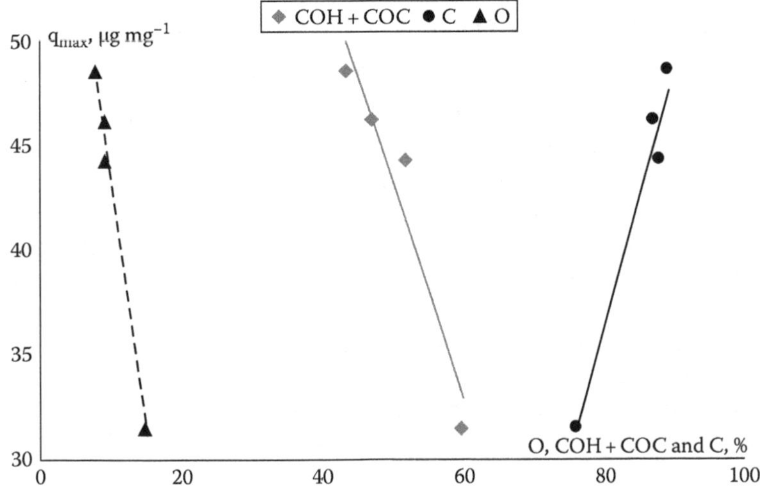

FIGURE 16.2 Influence of some physicochemical properties of activated carbons on total adsorption capacity (q_{max}). (Data from Durimel, A. et al., *Chem. Eng. J.*, 229, 239, 2013.)

than CLD (Dawson et al. 1979, Cabidoche et al. 2009), might have a higher affinity for the acidic carbon surface. It was proposed that adsorption mechanism of CLD should as well be governed by hydrophobic interactions with the AC graphene layer. Porous distribution did not influence CLD adsorption as the bagasse ACs prepared in the study of Durimel and coworkers contained significant amounts of both large micropores and mesopores, and their mean pore diameter was around 2 nm. According to the distance between the oxygen atom and the farthest chloride atom of CLD, the largest dimension of the molecule is 0.81 nm and CLD could enter large micropores and mesopores of all ACs.

16.3.2 Role of Temperature

The effect of temperature on CLD adsorption onto ACs was studied at different temperatures (20°C, 25°C, and 32°C), showing an increase of CLD adsorbed with the increase of temperature, which was characteristic of an endothermic process. Calculation of thermodynamic parameters such as standard free enthalpy ($\Delta G°$), enthalpy ($\Delta_{ads}H°$), and entropy ($\Delta S°$) showed negative free enthalpy values, suggesting that the adsorption of CLD onto AC surface is feasible and spontaneous. In addition, absolute value of $\Delta G°$ increases with temperature for all ACs. This result confirmed that CLD adsorption onto the ACs was favored at high temperature. Calculated values of enthalpy were close to 100 kJ mol^{-1}, for acidic samples, that might correspond to a chemisorption process. In addition, positive values of $\Delta S°$ suggested that entropy increases in the solid–liquid interface during the adsorption process, indicating that the degree of freedom of the adsorbed molecules increases too that would also indicate a decrease in the affinity between CLD and absorbent (Durimel 2010).

16.4 MOLECULAR MODELING STUDIES OF CHLORDECONE INTERACTIONS WITH AC SURFACE FUNCTIONAL GROUPS

The influence of surface group (SG) content over adsorption properties has been reported and studied to some extent both theoretically and experimentally for porous carbons, mainly AC and soot particles (McCallum et al. 1999, Oubal et al. 2010, Terzyk et al. 2011, Durimel et al. 2013, Enriquez-Victorero et al. 2014, Gamboa-Carballo et al. 2016). Then, modifying the SG content can lead to changes in the hydrophilic/lipophilic balance and so the modifications should guarantee the specific design purpose. In this regard, the computational chemistry can play a major role in designing specific modifications to obtain more efficient and specific adsorbents (Jensen et al. 2011). Recently, molecular modeling studies of lindane and CLD interactions with AC surface functional groups have been published (Durimel et al. 2013, 2015, Enriquez-Victorero et al. 2014, Gamboa-Carballo et al. 2016).

The results obtained with CLD will be discussed here. The description of the interactions between guest molecules and carbon surfaces is still challenging with current theoretical techniques. For this reason, the idealized carbon models based on small polycycles and their oxidized derivatives are a common choice (Cabaleiro-Lago et al. 2008, Picaud et al. 2008, Jenness and Jordan 2009, Oubal et al. 2010, Cho et al. 2013; Lazar et al. 2013, Njoku and Hameed (2011), Montero et al. (1998)).

In fact, naphthalene (Cabaleiro-Lago et al. 2008, Cho et al. 2013) and coronene (Jenness and Jordan 2009, Lazar et al. 2013) have repeatedly been employed as such models (Figure 16.3), and they were used to model CLD–AC interactions (Durimel et al. 2013, Gamboa-Carballo et al. 2016).

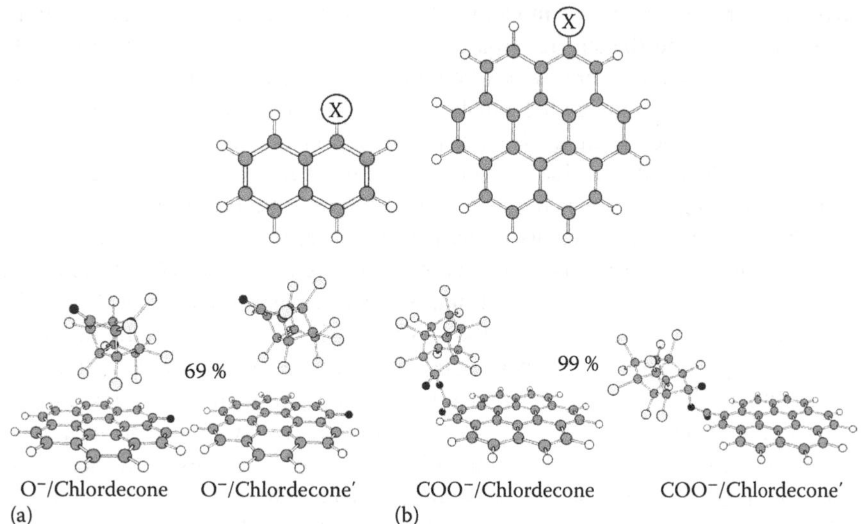

FIGURE 16.3 Activated carbon models consisting on (a) naphthalene (two rings model) and (b) coronene (seven ring model) and their oxidized forms, X = H, COOH, COO^-, OH, O^- (up). Two examples of distinctive minima structures and their populations (%) for the interaction between CLD and the AC model (bottom).

The use of such small models has both advantages and disadvantages, with the most obvious of the advantages being the savings in computational resources allowing either the application of relatively high-level calculations or a thorough exploration of the interaction space. Small carbon models such as naphthalene allow the use of most quantum chemical methods and the correct description of nonlocal correlations as well as the evaluation of zero point and thermal contributions to the enthalpy (Lazar et al. 2013) or the application of energy partition schemes to analyze binding characteristics. However, the dramatic reduction in size of these models with respect to the real structures can reduce the nonlocal contributions to the dispersion energies. In the specific case of ACs, the use of oxidized forms of polycycles completely neglects the influence of pore shape, defects, pore size, and pore connectivity but in return allows a quantum chemical description of the interactions with the adsorbed molecules. These models for AC do not take into account morphological or topological characteristics of the AC, but focuses on SG interaction with adsorbate. To evaluate the influence of acidic SGs over the adsorption of CLD onto AC, Durimel et al. (2013) and Gamboa-Carballo et al. (2016) used a methodology that combines semi-empirical methods and higher-level density functional theory. In order to verify the compliance of the conditions that favor the adsorption of CLD, theoretical calculations with multiple minima hypersurface (MMH) (Montero et al. 1998) procedures were applied. The AC is simulated by two simplest models: two and seven aromatic groups with different surface functional groups. The chemical changes produced in the functional groups at different pH were taken into account. The values of energy of association obtained by the MMH methodology demonstrate that the CLD–CLD intermolecular interactions are moderate (Durimel et al. 2013). In the case of studies concerning CLD and the surface functional groups of ACs, it was also found that the molecular interactions are weak when the oxygenated functional groups are not charged, being higher for COOH, which confirms the observed behavior of adsorption increase with increasing acidity groups in ACs (Durimel et al. 2013, Gamboa-Carballo et al. 2016). When, due to the pH of the medium, the functional groups are modified acquiring a negative charge, the interactions with the CLD increase appreciably. According to these results if the pH is acidic, then carboxyl and hydroxyl SGs will be in neutral form, and will not contribute significantly to CLD adsorption. On the other hand, when pH ≈ 7, estimations give a deprotonation of 3% for OH SG and about 90% for COOH, and then COOH SG will preferentially contribute to CLD adsorption onto AC at these conditions. These results are consistent with experimental findings by Durimel et al. (2013). At the same time, the interactions of CLD with the water molecules of the solvent will also increase with the variation of pH. This explains that CLD is in competition with the molecules of the solvent for adsorption sites, this being one of the conditions set by Giles et al. (1960) to account for the form of the adsorption isotherms (Durimel 2013). However, theoretical calculations suggested that if CLD molecule is associated with the SG, the molecules of water would not be able to compete for the adsorption sites (Gamboa-Carballo et al. 2015). As a result of MMH-PM7 calculations two distinctive interactions for the CLD-SGs of AC systems were obtained: interaction of the chlorines of CLD with the π-cloud of coronene (Cl···π-cloud), and donor–acceptor interaction between negatively charged oxygen of SGs (COO^- and O^-) and the electron-deficient carbonyl carbon of CLD, as shown in Figure 16.3 (Gamboa-Carballo et al. 2016). The bonding distances for O^-···CO interactions oscillated between 1.5 and 1.6 Å, which might correspond to covalent bonds, which suggests that chemical sorption may occur at neutral pH. This result is consistent with experimental findings (Durimel 2013). For the interactions Cl···π-cloud, optimization methods yielded some differences: While for PM7 the average interaction distances were about 2.7 Å, CAM B3LYP/6 31 + G(d,p) re-optimizations yielded an average of 3.6 Å. This result suggests that the semi-empirical Hamiltonian PM7 overestimates this kind of dispersive interactions, which may be the reason for variation in structures like COO^-/CLD/$(H_2O)_3$ and O^-/CLD/$(H_2O)_3$. However, in most of the cases, the structures conserved their geometry and the type of interaction. For that reason, it can be concluded that the semi-empirical Hamiltonian PM7 qualitatively described these complexes in terms of interactions types (Gamboa-Carballo et al. 2016). Finally, the studies performed using quantum theory of atoms in molecules allowed to characterize the topology of the electron density and its Laplacian at bond critical points and, hence, to describe intermolecular interactions. The interactions were mainly dispersive in nature (Van der Waals). For the interactions between water and CLD, the calculations showed frequently either closed shell hydrogen bonds with a high dispersive character or charge transfer H_2O···CO. Only a few structures with O^-···CO interactions have covalent bonds between the negatively charged oxygen in the SG and the electron-deficient carbon in CLD. The low electron density in these cases accounted for straight cylindrical single bonds, and density-based functions verified weak to strong covalent bonds. Once again, this result ratifies the idea of chemical sorption at neutral pH conditions (Gamboa-Carballo et al. 2016). Summarizing, the use of molecular modeling helps elucidate the nature of CLD interactions with surface functional groups of ACs using an inexpensive methodology. These results, together with the ongoing research to evaluate the role of basic groups in adsorption process, will allow deciding the best SGs for selecting the more adequate commercial ACs or for an efficient synthesis of new ACs for the adsorption of CLD from contaminated waters.

16.5 REGENERATION OF CHLORDECONE-CONTAMINATED ACTIVATED CARBONS

16.5.1 Thermal Treatment of Contaminated Activated Carbons

Thermal analyses of CLD-contaminated AC samples were carried out in order to obtain information on CLD thermal desorption from AC surface. Differential scanning calorimetry (DSC) and simultaneous thermal analyses (STA) were performed under an inert gas (nitrogen) to prevent reaction between the material and air. Thermogravimetric analyses were performed for raw ACs and CLD-contaminated ACs using a thermogravimetric balance. Samples were heated from 25°C to 700°C with a heating rate of 5°C min^{-1} under a nitrogen flow. Positive values of $\Delta H°$ were obtained, implying that desorption of CLD onto AC surface was an endothermic process, in agreement with previous data showing that CLD was strongly bound to AC surface and even covalently on the surface of AC sample exhibiting a high content of carboxylic groups. In addition, CLD desorption temperature was determined to be around 300°C. Moreover, from STA data, it was possible to obtain more information about CLD desorption by making the difference of the two graphics, considering that m_1 is the mass loss from the raw AC sample and m_2 the mass loss from the contaminated AC sample. The percentage of the initial mass desorbed was used to compare experiments from different initial masses. From the comparison of the graphics, it was possible to propose that the first decrease at 300°C was due to CLD desorption, while the second one might be linked to gasification of the AC sample due to chemical groups desorption (Durimel et al. 2013). For the four ACs tested, the first stage of the curve was clearly related to a mass loss, with an inflection point at 300°C. The second stage was first an increase of mass, as if the presence of CLD diminished the gasification of the AC. But then, at a higher temperature, another compound was desorbed. This phenomenon occurred for the three AC samples, sample prepared by water activation ($BagH_2O$), or with lower ratio of phosphoric acid/bagasse (Xp) (BagP0.5 and BagP1, with Xp = 0.5 and 1, respectively) while for the most acidic one BagP1.5 (Xp = 1.5) no increase of mass was observed during the gasification, confirming strong binding of CLD to the AC surface functional groups. Thermal studies have shown, as for adsorption in water, that CLD in gaseous phase desorption is an endothermic process. Desorption temperature obtained by DSC and STA for CLD is around 300°C, which is close to the melting point of the molecule (350°C).

16.5.2 Desorption by Solvent

CLD desorption study was carried out in different solvents—hexane, acetone, and ethanol—in order to determine the most appropriate one to desorb CLD from contaminated ACs. Desorption experiments were also performed in water considered as a control, allowing predicting the behavior of CLD in solution when contaminated ACs are in contact with "clean" water. Results showed that amount of CLD desorbed is higher for acidic bagasse AC. In water, the higher removal percentage of CLD initially adsorbed from AC surface was 12%, which was found for BagP1.5. For BagP0.5 and BagP1, removal percentages were calculated to be 4%. The polar solvents acetone and ethanol were both shown to have a very high affinity (Blanke et al. 1977, Bristeau et al. 2013) for CLD ($\log K_{OC} = 3.8$) and for AC surface allowing to solubilize higher amount of CLD molecule (between 30% and 45%) while for the less hydrophobic solvent hexane a low desorption rate (2.4%) were obtained. Acetone and ethanol molecules are able to occupy CLD binding sites for effective desorption of CLD from AC surface. On the other hand, hexanes that have a high affinity for hydrophobic molecules would therefore have less affinity with the carboxyl groups on the surface of ACs, which are the main CLD binding site.

The most acidic AC sample, BagP1.5, is the best AC for chemical desorption because it allows a fast and effective regeneration. This can be justified by BagP1.5 as it contains a high amount of mesopores and many carbonyl and carboxyl groups on its surface with high affinity for CLD, which may induce an exchange mechanism between CLD and the polar solvent molecules. Regeneration of contaminated ACs with ethanol would be a promising way in the French West Indies since this solvent is produced locally during sugarcane processing.

16.5.3 Vitamin B12–Mediated Chlordecone Degradation

Anaerobic bacteria are able to degrade a wide variety of aromatic compounds containing high numbers of chlorine atoms, using them as final electrons acceptor during a process called dehalorespiration (Sulfita et al. 1982, Holliger et al. 1997, Perez de Moza et al. 2014). Reductive dechlorination being such an important environmental microbial process, the question arises whether CLD is really refractory to microbial degradation (Dolfing et al. 2012). A study conducted by Orndorff and Colwell (1980) showed that only 16% of CLD was degraded to mono-hydro and di-hydrochlordecone under aerobic conditions. This poor degradability of CLD under aerobic conditions was recently confirmed (Fernandez-Bayo et al. 2013). On the other hand, *in vivo* CLD dechlorination was shown in methanogenic conditions using *Methanosarcina thermophila* species (Jablonski et al. 1996). CLD dechlorination was also achieved in the presence of vitamin B12 reduced by either $NaBH_4$ or acetoin (Schrauzer and Katz 1978).

In a recent work (Ranguin, 2015) using vitamin B12 (VB12 Co(III)) reduced upon Zn(0) or DTT (dithiothreitol) addition led to CLD removal immediately when the reductant was added. GC chromatogram of the reaction mixture shows the appearance of a unique CLD degradation intermediate at a retention time of 28.89 min. Mass spectra of this compound in MS/MS is shown in Figure 16.4 and it allows for

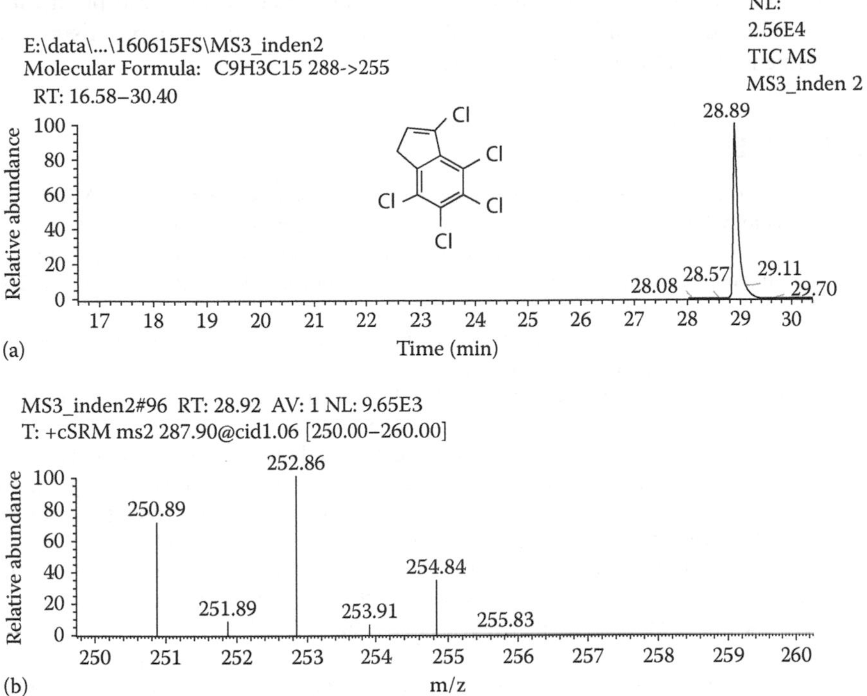

FIGURE 16.4 GC-MS chromatogram (a) and MS-MS spectra of pentachloroindene (b).

identification of a pentachloroindene intermediate (Ranguin 2015). Desorption using ethanol followed by vitamin B12–mediated reduction of CLD was proposed to be a promising way for treatment of CLD-contaminated ACs.

16.6 CONCLUSION

Adsorption of CLD in water is endothermic and this process corresponds to a heterogeneous process that occurs in the solid–liquid interface with diffusion of the solute occurring in a heterogeneous process in a complex matrix. The use of molecular modeling helps elucidate the nature of CLD's interactions with surface functional groups of ACs, using an inexpensive methodology. These results help determine the best SGs of the more adequate commercial ACs or for an efficient synthesis of new ACs for adsorption of CLD from contaminated waters.

Due to the presence of a high amount of oxygenated groups on the AC surface, higher amount of CLD is removed by polar solvents such as acetone and ethanol, than with hydrophobic solvents such as hexane. Ethanol is shown to be the best solvent for CLD desorption from different ACs and desorption kinetics correspond to a multistep diffusion process. Vitamin B12 reduced by zerovalent zinc is able to dechlorinate efficiently CLD, leading to the formation of a pentachloroindene dechlorination product that is identified by MS-MS spectroscopy. Desorption using ethanol followed by vitamin B12–mediated reduction of would be a promising way for treatment of CLD-contaminated ACs.

REFERENCES

Appleton T.J., R.I. Colder, S.W. Kingman, I.S. Lowndesand, and A.G. Read. 2005. Microwave technology for energy-efficient processing of waste. *Appl. Energy* 81: 85–113.

Bembnowska A., R. Pelech, and E. Milchert. 2003 Adsorption from aqueous solutions of chlorinated organic compounds onto activated carbons. *J. Colloid Interface Sci.* 265: 276–282.

Blanke R.V., M.W. Fariss, F.D. Griffith, and Jr. P. Guzelian. 1977. Analysis of chlordecone [Kepone] in biological specimens. *J. Anal. Toxicol.* 1: 57–62.

Borga K., G.W. Gabrielsen, and J.U. Skaare. 2001. Biomagnification of organochlorines along a Barents sea food chain. *Environ. Poll.* 113: 187–198.

Brender P., R. Gadiou, J.-C. Rietsch, P. Fioux, J. Dentzer, A. Ponche, and C. Vix-Guterl. 2012. Characterization of carbon surface chemistry by combined temperature programmed desorption with in situ x-ray photoelectron spectrometry and temperature programmed desorption with mass spectrometry analysis. *Anal. Chem.* 84: 2147–2153.

Bristeau S., L. Amalric, and C. Mouvet. 2013. Validation of chlordecone analysis for native and remediated French West Indies soils with high organic matter content. *Anal. Bioanal. Chem.* 406: 1073–1080. doi: 10.1007/s00216-013-7160-2.

Cabaleiro-Lago E.M., J. Rodríguez-Otero, and Á. Peña-Gallego. 2008. Computational study on the characteristics of the interaction in naphthalene···$(H_2X)_n$ = 1, 2 (X = O, S) clusters. *J. Phys. Chem. A* 112: 6344–6350.

Cabidoche Y.M., R. Achard, P. Cattan, C. Clermont-Dauphin, F. Massat, and J. Sansoulet. 2009. Long-term pollution by chlordecone of tropical volcanic soils in the French West Indies: A simple leaching model accounts for current residue. *Environ. Poll.* 157: 1697–1705.

Cho Y., S.K. Min, J. Yun, W.Y. Kim, A. Tkatchenko, and K.S. Kim. 2013. Noncovalent inter-actions of DNA bases with naphthalene and graphene. *J. Chem. Theory Comput.* 9: 2090–2096.

Dawson G.W., W.C. Weimer, and S.J. Shupe. 1979. Kepone—A case study of a persistent material. *The American Institute of Chemical Engineers (AIChE) Symposium Series* 75: 366–374.

Durimel A. 2010. PhD thesis. Study of remediation pathways of environments and materials contaminated by chlorinated pesticides, HCH and chlordecone, University of french West Indies and Guyana.

Durimel A., S. Altenor, R. Miranda-Quintana, P. Couespel Du Mesnil, U. Jauregui-Haza, R. Gadiou, and S. Gaspard. 2013. pH dependence of chlordecone adsorption on activated carbons and role of adsorbent physico-chemical properties. *Chem. Eng. J.* 229: 239–249.

Enriquez-Victorero C., D. Hernández-Valdés, A.L. Montero-Alejo, A. Durimel, S. Gaspard, and U. Jáuregui-Haza. 2014. Theoretical study of γ-hexachlorocyclohexane and β-hexachlorocyclohexane isomers interaction with surface groups of activated carbon model. *J. Mol. Graph. Model.* 51: 137–148.

EPA. 2009. *Toxicological Review of Chlordecone (Kepone).* U.S. EPA Publications, Washington, DC, EPA/635/R-07/004F.

Faur C., A. Cougnaud, G. Dreyfus, and P. Le Cloirec. 2008. Modelling the breakthrough of activated carbon filters by pesticides in surface waters with static and recurrent neural network. *Chem. Eng. J.* 145: 7–15.

Fernández-Bayo J.D., C. Saison, M. Voltz, U. Disko, D. Hofmann, and A.E. Berns. 2013. Chlordecone fate and mineralisation in a tropical soil (andosol) microcosm under aerobic conditions. *Sci. Total Environ.* 463–464: 395–403.

Figueiredo J.L., M.F.R. Pereira, M.M.A. Freitas, and J.J.M. Órfao. 1999. Modification of the surface chemistry of activated carbons. *Carbon* 37: 1379–1389.

Gamboa-Carballo J.J., K. Melchor-Rodríguez, D. Hernández-Valdés, C. Enríquez-Victorero, A.L. Montero-Alejo, S. Gaspard, and U. Jaúregui-Haza. 2016. Theoretical study of chlordecone and surface groups interaction in an activated carbon model under acid and neutral conditions. *J. Mol. Graph. Model.* 65: 83–93 (submitted).

Gaspard S., A. Durimel, N. Passé-Coutrin, T. Cesaire, and V. Jeanne-Rose. 2014. Activated carbons from plantae and marine biomass for water treatment. In *Biomass for Sustainable Applications-Energy Production and Storage and Pollution Remediation*, Sarra G. and Ncibi C. (eds.). Green Chemistry Series Book. Royal Society of Chemistry, London, U.K. ISBN 978-1-84973-600-8.

Giles C.H., T.H. MacEwan, S.N. Nakwa, and D.J. Smith. 1960. Studies in adsorption. Part XI. A system of classification of solution adsorption isotherms, and its use in diagnosis of adsorption mechanisms and in measurement of specific surface areas of solids. *J. Am. Chem. Soc.* 3973–3993.

Guzelian P.S. 1982 Comparative toxicology of chlordecone (Kepone) in humans and experimental animals. *Annu. Rev. Pharmacol. Toxicol.* 22: 89–113.

Holliger C., S. Gaspard, G. Glod, C. Heijman, R.P. Schwarzenbach, W. Schumacher, and F. Vazquez. 1997. Contaminated environments in the subsurface and bioremediation. *FEMS Microbiol. Rev.* 20: 517–523.

Jablonski P., D.J. Pheasant, and J.G. Ferry. 1996. Conversion of Kepone by Methanosarcina thermophile. *FEMS Microbiol. Lett.* 139: 169–173.

Jensen B., T. Kuznetsova, B. Kvamme, and Å. Oterhals. 2011. Molecular dynamics study of selective adsorption of PCB on activated carbon. *Fluid Phase Equilibr.* 307: 58–65.

Jenness G.R. and K.D. Jordan. 2009. DF-DFT-SAPT investigation of the interaction of a water molecule to coronene and dodecabenzocoronene: Implications for the water–graphite interaction. *J. Phys. Chem. C* 113: 10242–10248.

Kalderis D., D. Koutoulakis, P. Paraskeva, E. Diamadopoulos, E. Otal, J.O. Valle, and C. Fernández-Pereira. 2008. Adsorption of polluting substances on activated carbons prepared from rice husk and sugarcane bagasse. *Chem. Eng. J.* 144: 42–50.

Landau-Ossondo M., N. Rabia, J. Jos-Pelage, L.M. Marquet, Y. Isidore, C. Saint-Aime, M. Martin, P. Irigaray, and D. Belpomme. 2009. Why pesticides could be a common cause of prostate and breast cancers in the French Caribbean Island, Martinique. An overview on key mechanisms of pesticide-induced cancer. *Biomed. Pharmacother.* 63: 383–395.

Lazar P., F. Karlický, P. Jurecka, M. Kocman, E. Otyepková, K. Šafářová, and M. Otyepka. 2013. Adsorption of small organic molecules on graphene. *J. Am. Chem. Soc.* 135: 6372–6377.

Macarie H. 2012. Gibbs free energy of formation of chlordecone and potential degradation products: implications for remediation strategies and environmental fate. *Environ. Sci. Technol.* 46: 8131–8139.

Mall I.D., V.C. Srivastava, and N.K. Agarwal. 2007. Adsorptive removal of Auramine-O: Kinetic and equilibrium study. *J. Hazard. Mater.* 143: 386–395.

McCallum C.L., T.J. Bandosz, S.C. McGrother, E.A. Muller, and K.E. Gubbins. 1999. A molecular model for adsorption of water on activated carbon: Comparison of simulation and experiment. *Langmuir* 15: 533–544.

Montero L.A., A.M. Esteva, J. Molina, A. Zapardiel, L. Hernandez, H. Marquez, and A.A. Acosta. 1998. Theoretical approach to analytical properties of 2,4-diamino-5-phenylthiazole in water solution. Tautomerism and dependence on pH. *J. Am. Chem. Soc.* 120(1): 12023–12033.

Multigner, L., P. Kadhel, F. Rouget, P. Blanchet, and S. Cordier. 2016. Chlordecone exposure and adverse effects in French West Indies populations. *Environ. Sci. Pollut. Res.*, 23(1): 3–8.

Njoku V.O. and B.H. Hameed. 2011. Preparation and characterization of activated carbon from corncob by chemical activation with H_3PO_4 for 2,4-dichlorophenoxyacetic acid adsorption. *Chem. Eng. J.* 173: 391–399.

Orndorff S.A. and R.R. Colwell. 1980. Microbial transformation of Kepone. *Appl. Environ. Microbiol.* 39(2): 398–406.

Oubal M., S. Picaud, M.T. Rayez, and J.C. Rayez. 2010. A theoretical characterization of the interaction of water with oxidized carbonaceous clusters. *Carbon* 48: 1570–1579.

PerezdeMoza A., L. Laquitaine, M.C. Ncibi, and S. Gaspard. 2014. Microorganisms for soil treatment. In *Biomass for Sustainable Applications-Energy Production and Storage and Pollution Remediation*, Sarra G. and Ncibi C. (eds.), Green Chemistry Series Book. The Royal Society of Chemistry, London, U.K.

Picaud S., B. Collignon, P.N.M. Hoang, and Rayez J.-C. 2008. Adsorption of water molecules on partially oxidized graphite surfaces: A molecular dynamics study of the competition between OH and COOH sites. *Phys. Chem. Chem. Phys.* 10: 6998–7009.

Ranguin R. 2015. PhD thesis, Optimization of chlordecone quantification and setting up of a degradation process by hybrid materials activated carbon/cobalamin, University of french West-Indies, December 2015.

Rodríguez-Reinoso F. and R. Molina-Sabio. 1998. Textural and chemical characterization of microporous carbons. *Adv. Colloid Interface Sci.* 76–77: 271–294.

Rouquerol F., J. Rouquerol, and K. Sing. 1999. Adsorption by active carbons, In *Adsorption by Powders and Porous Solids*. Academic press, London, U.K.

Salman J.M., V.O. Njoku, and B.H. Hameed. 2011. Bentazon and carbofuran adsorption onto date seed activated carbon: Kinetics and equilibrium. *Chem. Eng. J.* 173(2): 361–368.

Schrauzer G.N. and Katz R.N. 1978. Reductive dechlorination and degradation of mirex and kepone with vitamin B_{12}. *Bioinorg. Chem.* 9: 123–143.

Sulfita J.M., A. Horowitz, D.R. Shelton, and J.M. Teidje. 1982. Dehalogenation: A novel pathway for the anaerobic biodegradation of haloaromatic compounds. *Science* 218: 1115–1117.

Tapiero H.T. and B.G. Nguyen. 2002. New Estrogens and environmental estrogens. *Biomed. Pharmacother.* 56: 36–42.

Terzyk A.P., P. Gauden, W. Zielínski, S. Furmaniak, R.P. Wesołowski, and K.K. Klimek. 2011. First molecular dynamics simulation insight into the mechanism of organics adsorption from aqueous solutions on microporous carbons. *Chem. Phys. Lett.* 515: 102–108.

Section VII

Management Approach

17 Chlordecone Contamination at the Farm Scale

Management Tools for Cropping System and Impact on Farm Sustainability

Florence Clostre, Magalie Lesueur Jannoyer, Jean-Marie Gaude, Céline Carles, Louise Meylan, and Philippe Letourmy

CONTENTS

17.1 INTRODUCTION

Soil pollution has a direct impact on farmers. In the case of soil contamination by chlordecone in the French West Indies, farmers must adapt their breeding and cropping systems in polluted fields. For example, breeding of free-ranged animals (poultry, cattle, pig, etc.) should be avoided in contaminated soils as soil intake can result in exceeding the regulatory threshold (Bouveret et al. 2013, Jondreville et al. 2014, Jurjanz et al. 2014). For some crops, cultivation in open field is possible without exceeding the maximum residue limit (MRL, 20 μg kg^{-1}) even in soils contaminated by chlordecone (Clostre et al. 2014, 2015, Dubuisson et al. 2007). But for other crops, such as root vegetables or cucurbits, a risk not to comply with regulation exists (Clostre et al. 2014, 2015).

For production sale, farmers have to meet the regulatory threshold in the final product. Regulatory thresholds are relevant for exposure management to protect consumers but they are difficult to enforce for farmers. Indeed, in the case of historical soil pollution, no tool had been developed to anticipate the compliance against MRL of the harvested organs based on soil analysis results. To build a decision tool for farm management, we used the data gathered concerning chlordecone uptake by crops. Our final objective was to combine the crop uptake with the MRL value to determine threshold values for soil contamination.

We also studied the impact of chlordecone soil pollution on management and sustainability of farms producing staple food (root vegetables) in Martinique with the aim to assess how farmers tried to adapt to this constraint and whether or not they succeeded.

17.2 THE RELATIONSHIP BETWEEN CHLORDECONE CONTENT IN SOILS AND IN CROP PRODUCTS: GROUPING INTO CATEGORIES

With the aim to design a simple tool for growers, we assessed if it was possible to make a classification of crop products according to their contamination potential. We classified crop products according to their estimated mean chlordecone content: nonuptakers or low-uptakers were below the limit of quantification (1 µg kg^{-1}), high-uptakers exceeded the MRL (20 µg kg^{-1}) and, in between, crop products were classified as mid-uptakers (1 < mean content < 20 µg kg^{-1}) (Table 17.1). We showed that this classification accounted for the main part of chlordecone content in crop products compared to other significant effects (Unpublished data).

The categories segregated different kinds of crops: The fruit tree products and the solanaceous constitute the majority of the "no-uptaker or low uptaker" category; cucurbits and leaf vegetables were mainly in the "mid-uptaker" category; and lastly the root and tuber vegetables were in the "high-uptaker" category.

The study of chlordecone uptake by crops showed that chlordecone content in plant organs was partly determined by soil content but between field and within field, its variability could be high. In the case of root vegetables, the relationship was linear (Clostre et al. 2015). For above-ground organs, the high variability in uptake caused the relationship to be poorer (Clostre et al. 2014).

TABLE 17.1
Classification of Crop Products in Three Uptake Categories

No- or Low-Uptakers	Mid-Uptakers	High-Uptakers
Air potato, ambarella, banana, bell pepper, cabbage, chayote, chili pepper, corn, eggplant, green bean, guava, lime, okra, papaya, passion fruit, pineapple, starfruit, tomato	Cucumber, lettuce, pumpkin, spring onion (leaves)	Dasheen, radish, spring onion (bulb), sweet potato, yam

Based on these results, we performed statistical analyses of chlordecone content in plant samples (consumed crop product) and in corresponding soil using a large dataset, including 741 data from 26 different crop products (Unpublished data). We showed that the crop product, the soil chlordecone content, and the type of soil (andosol, ferralsol, or nitisol) for each product had highly significant effect on the crop product contamination. This confirmed the relationship already observed for a range of organochlorines—including chlordecone—between the content in soil and in crop (Gaw et al. 2008, Mikes et al. 2009, Waliszewski et al. 2008, Zohair et al. 2006). These results were also consistent with other studies showing that chlordecone uptake by plants varies according to the type of soil and to the crop product (see Chapter 10). Similarly, for other organochlorines, pollutant uptake was shown to depend on the crop product (Gaw et al. 2008, Mattina et al. 2000, Namiki et al. 2015). And organochlorine bioavailability in soil and uptake by plants are known to depend on soil type (Alexander 2000, Harris and Sans 1967, Kiflom et al. 1999).

This first work provided information on the possibility of growing some crops on polluted soil without risking overlap of the regulatory threshold, regardless of the soil contamination, and on the high risk of growing roots and tubers on polluted soil. But a practical tool was needed to enlarge the crop options for growers according to the field chlordecone content.

17.3 FROM REGULATION TO PRACTICAL RECOMMENDATION: MAXIMUM RESIDUE LIMITS IN PRODUCT TRANSLATION IN SOIL CONTENT THRESHOLDS

We also demonstrated that compliance with regulation (MRL) was linked with the crop product and the uptake category (Unpublished data). The uptake category is thus relevant to be used when developing a simple tool for management by the farmers of cropping systems at the farm scale.

In a first work, an empiric approach was used to translate the MRL into threshold values for chlordecone contamination in soils. Indeed, vegetables did not exceed MRL when soil contamination remained below levels depending on the vegetable uptake potential (Clostre et al. 2011, Lesueur-Jannoyer et al. 2012). Based on the relationship between chlordecone content in soil and crop product, threshold values were determined for each uptake category using the maximum transfer observed.

The management tool was designed such as by reaching a compromise for simple and acceptable threshold values for farmers. In order to simplify the tool for farmers, and despite their known effect on chlordecone uptake (Clostre et al. 2014, 2015, Woignier et al. 2012), soil type and growing conditions were not taken into account. Indeed soil type is not always easy to assess, especially because of transitional soils, and growing conditions would be difficult to evaluate. Moreover, the threshold values used in the management tool were rounded to the easiest values to remember. We obtained the following operational threshold values: 0.1 and 1 mg kg^{-1} dry soil.

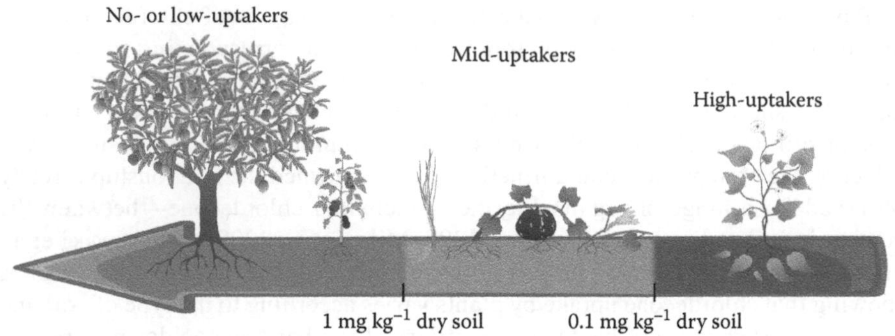

FIGURE 17.1 **(See color insert.)** Soil thresholds and arrow of the noncompliance risk according to crop uptake potential.

These threshold values led to three cases according to soil contamination (Figure 17.1):

1. Below 0.1 mg kg^{-1}: all crops can be cultivated
2. Between 0.1 and 1 mg kg^{-1}: recommendation to avoid cultivation of crops belonging to the high-uptaker category (root and tuber vegetables)
3. Above 1 mg kg^{-1}: recommendation to avoid cultivation of crops belonging to both the high-uptaker and mid-uptaker category (mostly root and tuber vegetables, cucurbits, and lettuces)

With this tool, farmers of the French West Indies can anticipate the contamination risk of crops and choose the most suitable crops for their cropping system based on the soil contamination of their fields. The overall aim was to allow farmers to keep cultivating their contaminated fields, even the "most" polluted.

This management tool has been integrated in the management programs of extension services for several years. In the French West Indies, the adoption of the tool by farmers led to a shift toward compliance of crop products with MRL (in the last few years according to the national control program (Direction de l'Alimentation 2015).

A more thorough statistical approach based on a predictive model of crop contamination with the chlordecone content in soil as a covariate is in progress and should allow determining thresholds according to the type of soil, the crop product, or the uptake category.

Contamination of soil by chlordecone led a part of the growers to modify their cropping system but it is difficult to assess the extent of the phenomenon. This is the reason why we took interest in the impact of soil pollution by chlordecone on farm sustainability.

17.4 IMPACT ON FARM SUSTAINABILITY: THE CASE OF ROOT VEGETABLE PRODUCTION

With the discovery of widespread contamination of foodstuffs and ecosystems in the French West Indies, regulatory action was taken. In 2003, an analysis of chlordecone content in soils was made mandatory prior to growing root crops (decree N°030725 of the regional Prefect of Martinique). MRL for chlordecone in crop products was set at 50 μg kg^{-1} in 2005 for some crop products, including root crops (ministerial order ECOC0500128A) and then the threshold was reduced to 20 μg kg^{-1} in 2008 by application of the "hygiene package" (ministerial order AGRG0816067A).

Root crop producers have been the most impacted among growers by the successive regulations regarding management of chlordecone contamination as root and tuber vegetables belong to the high-uptaker category. We studied farms producing staple food in Martinique with (24 farms) or without (10 farms) a part of their plots contaminated. Staple food production has been on the decline along with utilized agricultural land since the 1980s (DAAF Martinique 2003, 2012; Figure 17.2). Farmer interviews were conducted according to a methodology based on the work of Sutherland et al. (2012) and Wilson (2008), and farm trajectories were analyzed through a decisional model.

Among the farmers with contaminated plots, 9 farmers had plots contaminated over 1 mg kg^{-1} dry soil, and 10 had plots between 0.1 and 1 mg kg^{-1}. A very small number of farmers limited their production to root vegetables or to only a crop. Diversification was a generalized practice as staple food products were often a complementary activity to market gardening (cucumber, lettuce, etc.). Different triggers (opportunities or constraints) modified farm trajectories: land property, labor force, pests and diseases, equipment, and so on. Land property and pests were the most frequent constraints, followed by supply chain organization and labor force availability and cost (Figure 17.3). Chlordecone regulation was also quite frequently cited.

Farmers impacted by chlordecone adapted by modifying their crop rotations. Those with fields not contaminated by chlordecone moved the production of

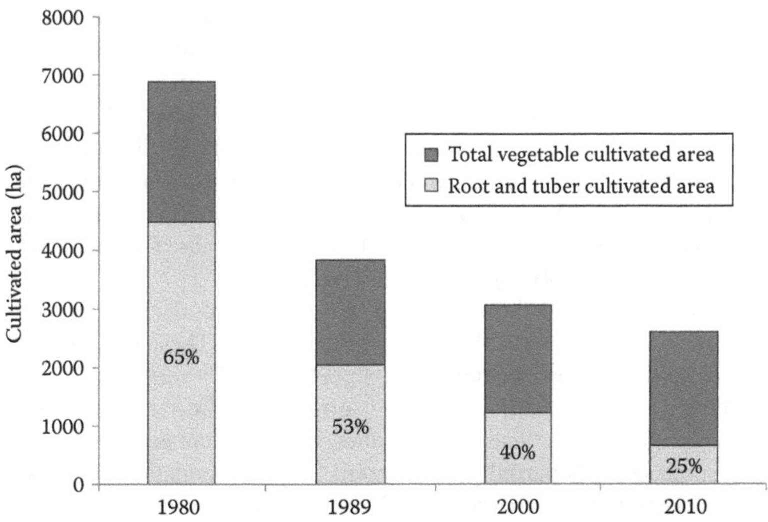

FIGURE 17.2 Cultivated area decline for the vegetable crops and the percentage of roots and tubers in Martinique. (Data from Ministry of Agriculture, Martinique, France, 2012.)

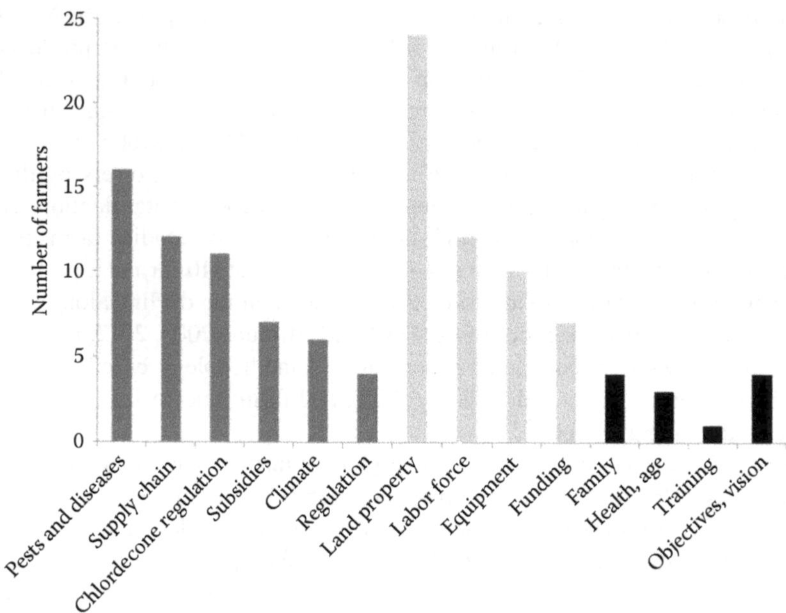

FIGURE 17.3 Factors that influence changes in farm management for root and tuber farming systems (in dark gray, farm external factors, in light gray, farm internal factors, in dark, farmer's characteristics).

chlordecone uptaker crops to chlordecone-free fields and limited their production on chlordecone polluted soils to crops posing no risk of exceeding regulatory threshold according to the decision tool presented earlier. Farmers whose fields were all contaminated above the recommended threshold of 0.1 mg kg^{-1} (dry soil) stopped growing root vegetables in favor of other crops. Thus, diversification of farming system was a factor of resilience and sustainability for small farms, which adapted and managed the sanitary risk for their products. This is consistent with Wilson (2007), who theorizes that most farmers attempt to remain within the scope of known successful management strategies. Major changes can occur when perceived constraints and/or opportunities are catalyzed by a trigger event, which will cause a reassessment of the farm and its management strategy (Sutherland et al. 2012).

Among the interviewed farmers, there was neither total crop orientation nor farm disappearance related to chlordecone soil pollution. These situations of major changes were due to other constraints or opportunities. Nevertheless, pollution by chlordecone was a compounding factor for already fragile farms by limiting the choice of crop productions and causing uncertainty or unnecessary precautions. Moreover, some growers partially reconverted to fruit production or soil-free production (greenhouse or hydroponics), which necessitated financial investment in new materials and also involved delay before production.

The stability of root vegetable farms with polluted fields could thus be obtained first through farming system diversification that allowed crop relocation or crop substitution, and second through the financial support for farming system restructuration (mostly diversification) or farm equipment modernization. To improve the process, technical and financial assistance could be provided specifically to these small farms to monitor their trajectory along time and support them with adapted training sessions.

17.5 CONCLUSION

The tools and analyses were set up in real conditions, with the integration of soil and climate specificities but also of farming system constraints. The simple and empirical tool to support farmers in their crop choice according to the field chlordecone content allowed farmers to anticipate the sanitary impact of polluted soil on their marketed products: limited risk could be taken at planting instead of compliance control at or after harvest.

In the case of root vegetable production, chlordecone soil pollution was not the main factor of the decline of cultivated areas, which was mainly explained by land property and the rise and difficulty in controlling new pests and diseases. In a polluted situation, farming system diversification and financial support guarantee farm stability and sustainability, but an effort could be made in the monitoring of the farm trajectories and in the technical support and training offered to these small farmers.

ACKNOWLEDGMENTS

Funding was provided by the French Chlordecone National Plan and Regional Health Agency of Martinique through the "JAFA" Family Garden Health Program, and by European Union under the FEDER programs. The authors thank the Directorate for Food, Agriculture and Forestry in Martinique for the data they provided.

REFERENCES

Alexander, M. 2000. Aging, bioavailability, and overestimation of risk from environmental pollutants. *Environmental Science & Technology* 34(20):4259–4265. doi: 10.1021/es001069+.

Bouveret, C., G. Rychen, S. Lerch, C. Jondreville, and C. Feidt. 2013. Relative bioavailability of tropical volcanic soil-bound chlordecone in piglets. *Journal of Agricultural and Food Chemistry* 61(38):9269–9274. doi: 10.1021/jf400697r.

Clostre, F., P. Cattan, J.-M. Gaude, C. Carles, P. Letourmy, and M. Lesueur-Jannoyer. 2015. Comparative fate of an organochlorine, chlordecone, and a related compound, chlordecone-5b-hydro, in soils and plants. *Science of the Total Environment* 532:292–300. doi: 10.1016/j.scitotenv.2015.06.026.

Clostre, F., M. Lesueur-Jannoyer, and B. Turpin. 2011. Impact des modes de préparation des aliments sur l'exposition des consommateurs à la chlordécone. Cirad, Martinique, France.

Clostre, F., P. Letourmy, and M. Lesueur-Jannoyer. 2015. Organochlorine (chlordecone) uptake by root vegetables. *Chemosphere* 118:96–102. doi: 10.1016/j.chemosphere.2014.06.076.

Clostre, F., P. Letourmy, B. Turpin, C. Carles, and M. Lesueur-Jannoyer. 2014. Soil type and growing conditions influence uptake and translocation of organochlorine (chlordecone) by *Cucurbitaceae* species. *Water, Air, & Soil Pollution* 225(10):1–11. doi: 10.1007/s11270-014-2153-0.

DAAF Martinique. 2003. Mémento de la statistique agricole. In *Agreste*. Direction de l'alimentation de l'agriculture et de la forêt de la Martinique, Fort-de-France, Martinique.

DAAF Martinique. 2012. Mémento de la statistique agricole. In *Agreste*. Direction de l'alimentation de l'agriculture et de la forêt de la Martinique, Fort-de-France, Martinique.

Direction de l'Alimentation, de l'Agriculture et de la Forêt de la Martinique. 2015. *Résultats des plans de contrôle dans les denrées végétales*. http://daaf972.agriculture.gouv.fr/spip.php?rubrique388. Accessed April 23, 2015.

Dubuisson, C., F. Héraud, J.-C. Leblanc, S. Gallotti, C. Flamand, A. Blateau, P. Quenel, and J.-L. Volatier. 2007. Impact of subsistence production on the management options to reduce the food exposure of the Martinican population to chlordecone. *Regulatory Toxicology and Pharmacology* 49(1):5–16. doi: 10.1016/j.yrtph.2007.04.008.

Gaw, S.K., N.D. Kim, G.L. Northcott, A.L. Wilkins, and G. Robinson. 2008. Uptake of ΣDDT, arsenic, cadmium, copper, and lead by lettuce and radish grown in contaminated horticultural soils. *Journal of Agricultural and Food Chemistry* 56(15):6584–6593. doi: 10.1021/jf073327t.

Harris, C.R. and W.W. Sans. 1967. Absorption of organochlorine insecticide residues from agricultural soils by root crops. *Journal of Agricultural and Food Chemistry* 15(5):861–863. doi: 10.1021/jf60153a022.

Jondreville, C., A. Lavigne, S. Jurjanz, C. Dalibard, J.-M. Liabeuf, F. Clostre, and M. Lesueur-Jannoyer. 2014. Contamination of free-range ducks by chlordecone in Martinique (French West Indies): A field study. *Science of the Total Environment* 493:336–341. doi: 10.1016/j.scitotenv.2014.05.083.

Jurjanz, S., C. Jondreville, M. Mahieu, A. Fournier, H. Archimède, G. Rychen, and C. Feidt. 2014. Relative bioavailability of soil-bound chlordecone in growing lambs. *Environmental Geochemistry and Health* 36(5):911–917. doi: 10.1007/s10653-014-9608-5.

Kiflom, W.G., S.O. Wandiga, P.K. Ng'ang'a, and G.N. Kamau. 1999. Variation of plant *p,p'*-DDT uptake with age and soil type and dependence of dissipation on temperature. *Environment International* 25(4):479–487. doi: 10.1016/S0160-4120(99)00005-7.

Lesueur-Jannoyer, M., P. Cattan, D. Monti, C. Saison, M. Voltz, T. Woignier, and Y.-M. Cabidoche. 2012. Chlordécone aux Antilles: évolution des systèmes de culture et leur incidence sur la dispersion de la pollution. *Agronomie Environnement & Sociétés* 2(1):45–58.

Mattina, M.I., W. Iannucci-Berger, and L. Dykas. 2000. Chlordane uptake and its translocation in food crops. *Journal of Agricultural and Food Chemistry* 48(5):1909–1915. doi: 10.1021/jf990566a.

Mikes, O., P. Cupr, S. Trapp, and J. Klanova. 2009. Uptake of polychlorinated biphenyls and organochlorine pesticides from soil and air into radishes (*Raphanus sativus*). *Environmental Pollution* 157(2):488–496. doi: 10.1016/j.envpol.2008.09.007.

Namiki, S., T. Otani, N. Seike, and S. Satoh. 2015. Differential uptake and translocation of β-HCH and dieldrin by several plant species from hydroponic medium. *Environmental Toxicology and Chemistry* 34(3):536–544. doi: 10.1002/etc.2815.

Sutherland, L.-A., R.J.F. Burton, J. Ingram, K. Blackstock, B. Slee, and N. Gotts. 2012. Triggering change: Towards a conceptualisation of major change processes in farm decision-making. *Journal of Environmental Management* 104:142–151. doi: 10.1016/j.jenvman.2012.03.013.

Waliszewski, S.M., O. Carvajal, S. Gómez-Arroyo, O. Amador-Muñoz, R. Villalobos-Pietrini, P.M. Hayward-Jones, and R. Valencia-Quintana. 2008. DDT and HCH isomer levels in soils, carrot root and carrot leaf samples. *Bulletin of Environmental Contamination and Toxicology* 81(4):343–347. doi: 10.1007/s00128-008-9484-8.

Wilson, G. 2007. *Multifunctional Agriculture: A Transition Theory Perspective.* CABI, Wallingford, CT.

Wilson, G.A. 2008. From "weak" to 'strong' multifunctionality: Conceptualising farm-level multifunctional transitional pathways. *Journal of Rural Studies* 24(3):367–383. doi: 10.1016/j.jrurstud.2007.12.010.

Woignier, T., F. Clostre, H. Macarie, and M. Jannoyer. 2012. Chlordecone retention in the fractal structure of volcanic clay. *Journal of Hazardous Materials* 241–242:224–230. doi: 10.1016/j.jhazmat.2012.09.034.

Zohair, A., A.-B. Salim, A.A. Soyibo, and A.J. Beck. 2006. Residues of polycyclic aromatic hydrocarbons (PAHs), polychlorinated biphenyls (PCBs) and organochlorine pesticides in organically-farmed vegetables. *Chemosphere* 63(4):541–553. doi: 10.1016/j.chemosphere.2005.09.012.

18 The *Family Gardens* Health Program

A Tool for Reducing the Exposure of Populations to Chlordecone in the French West Indies

Eric Godard and Guillaume Pompougnac

CONTENTS

18.1 INTRODUCTION

A number of measures have been taken since 1999 in Martinique and since 2000 in Guadeloupe to manage risks associated with chlordecone pollution. To limit consumers' exposure, the French Food Safety Agency (AFSSA) identified risky foods and laid down maximum residue limits (MRLs) for chlordecone. In its first assessment, the AFSSA underlined that the proposed regulation would not protect anyone who purchased food outside the commercial distribution channels: products for home consumption, gifts, and supplies from unregistered producers, for which other actions would be recommended. Previously, population groups at high risk of exposure had been identified by the Interregional Epidemiology Cell (Cire-AG), a local branch of the French Institute of Health Watch (InVS), to better target possible actions. This chapter describes this whole process.

18.2 CHARACTERIZATION OF POPULATION GROUPS AT HIGH RISK OF EXPOSURE TO CHLORDECONE THROUGH FOOD

The risk of exposure to chlordecone by the population of the French West Indies was evaluated in surveys of food consumption from 2005 to 2007 (AFSSA 2005, 2007). The risk for exceeding the toxicity reference value (TRV) concerned 1.9% of the population in Martinique and 1.3% in Guadeloupe (Blateau et al. 2011). The main results of the surveys can be summarized as follows:

- People living in a polluted area are likely to exceed the TRV.
- Children have a higher risk than adults.
- Consumption more than twice a week of root vegetables (e.g., sweet potatoes) or fish purchased from the short-circuit supply chain increases the risk of exposure.

To reduce the risk of exceeding the TRV, the AFSSA recommended not eating, more than twice a week, root vegetables grown in home gardens whose soil chlordecone concentration is unknown. In 2007, the AFSSA also recommended to respect the ban on fishing and to limit to four times a week the consumption of sea products whose origin was unknown. Family gardens polluted by chlordecone may be a source of high exposure through consumption of contaminated vegetables. These data justified the implementation of preventive actions targeting the populations concerned.

18.3 THE CONSTRUCTION OF JAFA PROGRAMS

The family gardens health program (JAFA) received a budget of €6.3 million (Chlordecone Interdepartmental Plan 2008–2010). It was piloted by the Directorates of Health and Social Development in Guadeloupe and Martinique (DSDS) and included the following actions:

- Identification of potential exposure to chlordecone through home surveys
- Regional and local information campaigns to raise awareness of the problem among the general population and the dissemination of recommendations
- A health education support program for people potentially exposed, to help them change their food habits, thereby reducing their exposure

One of the main challenges of the program was to avoid encouraging consuming industrial food with very high fat in a region with a high prevalence of obesity, diabetes, and hypertension. Another challenge involved was avoiding stigmatizing Creole family gardens, which are part of the historical heritage of the region.

18.4 PREPARATION OF SURVEYS FOR THE IDENTIFICATION OF THE POPULATIONS CONCERNED

18.4.1 Preparing Maps of the Plots to Be Surveyed by the Investigators

Polluted areas are more diffuse in Martinique, while in Guadeloupe, they are mainly located in the Basse-Terre part of the Island. In Martinique, the map of the risk of soil pollution by chlordecone, based on land occupation by banana plantations between 1970 and 1995 (Desprats et al. 2004) and updated by the Regional Directorate of Environment in 2007, was used along with cadastral maps, maps of built-up areas published by the National Geographic Institute (2002) and data from the 1999 population census conducted by the National Statistics Institute. This led to the identification of 17,887 plots to be inspected in 31 towns.

In Guadeloupe, the reference map showing potentially polluted soils in Guadeloupe was drawn by the Plant Protection Service of the Department of Agriculture and Forestry. This map is based on a land-use map showing the location of banana plantations between 1971 and 1993, supported by nearly 3,500 soil analyses. It distinguishes four levels of risk of soil contamination depending on the period of occupation of the banana monocrop. The two strongest risk levels, with a strip of 100 m around the plots, were taken into account. A total of 19,722 buildings were included in the survey in 11 towns in Basse-Terre.

18.4.2 Communication Prior to Field Surveys

Given the population's own representations of the problem and the lack of appropriate concrete information on the subject (INPES 2008), it was necessary to inform the population about the areas to be surveyed, to increase their knowledge about chlordecone, and to inform them about the preparation of the surveys to be conducted by the investigators. In Martinique, prior to the arrival of the investigators, information meetings for elected representatives and the general public were held in all the towns concerned, followed by meetings that were held in each neighborhood concerned. A total of 56 public meetings were attended by around 1,500 people between 2008 and 2010. In Guadeloupe, before the investigations started, 40 neighborhood meetings were organized in a fun and interactive way between 2009 and 2012, and were attended by a total of 1,220 people.

18.5 IDENTIFICATION OF POPULATIONS AT RISK OF HIGH EXPOSURE

After their training, the investigators conducted a door-to-door survey in each neighborhood. Soil tests were proposed if the home included a Creole garden where root vegetables were grown and eaten twice a week or more.

In Martinique, the investigators were made available by environmental protection associations (SEVE and Goutte d'Eau Lorrinoise) and coordinated by the Regional Education Committee for Health, which became the Regional Forum for Education and Promotion of Health (CRES-IREPS). The surveys were conducted between November 2007 and June 2010 in 31 municipalities. In late 2010, a diagnostic procedure was set up. All those who wished to know the soil pollution status of their family garden could get free sampling and get it analyzed. This procedure still exists, and the aim is to make it permanent. In Guadeloupe, the Pasteur Institute of Guadeloupe/Qualistat team conducted surveys and analyses between April 2009 and the end of December 2012 under the supervision of the IREPS of Guadeloupe. Both in Guadeloupe and Martinique, the risk of high exposure was assessed based on the assumption that only vegetables contributed to exposure to chlordecone. Figure 18.1 shows the investigation process that DSDS and IREPS prepared for Guadeloupe and Martinique.

When designing the program in Martinique, the threshold for soil contamination (i.e., the value that may lead to a chlordecone exposure exceeding the TRV via the consumption of root vegetables) was set at 250 $\mu g\ kg^{-1}$. This threshold was based on AFSSA recommendations on acceptable residue limits in food (50 $\mu g\ kg^{-1}$) and on the contamination susceptibility of the most sensitive plants (Cabidoche et al. 2006). However, the new MRLs established in 2008 (20 $\mu g\ kg^{-1}$) led to reduction of the soil contamination threshold to 100 $\mu g\ kg^{-1}$ of chlordecone.

Table 18.1 shows the achieved diagnoses at the end of the first semester of 2015. The consolidated numbers of households with soil sampled and potentially exposed households differ slightly from those presented by Vincent et al. (2011).

In Martinique, diagnostic surveys on request (911 analysis versus 1,401 initially) revealed new situations showing potential risks of exposure, and resulted in a nearly 60% increase in the number of homes concerned, that is, exceeding the threshold values.

In Guadeloupe, where no "diagnosis on request" was available, soil analyses highlighted proportionally much larger risks of exposure: 46% of the homes where a diagnosis was made have gardens containing soil that may produce root vegetables exceeding the MRLs, versus 21% in Martinique. Contamination levels were also higher; about 37% of homes in Guadeloupe whose garden soil was analyzed had a threshold of more than 250 $\mu g\ kg^{-1}$, versus 10% of homes in Martinique.

The assessment conducted by the Cire-AG on the basis of Escal, Calbas, and RESO surveys indicated that about 1,340 people in Martinique and 1,000 in Guadeloupe were likely to be exposed to chlordecone exceeding the TRV, especially by consuming

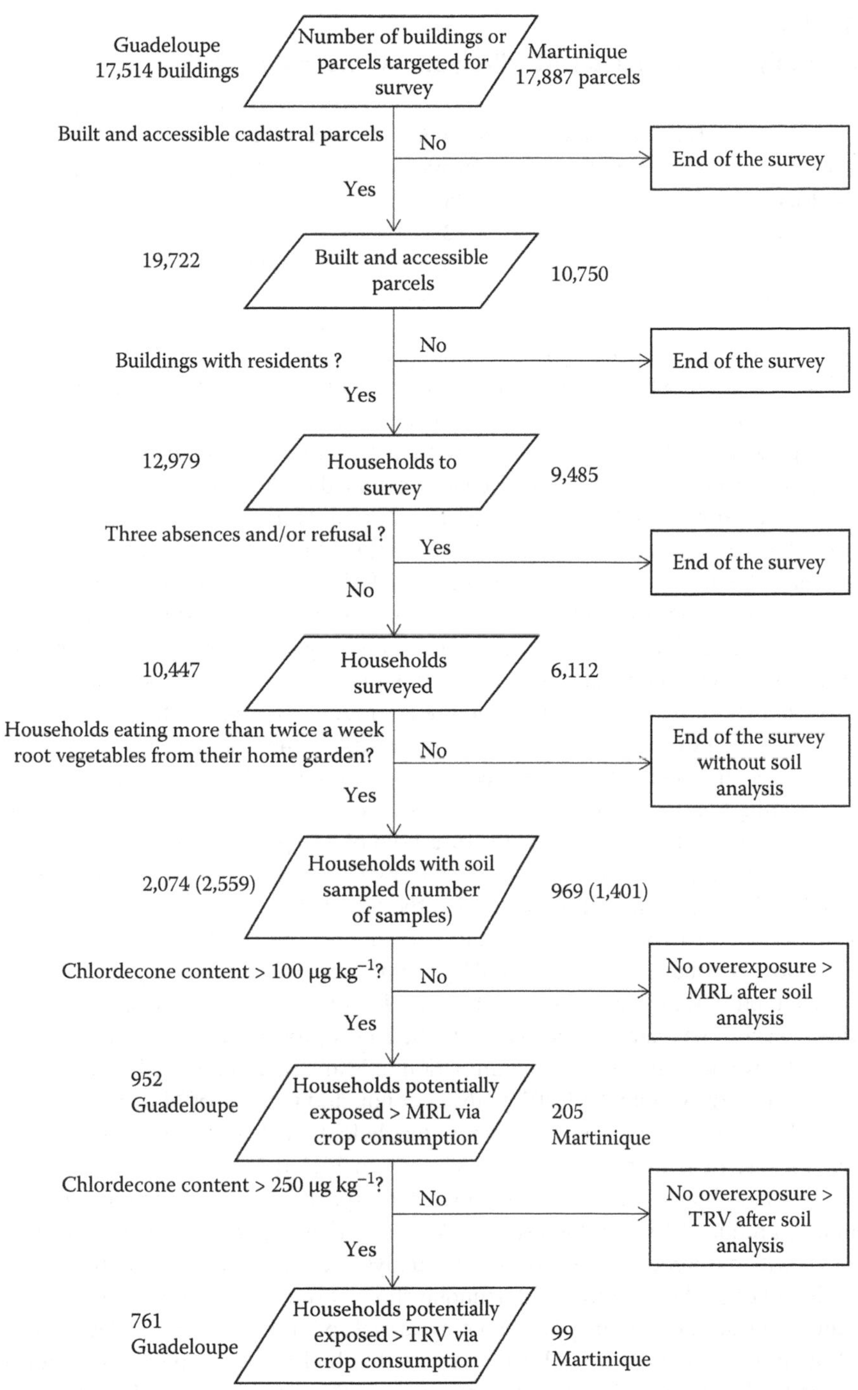

FIGURE 18.1 Investigation process.

TABLE 18.1
Soil Analyses Conducted (JAFA Program) at the End of 2015

Island	No. of Diagnoses of Homes	No. of Soil Analyses	No. of Homes with Soil CLD > = 100 μg kg^{-1}	No. of Homes with Soil CLD > = 250 μg kg^{-1}
Guadeloupe	2,074	2559	952	761
Martinique	1,632	2,312	316	160
Total	3,706	4,871	1,268	921

vegetables purchased from short-circuit supply chain (Blateau et al. 2011). The JAFA health program identified 761 homes (around 2000 people) in Guadeloupe and 160 homes (around 400 people) in Martinique whose garden soil was likely to produce root vegetables with a threshold of more than 50 μg kg^{-1}. Comparing results of the Cire-AG and of JAFA, one should conclude that the Cire-AG underestimated the risk in Guadeloupe and overestimated the risk in Martinique. Such a conclusion cannot be drawn since several factors, which the JAFA programs did not take into account, influence the risk of exceeding the TRV, as already discussed by Vincent et al. (2011). In addition, householders in nearly a third of Martinique homes could not be contacted during the first stage of the investigation, meaning the number of polluted plots and people at risk was certainly underestimated. Among the exposure factors not taken into account were products produced by family farms and fishery products, which were also identified as contributing to the risk of exceeding the TRV. Conversely, other factors led to overestimation of the population exposed to the risk of exceeding the TRV due to the consumption of root vegetables. First, we cannot be sure that when the soil exceeds the threshold of 250 μg kg^{-1}, the TRV is exceeded with the consumption of these vegetables grown on this soil; notably, research conducted by Clostre et al. (2011) showed that the transfer rates are generally lower than the upper-bound hypothesis (20%) and vary depending on the types of soils and vegetables. Second, reaching the TRV probably requires a much higher frequency of consumption than that recommended by AFSSA: eating the root vegetables only twice a week. The analyses performed by the JAFA program also revealed a trend in the predictive mapping of soil contamination in Martinique showing homes likely to be overexposed: Among the 1,401 analyses conducted in areas of Martinique considered to be at risk of pollution, 969 were in fact below the detection limit. Conversely, overexposure of homes located in areas not considered at risk on the map is nevertheless possible as only bananas grown for export were taken into account. But bananas produced for the domestic market and other crops were also occasionally treated with chlordecone. As a result, the analyses conducted for food producers by the Chamber of Agriculture in Martinique showed that 30% of polluted plots were located outside the area mapped as being at risk of soil pollution. In Guadeloupe, soil analyses conducted by JAFA on parcels that had not been identified as former banana fields revealed chlordecone residues in 65% of cases, with values exceeding 100 μg kg^{-1} in 39% of the plots (report by the IREPS Steering Committee JAFA 2012).

18.6 RETURN OF TEST RESULTS AND HEALTH RECOMMENDATIONS

How the results were returned to householders affected by soil testing was not the same in Guadeloupe and Martinique. In Martinique, the test results were posted to householders, with appropriate recommendations for the level of contamination of their soil. A home visit by a counselor was then proposed for all homes whose level of pollution exceeded 100 μg kg^{-1}, followed by two more visits to consolidate the expected behavioral changes. A single visit was proposed to householders who had requested a diagnosis. In Guadeloupe, results were returned by post to homes with soil pollution levels below 100 μg kg^{-1}. In the case of higher levels of pollution, the results were first presented and discussed in person by an advisor during a single home visit and then confirmed by post. In both cases, the objective was to avoid the dispersal of garden products and to help maintain a good nutritional balance by diversifying the household's supply of fruits, vegetables, and local starches. In Martinique, the letters presenting the results made the following recommendations for different levels of pollution:

- Results of soil analysis showing pollution between 100 and 250 μg kg^{-1}: Vegetables should not be sold but no limits were recommended concerning home consumption
- Results of soil analysis showing pollution between 250 and 1,000 μg kg^{-1}: Recommended that root vegetables should not be eaten more than twice a week
- Results of soil analysis showing pollution exceeding 1,000 μg kg^{-1}: Recommended stopping eating all root vegetables and cucurbits grown in the garden

Second, during three visits per household, a nutritionist and an agronomist examined individual actions that could reduce exposure, including the following:

1. Stopping or reducing consuming vegetables at risk, depending on the level of pollution
2. Replacing risky crops by less susceptible crops (tomato, pumpkin, etc.)
3. Relocating susceptible crops to less contaminated areas, using cultivation techniques that reduce exposure (e.g., adding soil) and diversify consumption by always eating a green vegetable with a root vegetable

By the end of the three scheduled visits (2014), 46 homes had not been visited by councilors. In Guadeloupe, the restitution phase of results to householders at risk of overexposure started in March 2010 and ended in December 2013. A total of 858 homes were visited. In all, 2,514 letters including results of the analyses were sent. The letter was slightly different from the letter sent in Martinique: When the level of soil contamination exceeded 100 μg kg^{-1}, it recommended limiting consumption to twice a week combined with thorough washing and peeling a thick outer slice of the vegetable; for contamination exceeding 1,000 μg kg^{-1}, it recommended stopping eating root vegetables. In Guadeloupe, the possibility of exceeding the MRL therefore prevailed over AFSSA recommendations based on the assessment of risk updated in 2007. Recommendations about crop diversification and food as well as the introduction of alternative cultivation techniques were the same on the two islands.

18.7 COMMUNICATION AND SUPPORT ACTIONS FOR THE PROGRAM

To promote food diversification and the consumption of fresh fruits and vegetables, local campaigns were conducted.

18.7.1 Collective Animations from Homes Identified as Being at the Risk of Exposure

In Martinique, agricultural and nutritional workshops were held for families referred by JAFA. In Guadeloupe, two collective concepts were created, the most advanced being the "JAFA Pilot Garden" (2010–2014) at homes where the level of contamination of the soil in the garden was between 100 and 1,000 μg kg^{-1}. These fun interactive events took the form of an "open house" in the garden of a householder with polluted soil. Practical ways of reducing exposure to chlordecone while continuing to cultivate the garden were presented such as growing a wider range of vegetables, promoting biodiversity and cultivation of the Creole garden, and encouraging food diversification. Fifty-eight percent of householders whose garden soil contained between 100 and 1,000 μg kg^{-1} chlordecone attended one of the 26 events organized. The same format will be used in 2016 for homes with highly polluted gardens (>1,000 μg kg^{-1}).

18.7.2 Mass Communication

General information on chlordecone pollution, its effects, and ways to protect oneself were broadcast to the public on radio and television. Efforts in Guadeloupe had a major impact. Thirty-two programs in French and Creole were broadcast in 2014, with questions asked by members of the general public and answers provided by researchers, and with scenes in Creole tackling problems of chlordecone pollution in everyday life. The same concept was used in Martinique in 2015. Two series were broadcast on television and shown in the cinema to publicize the JAFA program, including the opportunity to request a diagnosis, and the precautions to take with respect to products grown in family gardens.

18.7.3 Economic Support

In Martinique in 2010, an economic support program for families identified as potentially at risk of high exposure was implemented by an association that subsidized for the purpose. It took the form of a social grocery store that sold healthy root and tuber vegetables at 10% of the market price to households in the JAFA program to compensate for the reduction in consumption of garden products. In 2013, the association suggested creating a local exchange system in which JAFA households were invited to come and exchange services and safe products, whether products that are not sensitive to CLD soil contamination or CLD-free products grown in pollution-free gardens. Around one hundred families benefited from the distribution of home vegetables between 2010 and 2013, as very few householders actually went to the grocery store. The local exchange system, which replaced the social store where the first exchanges were organized in 2014, is still struggling to involve JAFA homes. In 2010, an experiment was set up in which "ready to plant" garden plots were made available in several municipalities. But it was not successful. People did not join for several reasons: reluctance to go somewhere else to work in the garden, worries the produce would be stolen or damaged, and so on.

18.8 THE SEARCH FOR ADDITIONAL RISK MANAGEMENT TOOLS

As Vincent (2011) wrote, to enable advisors to provide additional management tools to families, CIRAD Martinique was commissioned to conduct a series of studies on the effect of preparing vegetables on the final level of exposure to chlordecone. Research was conducted on different types of vegetables (dasheen, pumpkin, cucumber, sweet potatoes, yams, bulbiferous yams) and different compartments (skin, pulp, whole fruit) using several different modes of preparation (peeled, cooked, eaten raw, etc.). The study confirmed the favorable impact of peeling on the level of contamination of certain vegetables (pumpkin, sweet potato, dasheen) and that cooking had no favorable effect (Clostre et al. 2011, 2014). Knowledge was improved on the limits of contamination of different types of soil compatible with these crops (Clostre et al. 2011). The results of this research also led to the drafting of documents for distribution to householders so that vegetables can be safely grown depending on the level of soil pollution.

18.9 ACCOUNTING FOR THE RISKS INVOLVED IN BACKYARD FARMS

Two studies on ducks and laying hens revealed their high susceptibility to contamination via the soil and their food (Jondreville et al. 2013, 2014). A study was conducted in actual family farming conditions to analyze the contamination of products from these farms (poultry meat, liver, eggs) and to recommend preventive actions (see Chapter 11 for more information). The characteristics of animal housing and outside pens and the methods of feeding and watering were described, along with laying frequency and growth. The floors of the buildings and plots were analyzed. The thighs, abdominal fat, liver, and ovarian cluster of hens and roosters were analyzed, and, in some cases, the eggs laid by the hens too were analyzed. The results of the analysis revealed the risk of the MRL being exceeded when soil contamination exceeded 100 $\mu g\ kg^{-1}$. One possible way to keep contamination below the MRL is preventing the poultry having direct contact with the soil. Family farmers were then informed of the risk of contamination of their poultry and eggs in the case of soil pollution exceeding 100 $\mu g\ kg^{-1}$

and of the precautions to be taken: the minimum being to have the soil of the pens analyzed for pollution and to respect the recommendations outlined in an information brochure on animal housing, outside pens, and poultry feed.

18.10 PROGRAM EVALUATION

The first evaluation of the first 150 households who received the results of soil analysis (IPSOS 2012) was conducted in Guadeloupe in January and February 2012. It reported a very positive assessment of the support provided by JAFA by its beneficiaries, the efficiency of home support to encourage the implementation of recommendations, and the impact and benefits of participating in the activities organized within the program. However, it also reported the persistence of certain risk behaviors (failure to wash the vegetables or to peel a thick outside layer, continued consumption of root vegetables grown in contaminated gardens with no special precautions taken) and the fact that 50% of households raise animals for domestic consumption.

Another qualitative assessment, with five groups of people covered by the program, was conducted in 2013 and led by a sociologist of the Regional Health Observatory of Guadeloupe (ORSaG 2014). This evaluation highlighted a number of obstacles to the adoption of the changes proposed by the JAFA program: low credibility of the discourse of the authorities because they were held responsible for the pollution, questionable legitimacy of certain recommendations, the belief "they're not telling us everything" about the consequences of pollution, and deeply rooted values that may represent resistance to change, such as attachment to the garden (guarantee of independence and freedom) and "eat local food" to stay healthy and to affirm one's cultural identity.

Difficulties in accessing resources or the solutions proposed by the JAFA program were also reported: financial and technical inability to adapt products to soil pollution, inability to change their eating habits (by diversifying their diet), and difficulty in distinguishing the products to avoid when they don't come from one's own garden. Other obstacles to the changes proposed by the program were also identified: the effects of exposure to chlordecone are not sufficiently strong compared to other health problems, isolation, insecurity, lack of legitimacy accorded to official speeches, lack of perceived health benefits compared to the risk of loss of identity, and additional costs. All these points fail to identify the benefits to be had over the cost of accepting and applying the recommendations.

However, some favorable points also emerged from the evaluation: the feeling of being treated as a person, the friendly atmosphere of the collective events, access to practical information and knowledge tailored to their needs, awareness of the problems, and, in the case of preexisting doubt, the confirmation of the pollution of their soil being the first step toward change. On the other hand, the limited scope of the JAFA program was the subject of criticism: Other pesticides were not taken into account, the actions undertaken were judged to be too isolated, not extensive enough to effectively prevent exposure to chlordecone. The ORSaG report concluded: "The beneficiaries of the JAFA program call for more strict control of goods, a ban on roadside sales and promotion of a more environmentally friendly agriculture. As long as they feel there is no collective action for the benefit of the whole population, JAFA will be perceived as cauterizing a wooden leg". The respondents want more information about fishery products, medical treatment, and nutrition, and call for practical demonstrations rather than leaflets and brochures.

A report by the General Inspection Administration (Conseil Général de l'Environnement et al. 2011) commented on the high cost of JAFA programs given the number of people identified as being at high risk of exposure and who actually received support. JAFA programs implemented between 2008 and 2015 cost approximately €10 million. The cost of sampling and analyses accounted for less than 20% of the total sum. Unlike the "diagnosis on request" implemented in Martinique from the middle of 2010 on, the people we met were not necessarily asking for information or were systematic users of their garden, nor did they consume large quantities of products that represented a risk of exposure to chlordecone. If we exclude diagnosis on request, the programs were conducted for a given period of time to identify potentially overexposed people and help them reduce their exposure, whereas the use of family gardens and consumption of what is grown in them may change over time. In addition, knowledge of the contamination of family gardens resulting from the surveys conducted to date is far from exhaustive and the risk of exposure still exists for soils that have not been analyzed. The surveys did not cover the entire territory and contaminated soils can be found outside the areas on the contamination map. Finally, the transport of soil can transfer pollution to previously uncontaminated areas. One can thus question the effectiveness of the method used, the cost of which was very high compared to the number of homes involved, especially in Martinique, where the original program design does not take changing practices and behaviors into account. In comparison, diagnoses on request required a much lower budget but resulted in a 50% increase of detected situations where pollution potentially exceeded the TRV, at a cost of only €300,000. The diagnoses on request, free of charge for the householder, will likely meet future needs caused by changes in the use of the gardens, as well as the need to identify the level of risk for those who were not targeted by the JAFA program, at a lower cost than the original program. People who voluntarily request a diagnosis are aware of the problem of pollution, which should facilitate their acceptance and application of the recommendations. The 2015 information campaign on radio and TV in Martinique had a measurable effect, as it increased the number of requests for a diagnosis (the previous annual average was 150, which increased to 400 in 2015). As long as reliable information on risks continues to be available and the dissemination of food and agricultural recommendations also continues, diagnosis on request appears to be the best way to enable families who were unable to benefit from the first stage of the JAFA program, or those whose use and consumption of home garden products have changed, to manage their risk of exposure.

18.11 THE LIMITS OF JAFA PROGRAM IN REDUCING EXPOSURE

The JAFA program was designed to identify families with high exposure to chlordecone via consumption of vegetables grown in their garden and to provide them with tools to reduce their risk of exposure. The proposals for preventive measures were extended to family farms in 2015 and an awareness-raising campaign on the risks involved in purchasing vegetables from uncontrolled informal supply chains was organized. The diagnosis on request, which began in Martinique at the end of the investigations in 2010, is also available in Guadeloupe since January 2016. It offers households who did not meet the criteria for soil analyses the opportunity for soil diagnosis. This approach should be continued as it covers different situations connected with the use of family gardens: starting to cultivate a garden again, rearing livestock, adding soil from other sources, etc.

Visits to homes by counselors and collective activities undoubtedly succeeded in convincing some households with family gardens to change the species of edible plants they grow, what they eat, and their farming practices to protect themselves from exposure to chlordecone. But the evaluations conducted to date in Martinique by a mandated organization, and in Guadeloupe by polling institutes, and by the regional health observatory, agreed that it is extremely difficult to assess changes in behavior in the absence of a reliable indicator. Monitoring chlordecone levels in the population would certainly be a more reliable indicator of reduction in exposure, and would take all dietary sources of chlordecone into account. In this respect, the JAFA program cannot by itself respond to the problem of reducing dietary exposure to chlordecone. The AFSSA and Cire-AG assessments identified the supply and consumption of certain products acquired through informal distribution circuits as the main cause of high exposure. Although advice and guidance were given to affected households and the general population through its communication campaigns on the risks associated with these supply modes, the original objective of the JAFA program was not to address these last sources of the risk of exposure to chlordecone.

18.12 CONCLUSION AND PERSPECTIVES

The JAFA programs are concerned with the promotion of health: health being defined as "a state of complete physical, mental, and social wellbeing, not simply the absence of disease or infirmity" (WHO 1946). Health is the result of constant interaction between several determinants that can be grouped in three classes:

1. Individual determinants (genetic factors, age, gender, habits, and way of life, etc.)
2. Social determinants (community networks, families, etc.)
3. Environmental determinants (living conditions, socio-cultural, environmental and political conditions, etc.)

The problem of chlordecone needs to be considered in its local context:

- The historical pollution of soils by chlordecone in Guadeloupe and Martinique has resulted in distrust of the state and local authorities.
- The cultural dimension is crucial in the JAFA program because of the role of the Creole garden in the local cultural identity.
- The population already has knowledge and its own representations of the issue.

These aspects were known when the program was launched in 2009 and were confirmed in the qualitative assessment conducted by ORSaG in 2013. In this situation, a positive approach to the issue was preferred, especially in Guadeloupe, focusing on the benefits of the Creole family garden and on the quality of its products but including the constraints represented by pollution of the soil by chlordecone. This strategy, combined with the organization of fun interactive events in the neighborhoods based on local identity (language, cultural and culinary traditions, etc.), had some resonance with part of the population. In 2015, as part of the evaluation of the JAFA program aimed at increasing its efficiency, IREPS in Guadeloupe included the actions that had been undertaken in an ecological model to provide an overview of the program's targets and levels of intervention. The model showed that to increase the efficiency of the JAFA program and achieve the objective of reducing exposure to chlordecone, it will be necessary to work on each of the three determinants of adoption of positive health behaviors. Before 2015, significant effort was invested in identifying the individual situations but a little in the social and physical environment. Future JAFA activities in Guadeloupe should therefore focus on encouraging changes in behavior. Instead of doing something "for" the population, the aim is now to increase the capacity of the population to take charge of their own health. Through its actions, the "JAFA House" project should change the physical environment and its social determinants. JAFA House should be a friendly place that will

- Create and implement actions to reduce the impact of pesticides on health
- Create and strengthen intergenerational exchange and social ties between the beneficiaries and the public centered on the Creole garden
- Create and implement actions against chronic diseases

The JAFA House and associated educational garden should provide concrete practical examples of how to continue cultivating a Creole family garden polluted by chlordecone without using pesticides. Community research on alternative farming methods linked to agro-ecological farming techniques to reduce the use of inputs, and increase yields without contaminating crops should be implemented. The JAFA programs should continue to offer diagnoses on request, along with collective activities to promote crop diversification and cultivation techniques adapted to soil pollution. However, the JAFA program is only one action against high exposure to chlordecone (via informal distribution circuits and other sources mentioned earlier). In fact JAFA targets a specific population that is already very aware of the need for global action. A program with a wider scope is required to reduce exposure to chlordecone by members of the population who occasionally or mainly purchase food from informal production and distribution channels, including fishery products. This program could be an extension of the JAFA program. In addition, steps should be taken to better control potential exposure resulting from the consumption of local livestock, and regulations should be complemented by recommendations concerning the consumption of fishery products.

REFERENCES

AFSSA, 2005. Première évaluation de l'exposition alimentaire de la population martiniquaise au chlordécone: propositions de limites maximales provisoires de contamination dans les principaux aliments vecteurs. Rapport AFSSA, Maisons-Alfort, France, 39pp.

AFSSA, 2007. Actualisation de l'exposition alimentaire au chlordécone de la population antillaise. Evaluation de l'impact de mesures de maîtrise des risques. Rapport AFSSA, Maisons-Alfort, France, 48pp.

Blateau A., Flamand C., Pedrono G., Ségala C., and P. Quénel, 2011. Caractérisation des groupes de population à risque d'exposition élevée vis-à-vis du chlordécone via l'alimentation, Guadeloupe et Martinique, 2003–2009. *BEH* 3-4-5 février 2011, pp. 30–34.

Cabidoche Y.M., M. Jannoyer, and H. Vannière, 2006. Conclusions du Groupe d'Etude et de Prospective "Pollution par les organochlorés aux Antilles—aspects agronomiques". Cirad-FLOHR, Montpellier, Inra Petit-Bourg, Guadeloupe, France, 66pp.

Clostre F. and M. Lesueur-Jannoyer, 2011. Tests d'alternatives culturales sur sols contaminés par la chlordécone. Document de synthèse Cirad-SEVE, Martinique, France, 27pp.

Clostre F., M. Lesueur-Jannoyer, and B. Turpin, 2011. Impact des modes de préparation des aliments sur l'exposition des consommateurs à la chlordécone; Volet recherche du programme Jafa, rapport final, Cirad Martinique, France, 147pp.

Clostre, F., P. Letourmy, L. Thuriès, and M. Lesueur-Jannoyer, 2014. Effect of home food processing on chlordecone (organochlorine) content in vegetables. *Science of the Total Environment* 490: 1044–50.

Conseil Général de l'Environnement et du Développement Durable, Inspection Générale des Affaires Sociales, Conseil général de l'Alimentation de l'Agriculture de la Ruralité et des espaces Ruraux, Inspection Générale de l'Administration de l'Education Nationale et de la Recherche, 2011. Rapport d'évaluation des plans d'action chlordécone aux Antilles (Martinique, Guadeloupe). Paris, France, 104pp.

Desprats J.F., J.P. Comte, and Ch. Chabrier, 2004. Cartographie du risque e pollution des sols de Martinique par les organochlorés. Rapport phase 3—BRGM RP53262FR, 6 cartes, 23 pp.

INPES, 2008. Etude sur les connaissances, les perceptions et les comportements des populations de Martinique et de Guadeloupe vis-à-vis de la chlordécone. Saint-Denis, France, 45pp.

Ipsos Antilles, 2012. Rapport d'évaluation du programme Jafa auprès des 150 1er foyers ayant bénéficié d'un accompagnement individuel. Baie Mahaut, Guadeloupe, France, 37pp.

Jondreville, C., C. Bouveret, M. Lesueur-Jannoyer, G. Rychen, and C. Feidt. 2013. Relative bioavailability of tropical volcanic soil-bound chlordecone in laying hens (*Gallus domesticus*). *Environmental Science and Pollution Research* 20(1):292–299.

Jondreville, C., A. Lavigne, S. Jurjanz, C. Dalibard, J.-M. Liabeuf, F. Clostre, and M. Lesueur-Jannoyer. 2014. Contamination of free-range ducks by chlordecone in Martinique (French West Indies): A field study. *Science of the Total Environment* 493:336–341. doi: 10.1016/j.scitotenv.2014.05.083.

Observatoire Régional de la Santé de Guadeloupe (ORSaG), 2014. Evaluation qualitative du programme Jafa auprès des foyers bénéficiaires des actions. Rapport final. Basse-Terre, Guadeloupe, France, 32pp.

Vincent J., D. Camy, G. Thalmensi, M. Julien, M. Ledrans, P. Quénel, A. Blateau, and E. Godard, 2011. Le programme de santé des jardins familiaux en Martinique. Environnement Risque Santé vol. 10, n°5: 395-403 WHO, 1946. Constitution of the World Health Organization. Off Rec. Wld Hlth Org. 2, 100 doi: http://www.who.int/governance/eb/who_constitution_en.pdf.

19 The Challenge of Knowledge Representation to Better Understand Environmental Pollution

Philippe Cattan, Jean-Philippe Tonneau, Jean Baptiste Charlier, Laure Ducreux, Marc Voltz, Jean Pierre Bricquet, Patrick Andrieux, Luc Arnaud, and Magalie Lesueur Jannoyer

CONTENTS

19.1 INTRODUCTION

Assessing the impact of pollution in the environment basically supposes two things: first that we are aware of the risk of pollution and second that we have sufficient knowledge about pollutant dispersion. This poses two questions: The first one is about how to build a representation of the risk of pollution that determines the triggering of survey campaigns; the second one deals with our understanding of the pollution processes that determines where and how to act. Representation and understanding are dynamic processes that feed each other. In the case of chlordecone, their low evolution explains that, although the chlordecone impact on the environment has been evidenced since the end of the 1970s (Kermarec 1979, Snegaroff 1977), the authorities began the broad assessment of pollution only from the year 2000. Moreover, this assessment did not involve all compartments of the environment but was progressively extended: water in 2000; then soils, plants, and aquatic animals; and finally, terrestrial animals in 2008. Could this delay have been shortened? This chapter relates possible ways currently being tested that would have improved this sequence.

19.2 BUILDING A REPRESENTATION OF POLLUTION PHENOMENON

A good research is based on the definition of a clear research problem. Grasping what is perceived as the "problem" allows the scientist to identify needs for additional knowledge, which will justify and structure the research. Nonetheless, it is important to stress that the understanding of the issues at stake remains partial and highly dependent on existing knowledge. A clearly defined problematic is even more important in the case of action-research projects involving different partners. These approaches are based on the observation that innovations are poorly considered in stakeholders' daily practices. This is especially true for chlordecone, a widely used insecticide in the French West Indies. It took 7 years for practitioners to enforce mitigation processes after the field treatments ended in 1993, and knowing that impact on the environment has been evidenced since the end of the 1970s. The actors did not comprehend the urgency of the situation and/or chose to disregard it. Furthermore, they did not show signs of social mobilization to "address the issue" and even the social demand for research on the subject remained low.

For many scientists, the lack of innovation use is related to drawbacks in expert and scientific communications (Baulieu 2003). This viewpoint supports that the key for transmission and diffusion of knowledge is to create the necessary conditions for the adaptation and adoption of solutions (Benor et al. 1984, Schutz 1964). Other authors argue for a more complex process (Bouilloud 2000, Ghora-Gobin 1993). Knowledge is useless if a given actor cannot use them in given situations and thus knowledge becomes relevant when translated into practice (Klein 2001). The challenge is then to develop "capacities to adapt" so that societies can cope with change (Folke et al. 2003).

Action-research aims to develop this capacity to adapt. "Action research" and "innovation in partnership" are terms that characterize research methods that bring together researchers and stakeholders (mainly producers, technicians, decision makers, and managers) to promote the co-construction of not only technical but also organizational and institutional innovations. However, such an effort is not easy because, when faced with a challenge, different actors tend to read the same situation differently, due to multiple perceptions. The actors often have competing interests, have little coordination between themselves, and lack communications and operational attributes.

How can research be organized and conducted in such conditions? What mechanism can, simultaneously, co-construct a shared vision of a given problem (understanding), identify existing knowledge, including that of the actors, identify knowledge requirements, and plan actions, including those of research. Using experiments conducted within research project on land degradation (Balestrat et al. 2011) and river pollution (Lalande et al. 2014), we propose a modeling approach to address this lack of shared understanding of a common problem. Our methodological choice is the DPSIR model, which stands for Drivers, Pressures, States, Impacts, and Responses. It is a tool that supports participatory processes and intends to foster a shared and systemic analysis of an issue, along with the creation of action plans, and the co-production of necessary knowledge and skills.

In this chapter, we propose to first recall the model's underlying concepts and describe its role in the scientific process. We then present the mechanism that was designed in the case of chlordecone pollution in the French West Indies.

19.2.1 Research and Representation Model

Scientific approaches, whether they are hypothetical-deductive or inductive, always stress the importance of the model (Kuhn 1957, Nersessian 1999). A model is defined by Hagget as "a schematic representation of reality" (Hagget, cited by Ferras 1998). The model is an explicit representation that offers an integrated view of the issues involved by linking its various constitutive elements through semantic or causal relationships. It retains a certain level of complexity while providing an organized and simplified synoptic view of reality, which is accessible to all the actors involved in its development.

Based on literature review (Bateson 1984, de Rosnay 1975, Le Moigne 1977, Morin 2005, Simon 1974, von Bertalanffy 1950), the activity of modeling consists in describing a complex reality in a simplified way by retaining only those elements considered important so that known or assumed laws can be applied on these modeled elements in order to arrive at explanations or predictions. The models are constructed from abstractions, analogical reasoning, experiments, and analyses of a limited number of cases (Nerssessian 1999). A model can be digital. It is then expressed in the form of equations implemented within a computer application that can be used to manipulate the model and perform simulations. A model can also be discursive. Its quality is then directly dependent on the coherence of the underlying reasoning. A model can also be analogical and take the form of physical models of varying complexity, sketches, or other representations.

Participatory modeling can be used to support a group of actors (a set of individuals or legal entities who decide to organize themselves into a community in order to "seize" and address an important issue). The challenge is to integrate multiple viewpoints with different perceptions, interests, perspectives of actions, locations, territorial resources, time steps, and periods. Armatte (2005) stresses the importance of highlighting and integrating multiple perspectives within the model. The model tends to reshape the way of thinking of the different actors in defining the problem (Joerin and Rondier 2007). As an intermediate object (Vinck 1999), the model allows the actors to exchange viewpoints and thus constitutes an "aid" to collective learning processes (Lemoisson and Passouant 2012). Co-construction allows the actors to share the analysis of the situation, define common actions, and identify indicators to assess these actions (Gilbert and Boutler 2000).

19.2.2 The DPSIR Model: A Tool for Understanding, Sharing, and Planning

Describing a phenomenon is not sufficient to understand it. Hence, it is crucial to identify the structural processes responsible for the state of the phenomenon and the factors that will affect those processes. Modeling a phenomenon can thus facilitate its understanding, especially if it seeks to identify cause-and-effect relationships. The DPSIR model fulfils this criteria since it is designed to reveal these linkages.

In the case of chlordecone pollution in the West Indies, we used three DPSIR models: the model of the dynamics, the model of action, and the model of observation (Figure 19.1).

We used the model of the dynamics to understand the issue. The DPSIR model helps in apprehending the complexity of the mechanisms and dynamics of the chlordecone phenomenon. The driving forces (D) are considered structural changes (economic and social) external to the territory but with an impact on its dynamics. These driving forces produce pressures (P), a set of changes (degradations or improvements) in the state (S) of the level of resources, activity systems, and territorial systems, taken as a whole. The effect of these variations generates impacts (I), which can be environmental, social or economic, and which are perceived and accepted at varying degrees by society. The responses (R) are public measures or private behaviors adopted in response to the driving forces, pressures, and impacts. This model

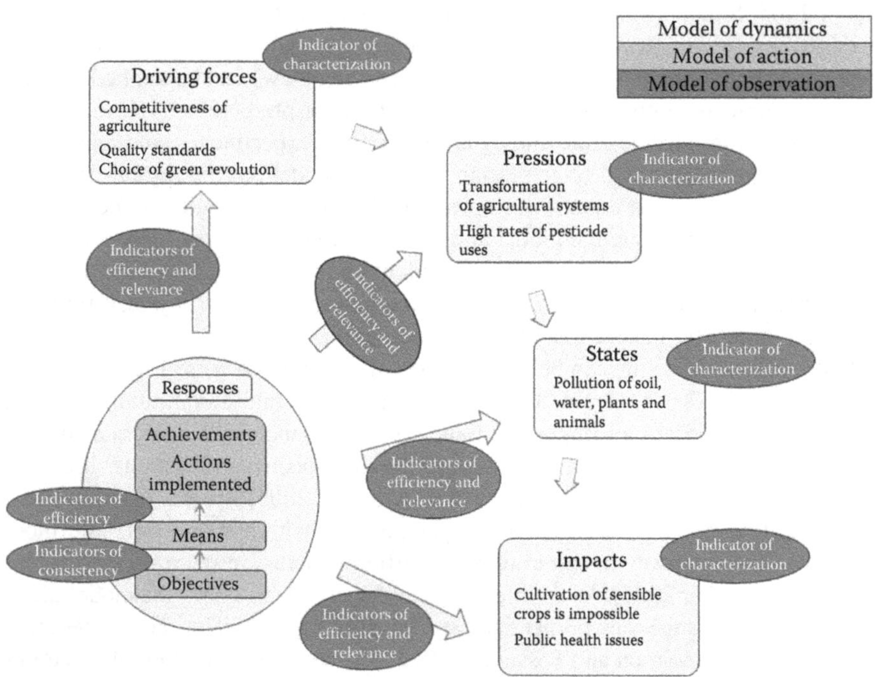

FIGURE 19.1 Schematic organization of the DPSIR model.

of the dynamics relies on the existing body of knowledge, and encompasses both scientific knowledge and the actors' "profane"* knowledge.

The model of action stems from the model of the dynamics. It specifies the most relevant actions to be undertaken. These involve implementing innovations (through training mechanisms and experiments in the real environment), testing and adapting inventions, and undertaking research. The actors' practices and responses are analyzed in relation with the appropriateness of their response regarding certain phenomena and dynamics. During the modeling process, current practices and research requirements are discussed and, as the understanding of the system evolves, potentially redefined. The model of action is used prospectively to devise responses to scenarios based on possible developments. It allows the actors to co-design action plans and implementation strategies. Actions to respond to questions identified from observed phenomena can be chosen with the help of simulations made possible by this model. The co-design of action plans includes identifying indicators to monitor actions, to indicate the state of different components of the system, and to measure impacts.

Built to complement the model of action, the model of observation specifies in detail the observations that are required to generate the system of indicators of the model of action. This step is also used to evaluate the nature of existing data and plan the acquisition of new data needed to estimate indicators. Datasets are also used for consultation, facilitation, and brainstorming workshops, and provide stakeholders with adequate information to meet their particular requirements and facilitate their decision-making processes.

19.3 INCREASING OUR UNDERSTANDING OF THE POLLUTION PROCESSES

Our understanding of pollution processes depends on the objective information we can gain from the observation of our environment. The key element to acquire this objective information is undoubtedly the measurement device. Two main targets can be assigned to it. The first one is to assess the pollution state of the different compartments of the environment. The second one is to gain knowledge in order to build an effective representation of the pesticide fate in the environment. Water is the main vector of pesticide fate, it thus requires focusing on hydrological units in which pesticide fluxes converge toward a main outlet. In this section, we give some indications about the guideline of monitoring pesticide fluxes in the environment from the plot scale (the scale of pesticide application) to the catchment scale (the scale integrating all the environmental compartments). An example of catchment monitoring in the Observatory of Pollution from Agriculture and Chlordecone (OPA-C), in the French West Indies, is given to illustrate the study case of diffuse pollution in tropical volcanic areas.

19.3.1 Guidelines of Monitoring Pesticide in the Environment

The main processes involved in chlordecone dispersion are presented Figure 19.2, which represents the hydrological and contamination processes and the environmental

* Ordinary or lay knowledge, based in common sense and embedded in life experience.

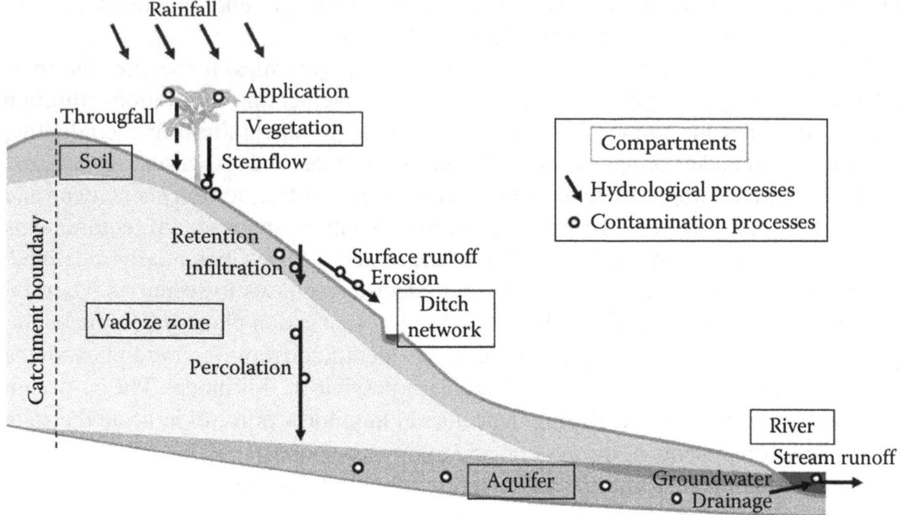

FIGURE 19.2 Main hydrological processes driving soil and water contamination by pesticides at the catchment scale.

compartments they are related to. Since chlordecone pollution takes place in the context of tropical climate in volcanic islands, particular attention must be paid to the following characteristics: high rainfall amount (2–4 meter by year), cultivation of plants like banana providing high rainfall redistribution by plant canopy (Cattan et al. 2007a), and soils like andosol with high infiltration rate (hydraulic conductivity at saturation that yields more than 60 mm h^{-1}), so aquifers play a major role in the hydrological functioning at catchment scale (Charlier et al. 2008).

Table 19.1 incorporates the essentials elements of Figure 19.2, and gives information about the means to characterize each process and help in pollution monitoring. For chlordecone, researchers have focused on soil retention process owing to the particular structure of allophane soils. However, there is a possibility that this focus also reflects the facility to study this process in this compartment of the environment. Maybe if we relate to the number of references in the FWI, local studies in laboratory (e.g., soil retention) are easier to conduct than local studies in situ (e.g., rainfall partitioning by canopy, percolation in vadose zone), which themselves are easier to conduct than studies at the plot (soil erosion, runoff) or larger scale (ditch network, aquifers, and rivers). Whatever, it appears that the usual monitoring at the outlet of a river remains insufficient to identify the relevant pollution issues and then pilot an action plan (where and how to act). Additional monitoring focusing on the process of interest has to be conducted. Figure 19.3 illustrates which type of monitoring could be implemented according to plot, subcatchment, and catchment scale.

TABLE 19.1
Main Hydrological Processes Driving Contamination Processes, and Corresponding Monitoring Scheme to Be Set Up

Compartment	Process	Variable	Variation Factor	Monitoring of Pollution	Complementary Monitoring to Improve the Functioning Knowledge	References in FWI
Vegetation	Rainfall partitioning by canopy	LAI, funneling ratio	Plant type	Stemflow measurement		Cattan et al. (2007a, 2009), Charlier (2009b)
Soil	Soil retention	K_{oc} (soil organic carbon–water partitioning coefficient)	Soil type Material age Agricultural practices	Batch experiment	Organic matter sampling	Cabidoche et al. (2009), Levillain et al. (2012), Woignier (2012a,b), Woignier et al. (2013), Fernandez Bayo et al. (2013)
Soil	Runoff/infiltration partitioning	Soil hydraulic conductivity	Soil Soil tillage Surface state/land cover	Infiltration tests Rainfall simulator and runoff plots	Chemical analysis in water at the plot scale	Cattan et al. (2006)
Soil	Erosion (transportation of CLD in surface water either in dissolved form or adsorbed on eroded soil particles)	Soil structural stability	Soil type Surface state/land cover elevation	Rainfall simulator and runoff plots Sampling during flood events	Mineralogy of sediments	

(*Continued*)

TABLE 19.1 (*Continued*)

Main Hydrological Processes Driving Contamination Processes, and Corresponding Monitoring Scheme to Be Set Up

Compartment	Process	Variable	Variation Factor	Monitoring of Pollution	Complementary Monitoring to Improve the Functioning Knowledge	References in FWI
Ditch network	Surface water drainage	Hydrological connectivity of plots	Network geometry	Gauging station at the sub-basin scale	Physico-chemical monitoring	Charlier (2009a)
Vadose zone	Percolation from soil toward aquifers	Soil structure permeability	Soil type Geological formation	Column experiments; lysimeter; shallow piezometer	Chemical analysis in groundwater	Cattan et al. (2007b), Sansoulet et al. (2008)
Aquifer	Groundwater drainage	Residence time	Geology Material age hydraulic properties	Tracer	Physico-chemical monitoring; chemical analysis of groundwater; water datation	Gourcy et al. (2009), Charlier et al. (2009a)
River	Stream water drainage	Hydrological connectivity of sub-catchments	Slope, river bed geometry, surface/groundwater interactions	Gauging station at the basin outlet	Physico-chemical monitoring; chemical analysis of groundwater; water datation	

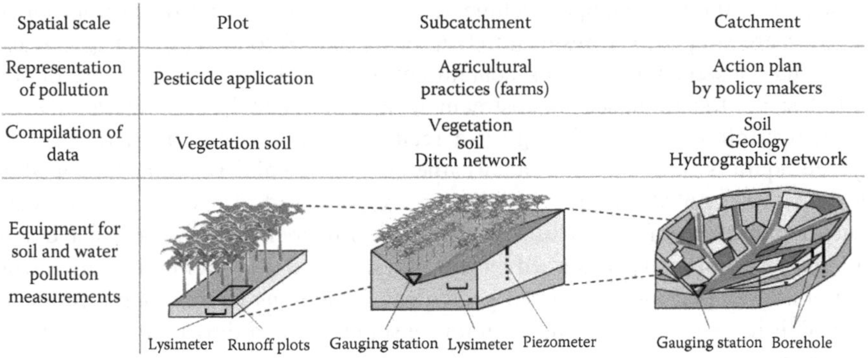

Spatial scale	Plot	Subcatchment	Catchment
Representation of pollution	Pesticide application	Agricultural practices (farms)	Action plan by policy makers
Compilation of data	Vegetation soil	Vegetation soil Ditch network	Soil Geology Hydrographic network
Equipment for soil and water pollution measurements	Lysimeter Runoff plots	Gauging station Lysimeter Piezometer	Gauging station Borehole

FIGURE 19.3 Compilation of data and corresponding monitoring to measure pollution from plot to catchment scale.

19.3.2 Example of the Observatory of Pollution from Agriculture and Chlordecone (OPA-C)

Besides other processes, the factors of variations define different sets of interaction and conditions of pesticide dispersion. For the OPA-C observatory, this led to identification of two complementary catchments, one in Guadeloupe and the other one in Martinique (see Figure 19.4).

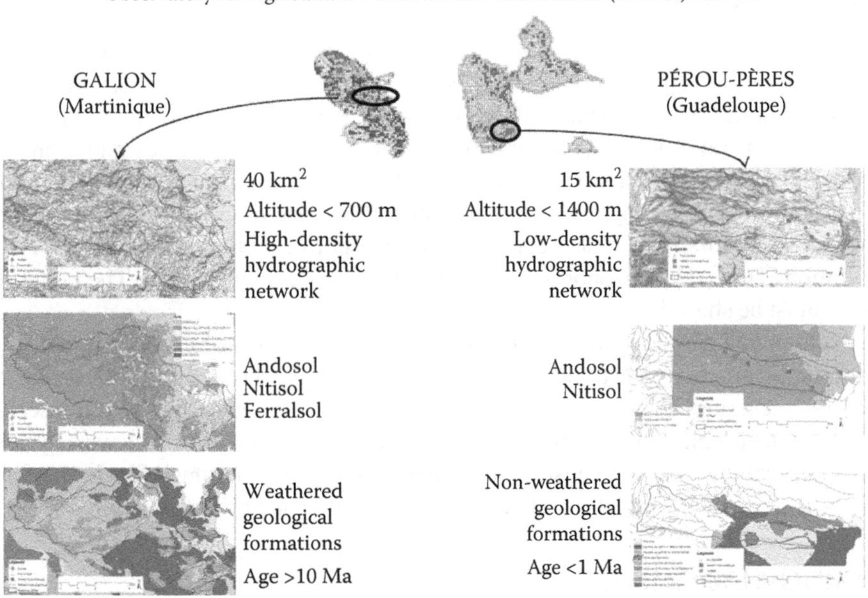

FIGURE 19.4 **(See color insert.)** Illustration of the main environmental characteristics of the Observatory of Pollution from Agriculture and Chlordecone (OPA-C).

Briefly, in the Guadeloupe catchment, hypotheses on the pesticide transport to the river are supposed to result in: (1) a weak runoff contribution on permeable soils, and consequently of weak contamination by chlordecone adsorbed on soil particles; (2) the influence of upstream head-catchment in noncultivated zones (tropical rainforests); and (3) high and relatively quick contribution from multi-layered aquifers developed into nonweathered geological formations having short residence time. In Martinique, hypotheses are supposed to result in: (1) a greater contribution of runoff and subsurface runoff on deeper and less permeable soils and greater chlordecone adsorbed on soil particles and (2) lower contribution from aquifers developed into weathered geological formations having longer residence time. This helps differentiate the monitoring schemes according to these contrasted hypotheses:

- Guadeloupe:
    - Monitoring of runoff in two geomorphologically different rivers (one having a large head-water catchment in the rainforest, and the other one originated from the cultivated area)
    - Monitoring groundwater levels in boreholes of various depth to intercept different aquifer layers (20 < depth < 80 m)
- Martinique:
    - Monitoring of runoff in several stations along the same river leading to measure flows at various scales
    - Monitoring water levels in several "shallow" boreholes (depth <20 m)

19.4 CONCLUSION

Assessing the impact of pollution, notably those of chlordecone, involves both a human issue and a physical issue. Both, inseparably tied, feed each other: the representation of the risk of pollution determines where and how to act, and to measure; measurement modifies our understanding of the pollution processes and then our representation of pollution risk. Consequently, no management of pollution could exist without considering each of these issues. The case of chlordecone shows that the usual investment in measurement is insufficient to encompass all dimensions of pollution. This leads to abnormal delay in evidencing contamination. The pollution issue must be shared.

Human and physical issues interact, thus assessing the impact of pollution is a continuous process, not a one-shot action. This process lasts with the problem of pollution and implies a collective investment: time investment to modify representation and action plan; monetary investment that must be budgeted on several years; and political investment to ensure that general interest is accounted for. In the final analysis, the state of pollution reflects the efficiency of social organization.

ACKNOWLEDGMENTS

Funding was provided by the French Chlordecone National Plan, the French National Research Agency ("Chlordexco" project), and the 2007–2013 FEDER fund.

REFERENCES

Armatte, M. 2005. Lucien March: Statistiques sans probabilité. *Journal électronique d'histoire des probabilités et de la statistique* 2(1):1–19.

Balestrat, M., E. Barbe, J.P. Chéry, P. Lagacherie, and J.P Tonneau. 2011. Reconnaissance du patrimoine agronomique des sols: une démarche novatrice en Languedoc-Roussillon. *Norois* 221:83–96.

Bateson, G. 1984. *La nature et la pensée.* Paris, France: Editions du Seuil.

Baulieu, E.-E. 2003. La science est progrès. Allocution du 14 janvier 2003, 14h30 France. http://www.academie-sciences.fr/activite/conf/exposeBaulieu_140103.pdf. Accessed April 16, 2008.

Benor, D., M.W.P. Baxter, and J.Q. Harrison. 1984. *Agricultural Extension: The Training and Visit System.* Washington, DC: World Bank.

Bouilloud, J.P. 2000. Sciences sociales et demande sociale. Pour une méthodologie de la réception. *Sciences de la société* 49:167–178.

Cabidoche, Y.M., R. Achard, P. Cattan, C. Clermont-Dauphin, F. Massat, and J. Sansoulet. 2009. Long-term pollution by chlordecone of tropical volcanic soils in the French West Indies: A simple leaching model accounts for current residue. *Environmental Pollution* 157(5):1697–1705. doi: 10.1016/j.envpol.2008.12.015.

Cattan, P., F. Bussière, and A. Nouvellon. 2007a. Evidence of large rainfall partitioning patterns by banana and impact on surface runoff generation. *Hydrological Processes* 21(16):2196–2205. doi: 10.1002/hyp.6588.

Cattan, P., Y.M. Cabidoche, J.G. Lacas, and M. Voltz. 2006. Effects of tillage and mulching on runoff under banana (*Musa* spp.) on a tropical Andosol. *Soil and Tillage Research* 86(1):38–51. doi: 10.1016/j.still.2005.02.002.

Cattan, P., S.M. Ruy, Y.M. Cabidoche, A. Findeling, P. Desbois, and J.B. Charlier. 2009. Effect on runoff of rainfall redistribution by the impluvium-shaped canopy of banana cultivated on an Andosol with a high infiltration rate. *Journal of Hydrology* 368(1–4):251–261. doi: 10.1016/j.jhydrol.2009.02.020.

Cattan, P., M. Voltz, Y.M. Cabidoche, J.G. Lacas, and J. Sansoulet. 2007b. Spatial and temporal variations in percolation fluxes in a tropical Andosol influenced by banana cropping patterns. *Journal of Hydrology* 335(1–2):157–169. doi: 10.1016/j.jhydrol.2006.11.009.

Charlier, J.-B., P. Cattan, R. Moussa, and M. Voltz. 2008. Hydrological behaviour and modelling of a volcanic tropical cultivated catchment. *Hydrological Processes* 22(22):4355–4370. doi: 10.1002/hyp.7040.

Charlier, J.-B., P. Cattan, M. Voltz, and R. Moussa. 2009a. Transport of a nematicide in surface and groundwaters in a tropical volcanic catchment. *Journal of Environmental Quality* 38(3):1031–1041. doi: 10.2134/jeq2008.0355.

Charlier, J.B., R. Moussa, P. Cattan, Y.M. Cabidoche, and M. Voltz. 2009b. Modelling runoff at the plot scale taking into account rainfall partitioning by vegetation: Application to stemflow of banana (*Musa* spp.) plant. *Hydrology and Earth System Sciences* 13(11):2151–2168. doi: 10.5194/hess-13-2151-2009.

de Rosnay, J. 1975. *Le Macroscope: Vers une vision globale.* Points, Seuil, Paris, France.

Fernandez Bayo, J., C. Saison, C. Geniez, M. Voltz, H. Vereecken, and A.E. Berns. 2013. Sorption characteristics of chlordecone and cadusafos in tropical agricultural soils. *Current Organic Chemistry* 17(24):2976–2984. doi: 10.2174/13852728113179990121.

Ferras, R. 1998. *Les modèles graphiques en géographie, Collection géo-poche.* Paris, France: Economica-Reclus.

Folke, C., J. Colding, and F. Berkes. 2003. Building resilience and adaptive capacity in social-ecological systems. In *Navigating Social-Ecological Systems: Building Resilience for Complexity and Change*, Berkes, F., Colding, J., and Folke, C. (eds.). Cambridge, U.K.: Cambridge University Press, pp. 352–387.

Ghora-Gobin, C. 1993. Crises de la ville et limites de la connaissance théorique. Pour une conceptualisation de la mise en œuvre. *Sciences de la société* 30:171–180.

Gilbert, K.J. and J.C. Boutler. 2000. *Developing Model in Science Education.* Dordrecht, the Netherlands: Kluwer Academic Publisher.

Gourcy, L., N. Baran, and B. Vittecoq. 2009. Improving the knowledge of pesticide and nitrate transfer processes using age-dating tools (CFC, SF6, 3H) in a volcanic island (Martinique, French West Indies). *Journal of Contaminant Hydrology* 108(3–4):107–117.

Joerin, F. and P. Rondier. 2007. Les indicateurs et la décision territoriale: Pourquoi? quand? comment. In *Les indicateurs socio-territoriaux et les métropoles*, Sénécal, G. (ed.). Québec, Canada: Presses de l'Université Laval, pp. 9–36.

Kermarec, A. 1979. Niveau Actuel de La Contamination Des Chaînes Biologiques En Guadeloupe: Pesticides et Métaux Lourds. Guadeloupe, France: Duclos.

Klein, É. 2001. Les vacillements de l'idée de progrès. In Peut-on encore croire au progrès?, Bourg D., Besnier J.M. (Eds), Paris: PUF, p. 67–77.

Kuhn, T.S. 1957. *The Copernican Revolution: Planetary Astronomy in the Development of Western Thought.* Cambridge, U.K.: Harvard University Press.

Lalande, N., F. Cernesson, A. Decherf, and M.-G. Tournoud. 2014. Implementing the DPSIR framework to link water quality of rivers to land use: Methodological issues and preliminary field test. *International Journal of River Basin Management* 12(3):201–217. doi: 10.1080/15715124.2014.906443.

Le Moigne, J.L. 1977. *La théorie du système général: Théorie de la modélisation.* Paris, France: Presses universitaires de France.

Lemoisson, P. and M. Passouant. 2012. Un cadre pour la construction interactive de connaissances lors de la conception d'un observatoire des pratiques territoriales. *Cahiers Agricultures* 21(1):11–17. doi: 10.1684/agr.2012.0538.

Levillain, J., P. Cattan, F. Colin, M. Voltz, and Y.-M. Cabidoche. 2012. Analysis of environmental and farming factors of soil contamination by a persistent organic pollutant, chlordecone, in a banana production area of French West Indies. *Agriculture, Ecosystems & Environment* 159:123–132. doi: 10.1016/j.agee.2012.07.005.

Morin, E. 2005. *Introduction à la pensée complexe*, *Essais*: Point Seuil, Paris, France.

Nersessian, N. 1999. Model-based reasoning in conceptual change. In *Model-Based Reasoning in Scientific Discovery*, Magnani, L., Nersessian, N., and Thagard, T. (eds.). New York: Springer, pp. 5–22.

Sansoulet, J., Y.-M. Cabidoche, P. Cattan, S. Ruy, and J. Šimůnek. 2008. Spatially distributed water fluxes in an andisol under banana plants: Experiments and three-dimensional modeling. *Vadose Zone Journal* 7(2):819–829. doi: 10.2136/vzj2007.0073.

Schutz, T.W. 1964. *Transforming Traditional Agriculture.* New Haven, CT: Yale University Press.

Simon, H.A. 1974. *La Science des systèmes: Science de l'artificiel.* Paris, France: Épi.

Snegaroff, J. 1977. Les résidus d'insecticides organochlorés dans les sols et les rivières de la région bananière de Guadeloupe. *Phytiatrie-Phytopharmacie* 26:251–268.

Vinck, D. 1999. Les objets intermédiaires dans les réseaux de coopération scientifique: Contribution à la prise en compte des objets dans les dynamiques sociales. *Revue française de sociologie* 40(2):385–414. doi: 10.2307/3322770.

von Bertalanffy, L. 1950. *Théorie générale des systèmes.* Paris, France: Dunod.

Woignier, T., F. Clostre, H. Macarie, and M. Jannoyer. 2012a. Chlordecone retention in the fractal structure of volcanic clay. *Journal of Hazardous Materials* 241–242:224–230. doi: 10.1016/j.jhazmat.2012.09.034.

Woignier, T., P. Fernandes, M. Jannoyer-Lesueur, and A. Soler. 2012b. Sequestration of chlordecone in the porous structure of an andosol and effects of added organic matter: An alternative to decontamination. *European Journal of Soil Science* 63(5):717–723. doi: 10.1111/j.1365-2389.2012.01471.x.

Woignier, T., P. Fernandes, A. Soler, F. Clostre, C. Carles, L. Rangon, and M. Lesueur-Jannoyer. 2013. Soil microstructure and organic matter: Keys for chlordecone sequestration. *Journal of Hazardous Materials* 262:357–364. doi: http://dx.doi.org/10.1016/j.jhazmat.2013.08.070.

Section VIII

Conclusion

20 Crisis Management of Chronic Pollution by Chlordecone
Conclusions

Philippe Cattan, Florence Clostre, Thierry Woignier, and Magalie Lesueur Jannoyer

CONTENTS

By observing the chlordecone crisis in the French West Indies, a number of conclusions can be drawn. Globally, two main considerations stand out. The first concerns the main driver of pollution that led to an intensive use of chlordecone around the 1980s (Figure 20.1) despite events that may have warned the stakeholders concerned: the Hopewell disaster and the ban on the use of Kepone (chlordecone) in the United States, and the Kermarrec report in the French West Indies. This may reflect the production-based viewpoint that prevailed in the society during this period. Likewise, the clear change that occurred until 2000 may be related to the emergence of the environmental viewpoint that progressively settled in all strata of society: research with first studies on pesticide fate in 1994 for the French West Indies, administration that increased control on environmental states and agricultural practices, and finally farmers drafting technical specifications for a better land use. Thus, the chlordecone crisis could not be restricted to the agricultural activity alone; it also involves the evolution of society. The second consideration is probably that science is not able to bring solution to all threats originating from human activities. In the case of chlordecone, no remediation is today technically effective and thus we have to live with the pollution.

How can we do this? The chlordecone case helps draw a scheme of pollution management and we identified five main stages.

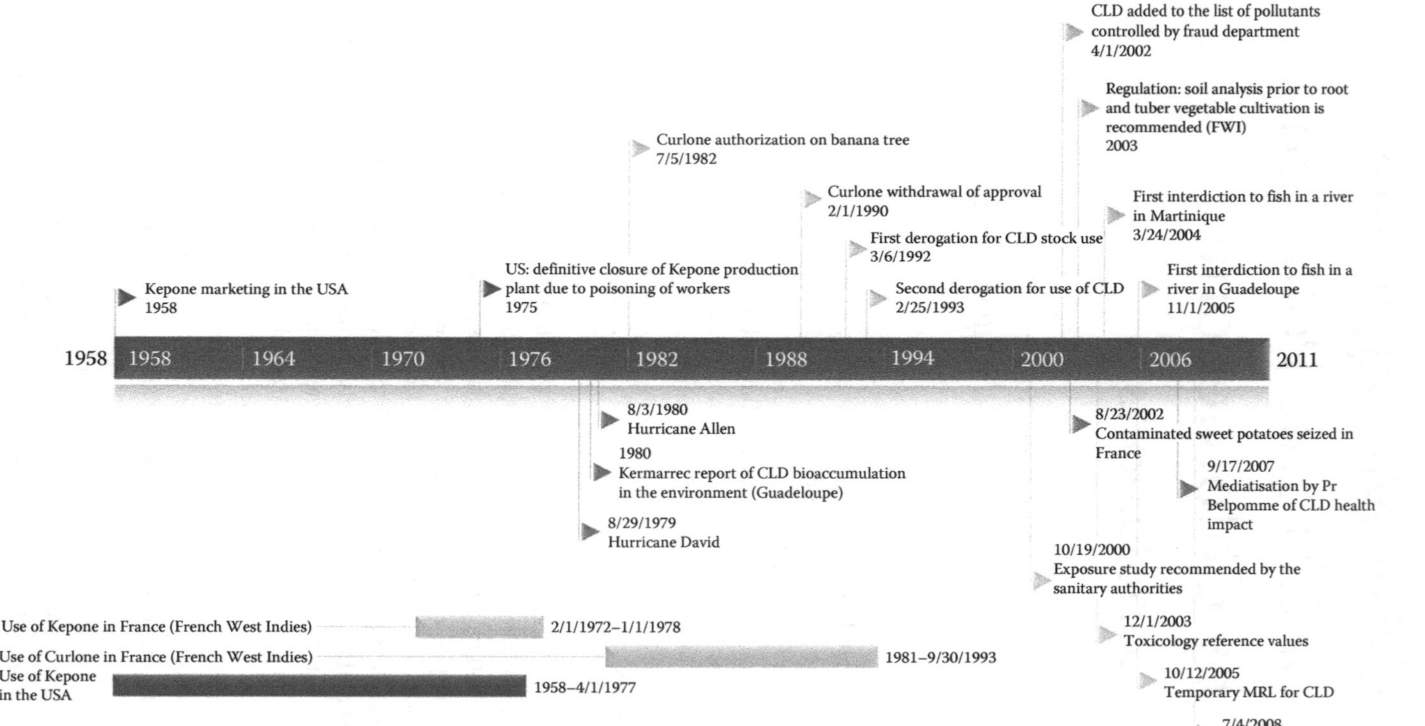

FIGURE 20.1 Timeline of chlordecone (CLD) use and pollution management.

20.1 FIRST STAGE: HIGHLIGHTS OF POLLUTION

Two main elements have led to pollution awareness:

1. *Fresh ideas*: At the beginning of the chlordecone crisis, an external viewpoint (Fagot and Mestre report) on the local management of water resources made the local authorities aware of environmental pollution. This external perspective occurred again with Pr Belpomme bringing controversy and preceding the establishment of the National chlordecone Plan in 2009. Undoubtedly, these external involvements have titled the local controversies to the environmental side.
2. *Evidence of pollution*: The first measurements in some compartments of the environment provided the necessary information. It was first the case for water spring, then for plants, soil, and animals. Only few measurements were necessary at this stage and they were sufficient to highlight the problem and trigger further development.

20.2 SECOND STAGE: BUILDING OF STANDARDS

This stage began early and resulted in toxicology reference values or contamination thresholds. It was the necessary step to decide what was good or bad and where to act. It was a difficult task. For chlordecone, it took 5 years from the determination of toxicology reference values in 2003 to those of maximum residue limits in 2008. And, as thresholds were based on compromises (technical, political, sanitary, etc.), references for meat are still under discussion.

20.3 THIRD STAGE: BUILDING TOOLS AND METHODS

This stage was concomitant with stage 2. It was a preliminary stage before characterizing chlordecone pollution on a large scale in various environments. It consisted of establishing protocols for sampling (which depth of soil? which river courses? which parts of plant or animal? etc.); extraction processes for different matrices (water, soil, plant and animal tissues, sometimes in very low quantities); analytical processes providing accurate and faster results; and building of indicators like sentinel species in various conditions. In chlordecone's case, everything was to be built owing to the lack of references for this molecule.

20.4 FOURTH STAGE: UNDERSTANDING THE POLLUTION PROCESSES AND ASSESSING THE IMPACTS

This stage aimed to build a representation of the pollution processes. It tended to answer questions about the origin of pollution, its duration, its impact on human health and environment, and its management. This stage involved long-term studies since it aimed to determine long-term effects of pollution on the development of living organisms. The most emblematic study here dealt with children's development.

For this kind of long-term effect, long series of measurements are necessary. This was accounted for only from 2008 with the establishment of National Chlordecone Plan and the beginning of surveys on human beings, animals, and environment. Thus, the delay to begin data acquisition (2008, i.e., 8 years after the pollution was highlighted) hindered the capacity to build models and also to anticipate the pollution evolution. This is of major concern.

These studies relative to this fourth stage deeply modified the representation of pollution: evidence of real dangers for human beings from very weak concentrations and high accumulation of chlordecone in specific organs or animal parts (liver of cow, shells of shrimp, etc.), in some cases possible depuration, and connection between all the compartments of the environment. Accordingly, the pollution management was also deeply modified. Prohibition of consumption was the first rule, then methodological solutions arose with the increase of knowledge, management of cattle in contaminated areas, and adapting crops to the level of soil contamination. Thus, the simple characterization of pollution that occurred in the first stage was not sufficient and efficient to manage pollution. Research plays a key role here to give a relevant representation of the pollutant fate, which may result in management actions.

20.5 FIFTH STAGE: MANAGEMENT OF THE POLLUTION

Surely, following the usual scheme underlying industrial development, the authorities expressed great hope in finding a technical solution to the chlordecone crisis: Science will identify the miracle solution that will lead to removal of chlordecone from soils and consequently from all environmental compartments. Unfortunately, it was not so simple. Studies on remediation did not give the expected results, and today no solution appears in the short term: Inhabitants have to live with the pollution. However, organizational solutions exist regarding consumer protection, change of activities, and implementation of rules. This involves developing monitoring policies and information systems to better produce adequate information and share it to improve the understanding of the pollution. Presently, such information systems are being developed by various actors (notably local water agencies) and represent the memory of environmental states. However, their stable financing in the long-term course (10–15 years) is not secured.

20.6 WHAT'S NEXT?

Since pollution may continue for decades, what can be expected? The current poor involvement of inhabitants on this subject (French ministries have funded most of the actions on chlordecone; the involvement of regional institutions remains weak and there are no private funding), the short duration of successive chlordecone plans (3 years each), and the lack of long-term thinking on pollution management make us fear that this issue will die with the departure of the few people in research institutes and administration, who brought and kept the problem in limelight. Nevertheless, the cost is high in terms of economic impacts, and of funding for chlordecone plans, researchers, pollution managers, and officers to remediate the consequences of a former decision of a molecule application, and the process is very long as it involves different stages we described previously. In the future, it is expected that this chlordecone crisis should give pause for thought and enlighten future decisions related to human activity.

20.7 TO CONCLUDE

It appears that management of chlordecone pollution necessitated building a complex system involving different actors, different thoughts, and different technologies. It entails high costs that only developing countries may have to bear. Since this kind of pollutant was disseminated worldwide, this may become a main concern for the poorest countries. This raises the question of responsibility for pollution and who pays: the governments who allowed the use? The sellers who disseminated the pollutant around the world? The users who applied it? Those who collectively agreed on the use and promoted tacitly this type of development?

In fact, the phenomenon of pollution is a result of society's functioning. From this viewpoint, the evolution of practices highly depends on changes of thoughts and on representations. Chlordecone crisis shows that information flow through information systems, transparency policies, and mobilization of external points of view are determinant factors for such a progress. It also shows that opposition to transparency was, and still is, numerous in all society strata, including research. This is the main challenge we have to face today.

Index

J

K

L

M

N

O

Milton Keynes UK
Ingram Content Group UK Ltd.
UKHW021622071024
449327UK00020BA/1155

9 780367 658373